Green Energy and Technology

Climate change, environmental impact and the limited natural resources urge scientific research and novel technical solutions. The monograph series Green Energy and Technology serves as a publishing platform for scientific and technological approaches to "green"—i.e. environmentally friendly and sustainable—technologies. While a focus lies on energy and power supply, it also covers "green" solutions in industrial engineering and engineering design. Green Energy and Technology addresses researchers, advanced students, technical consultants as well as decision makers in industries and politics. Hence, the level of presentation spans from instructional to highly technical.

Indexed in Scopus.

Indexed in Ei Compendex.

Kibret Mequanint • Ababay Ketema Worku
Muluken Zegeye Getie
Zerihun Getahun Workineh
Editors

Sustainable Development Research in Materials and Energy

Advancement of Science and Technology

Springer

Editors
Kibret Mequanint
Department of Chemical
and Biochemical Engineering
University of Western Ontario
London, ON, Canada

Ababay Ketema Worku
Bahir Dar Institute of Technology
Bahir Dar University
Bahir Dar, Ethiopia

Muluken Zegeye Getie
Bahir Dar Institute of Technology
Bahir Dar University
Bahir Dar, Ethiopia

Zerihun Getahun Workineh
Information Technology and Communication
Pompeu Fabra University
Barcelona, Barcelona, Spain

ISSN 1865-3529 ISSN 1865-3537 (electronic)
Green Energy and Technology
ISBN 978-3-031-69862-0 ISBN 978-3-031-69860-6 (eBook)
https://doi.org/10.1007/978-3-031-69860-6

This Springer imprint is published by the registered company Springer Nature Switzerland AG
The registered company address is: Gewerbestrasse 11, 6330 Cham, Switzerland

Contents

Part I
Materials Today

Chapter 1
Evaluation and Optimization of Sustainable Ecofriendly Bio-based Micro-composites for Smart Packaging Application Using Micro-Clay Particles as Reinforcement

Asmare Tezera, Desalegn Abera, Zenamarkos Bantie, Behailu Bisenebt, Assefa Beyene, Ejigayehu Deslaegn, Tesfa Nega, Mequannt Demeke, Surafiel Aregahegn, Tessafa Abrham, and Gashaw Arega

1.1 Introduction

The increasing concern over environmental pollution caused by nonbiodegradable plastics has led to a growing interest in the development of biodegradable alternatives. Starch-based plastics have emerged as a promising solution due to their renewability and biodegradability. However, the current quality of starch-based plastics is insufficient to replace traditional plastics on a large scale. Thus, there is a need to enhance the mechanical properties and degradation rate of starch-based biodegradable plastics. Starch-based biodegradable plastic is a promising alternative to traditional plastics, as it is made from natural and renewable resources and can be decomposed by natural processes. However, the properties of starch-based biodegradable plastic, such as its strength and durability, need to be improved to meet the requirements for various applications [1]. One potential solution is to use clay particles as a filler to improve the properties of the material. The production and disposal of traditional plastics have significant harmful effects on the environment, leading to pollution, resource depletion, and climate change. Biodegradable plastics

A. Tezera (✉) · D. Abera · Z. Bantie · B. Bisenebt · E. Deslaegn · T. Nega · M. Demeke
S. Aregahegn · T. Abrham
Faculty of Chemical & Food Engineering, Bahir Dar Institute of Technology, Bahir Dar University, Bahir Dar, Ethiopia

A. Beyene
Faculty of Industrial & Mechanical Engineering, Bahir Dar Institute of Technology, Bahir Dar University, Bahir Dar, Ethiopia

G. Arega
Department of Mechanical Engineering, Technology Institute, University of Gonder, Gonder, Ethiopia

K. Mequanint et al. (eds.), *Sustainable Development Research in Materials and Energy*, Green Energy and Technology,
https://doi.org/10.1007/978-3-031-69860-6_1

are a promising alternative to traditional plastics, as they can be decomposed by natural processes without leaving harmful residues or microplastics [2]. Mango seed kernel- and cassava starch-based biodegradable plastic is particularly attractive as it is made from natural and renewable resources, which reduces the reliance on fossil fuels and supports local agriculture. Additionally, the use of waste materials, such as mango seed kernels, can help to reduce waste and create value from what would otherwise be discarded. The mango, which belongs to the Anacardiaceae family, is a tropical fruit that can be found across various tropical and subtropical regions [3]. While the flesh of the mango fruit is typically eaten fresh or processed into jams, juices, and other products, the seed kernel of the mango also has plenty of potential. Despite containing a significant amount of starch, these valuable seeds are often carelessly discarded, causing significant environmental pollution [4]. Creating starch-based film using organic waste such as mango seeds is an accessible way to reduce environmental issues caused by traditional petroleum-based plastics. Currently, packaging materials for food and dry goods are made from a mix of petroleum-based plastic, metals, glass, paper, and board, often in combinations. While plastic is popular for its long shelf-life properties, it's not easily biodegradable and is contributing to environmental issues. Burning and burying plastic waste is not a viable solution, and this has led to interest in creating degradable bioplastics [5]. The excessive use of nonbiodegradable plastics has contributed significantly to environmental pollution and the depletion of natural resources. To address this issue, the development of starch-based biodegradable plastics has gained attention. However, the mechanical properties and degradation rate of these plastics are still limited. Therefore, there is a need to find an effective solution to improve the mechanical properties of starch-based biodegradable plastics, while also enhancing their degradation rate.

The research on starch-based biodegradable plastic holds significant importance in today's world. With the growing concern for environmental sustainability and the need to reduce plastic waste, exploring alternative materials is crucial. Starch-based biodegradable plastics offer a promising solution as they are derived from renewable resources and have the potential to degrade naturally, minimizing their impact on the environment. By conducting research on these materials, we can gain valuable insights into their properties, behavior, and potential applications. This knowledge can drive advancements in the development of sustainable packaging solutions, contributing to a more eco-friendly and circular economy. Ultimately, this research plays a vital role in promoting a greener future and addressing the global challenge of plastic pollution. Over 67 million tons of packaging waste have been generated in recent years, leading to serious environmental concerns. Furthermore, challenges facing the economy and safety have encouraged scientists to research degradable polymers as a substitute for petrochemical-based polymers. These degradable polymers, or bioplastics, offer a much more eco-friendly solution compared to conventional plastics which pose a range of environmental issues. To address this challenge, researchers have proposed a low-cost and renewable substrate that can address waste generated in agriculture. This approach suggests using starch from agricultural waste as a biodegradable polymer for developing bioplastics due to its abundant sources and performance, and it is suitable as a material for producing low-cost bioplastics [6].

The term "polymer" comes from the Greek words "poly" meaning many and "meros" meaning parts. Scientists often use the term "macromolecule" in place of polymer to denote huge molecules. The properties of plastics are determined by the long chains of atoms that make up the polymers and the bonds within and between these chains. However, not all polymers can be considered plastics. Plastics are modified polymers or macromolecules that are reinforced with additives to create a mechanically strong, dimensionally stable material. In the past 60 years, synthetic polymers have seen significant growth, becoming a major area of focus in polymer science. Researchers are exploring biodegradable polymers made from renewable resources as a potential replacement for synthetic plastics. The current ASTM definition of biodegradable polymers describes them as degradable due to the action of naturally occurring microorganisms such as bacteria, fungi, and algae. Recent research has shown that biodegradable bioplastics made solely from starch can be thermally unstable and lack mechanical strength due to the absence of reinforcing materials could not give any longer service. In order to improve the thermomechanical properties of the film in addition to extend its shelf life and service, natural pottery clay is the best solution to overcome such shortcomings [7].

The clay layers also induce changes in the molecular mobility of the polymer matrix, which may enhance its thermal and mechanical properties as well as oxidation stability, among other advantages [8]. Mango seed kernel- and cassava root starch-based biodegradable plastic is a type of plastic that is made from natural and renewable resources. This type of plastic can be decomposed by bacteria and other natural processes, which makes it a more environmentally friendly alternative to traditional plastic. The process of making this biodegradable plastic involves extracting the starch from waste mango seed kernels and mixing it with cassava starch. The mixture is then processed with clay into a microplastic material that can be molded into various shapes and forms. The benefits of using this type of biodegradable plastic include reducing the amount of waste that ends up in landfills and the ocean, as well as reducing the environmental impact of plastic production. Additionally, this type of plastic can be produced using locally sourced, renewable resources, which can help to support local economies. Generally mango seed kernel- and cassava root starch-based biodegradable plastic represents an important innovation in the field of sustainable materials and has the potential to help address some of the environmental challenges associated with traditional plastic production and disposal [9, 10].

The aim of this research is to optimize the properties of starch-based biodegradable plastic using clay particles as a filler. Specifically, the research aims to investigate the effect of clay particle concentration on the mechanical properties of the material and to evaluate the biodegradability of the material with clay particles as a filler and to compare the properties of the material with and without clay particles to identify the benefits and limitations of using clay as a filler in starch-based biodegradable plastic.

Fillers are often used in the production of biodegradable plastics to improve their mechanical, thermal, and barrier properties. These properties can be critical for various applications, such as packaging, agriculture, and medical products. Fillers can also be used to reduce the cost of biodegradable plastic production by reducing the amount of expensive biodegradable polymers needed to achieve the desired properties [11]. In addition, fillers can provide other benefits, such as increasing the stiffness, reducing shrinkage, improving the printability, and enhancing the biodegradability of the material. Clay particles are a common type of filler used in biodegradable plastics due to their low cost, wide availability, and ability to improve mechanical properties, such as strength and toughness [12]. They can also improve the thermal and barrier properties of the material, such as moisture and gas permeability. In general, the use of fillers can help to optimize the properties of biodegradable plastics and make them more suitable for various applications [13].

1.2 Materials and Methods

1.2.1 Materials and Chemicals

The starch extracted from mango seed kernel and cassava root and the mango fruit waste obtained from the nearby city Bahir Dar and the cassava root were brought Hawassa northern region of Ethiopia, and the clay was locally available. Glycerol, a necessary component, can be easily obtained from the nearby market. Mango seeds were procured from local juice traders in Bahir Dar city. The experiment mainly utilized distilled water, glycerol obtained from the nearby market, sodium metabisulfite as an antioxidant, liquid sulfur dioxide as an extracting agent, and micro-pottery clay as a filler. Additionally, chemicals such as sodium hydroxide, methanol, ethanol, iodine solution, KI, acetic acid, hydrochloric acid, sulfuric acid, and sodium chloride were also used.

1.2.2 Equipments Used

The main equipments used during the experiment was balance (JF-2004), peeler, miller, sieve, bath (TBS451PA), cylinder, muslin cloth, oven (DHG-9023A 20 L), flasks, and a universal testing machine (UTM 100ST.

1.2.3 Methods of Starch Extraction

The waste mango seed kernel collected from the nearby city and the cassava root were collected form Hawassa, washed with clean water, shredded, peeled, and then

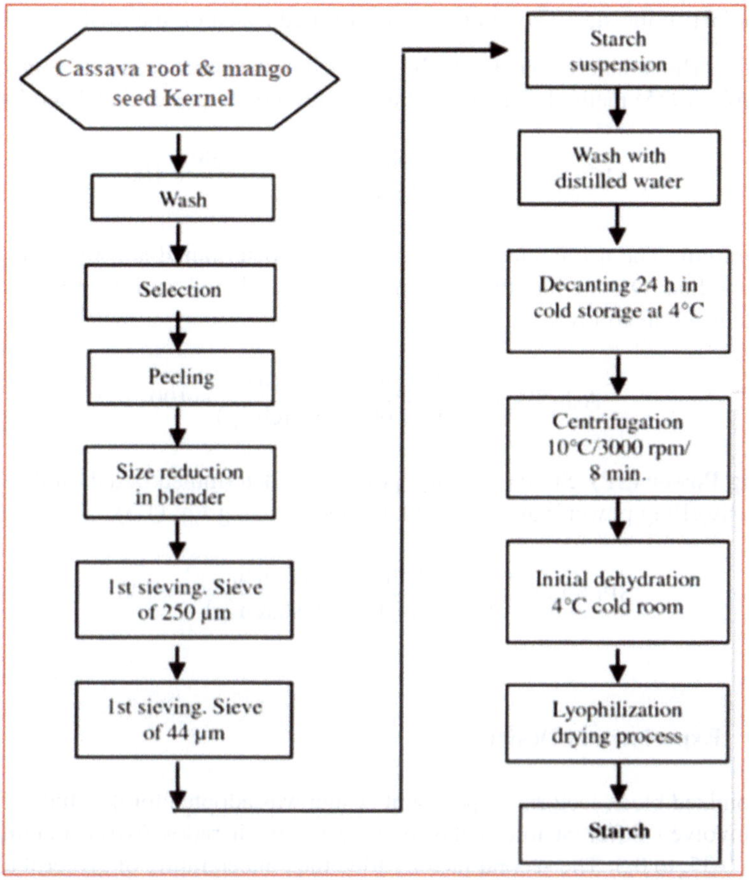

Fig. 1.1 Starch extraction process

fed into the cutter mixer. Once the dough is formed, the sediment was separated from the slurry muslin cloth and washed many times. After the second settling, the starch sediment was dried using an oven dryer with a temperature of 100°C for 24 h so as to remove the moisture. The extraction process is shown in Fig. 1.1. Finally, the dried starch was ground and sieved within the range of 0.5–0.125 µm.

1.2.4 Characterization of Starch

Confirmation Test The confirmation test of starch was carried out using iodine solution. Iodine solution was used to identify whether the extracted one is really starch or not [14].

pH The pH of the starch solution was determined based on the method of [15].

Moisture The moisture content (MC) presented in the starch was determined based on AOAC (1975) method. The percentage of the moisture was calculated by

$$\text{MC }(\%) = \frac{\text{Initial weight} - \text{Final weight}}{\text{Initial weight}} * 100 \tag{1.1}$$

Ash Content The ash content of the starch was determined according to AOAC (1990) method. On cooling, the resulting ash was weighed, and the percentage ash was calculated as follows:

$$\text{Ash}(\%) = \frac{\text{Weight of Ash (g)}}{\text{Weight of dry starch (g)}} * 100 \tag{1.2}$$

Swelling Power (SP) Starch swelling power was determined based on the method of [16]. Swelling power was calculated as follows using Eq. (1.3):

$$\text{SP}(\%) = \frac{\text{Weight of the wet sediment (g)}}{\text{Weight of the dry starch(g)}} * 100 \tag{1.3}$$

1.2.4.1 Experimental Design

A randomized block factorial experiment design was adopted for the study. The first factor involved different micro-clay particle to starch ratios (w/w), ranging from 0, 0.25, 0.35, to 0.5. The second factor varied the concentration of glycerol to starch (v/w), with ratios of 0, 1.25, 1.5, and 1.75. The investigation aimed to analyze the combined and individual effects of micro-clay particle and plasticizer on the film's overall properties.

1.2.4.2 Development Microplastic Film

To begin the experiment, a solution containing 5 grams of starch (2.5 g each) and 100 milliliters of distilled water was prepared. The mixture was stirred manually while gradually increasing the temperature between 55, 65, 75, and 85°C at a heating rate of 10°C per minute to induce gelatinization. Micro-clay particle was added to the solution at concentrations of 0, 0.25, 0.35, and 0.5 w/v along with glycerol at concentrations (w/v) of, 1.25, 1.5, and 1.75. The resulting film-forming solutions were cast into glass Petri dishes and allowed to dry for 6 h at a constant temperature around 33°C. Finally, the films were peeled off after reaching a constant weight.

1.2.4.3 Product Characterization

Thickness
To measure the film thickness, a digital micrometer was used, and three random locations were selected to take measurements for each film. The obtained values were then averaged to yield a representative measurement [17].

Transmittance and Transparency
The film's transparency was measured using a UV spectrophotometer at a wavelength of 620 nm. Rectangular pieces of the film were shaped and placed inside the spectrophotometer cell at the same wavelength. For the micro-composite films, strips measuring 1 by 3 cm were cut and placed in the test cell of the spectrophotometer, with air serving as the reference. Transparency was expressed as transmittance units per thickness unit [18].

$$\text{Transparency}(\%\text{T}) = \frac{-\text{LogT620nm}}{\text{X}} \tag{1.4}$$

where T620 is the transmittance at 620 nm and X is the film thickness (mm).

Moisture Absorption (MA)
To determine the moisture content of the film, the standard vacuum oven method (AOAC, 1995) was employed. This involved cutting individual 3 × 3 cm pieces of the film and weighing them. The film pieces were then dried in an oven at 105°C for 24h until they reached an equilibrium weight. The amount of moisture absorbed by the film was calculated using Eq. (1.5).

$$\text{MA}(\%) = \frac{\text{Initial weight} - \text{Final weight}}{\text{Initial weight}} * 100 \tag{1.5}$$

Tensile Strength (TS)
The tensile strength of the film was analyzed using the universal testing machine in compliance with the ASTM D-882-9 standard. To determine the tensile strength and % elongation, a specimen in the shape of a dumbbell was utilized. The initial grip separation and cross-head speed were set at 20 mm and 100 mm/min, respectively. The caliper micrometer was utilized to measure the width and thickness of the film.

$$\text{TS (MPa)} = \frac{\text{Force(N)}}{\text{Area(Width in mm} * \text{Thickness in mm)}} \tag{1.6}$$

Biodegradability
The film's biodegradability was assessed through a soil burial method. The film was cut into small pieces measuring 3 × 3 mm and buried 80 mm deep in the soil for a duration of 2 weeks. Prior to burial, the initial weight of the film (W1) was noted. After the 2-week period, the film pieces were removed from the soil, washed with distilled water to remove any soil particles, and dried on filter paper. The samples were then placed in a desiccator for 48 hours until they reached a constant weight.

The final weight of the samples (W2) was recorded, and the degree of degradability (%W) was calculated using the formula below [19]:

$$W(\%) = \frac{(W1 - W2) \times 100}{W1} * 100 \tag{1.7}$$

Coulorance of the Films

The Minolta CR300 Series digital color meter was used to determine the color of the films. The L*, a*, b* method was used to obtain color values ranging from L = 0 (black) to 100 (white) for lightness (L), red-green chromaticity (a), and yellow-blue chromaticity (b). To measure the color, film specimens were placed on a white standard plate with specific L*, a*, and b* values. Three readings were taken from different areas of each micro-film, and Eqs. (1.8) and (1.9) were used to calculate the total color difference (E) and whiteness (WI) indexes of the samples, respectively.

$$\Delta E = \sqrt{(L* - L)^2 + (a* - a)^2 + (b* - b)^2} \tag{1.8}$$

$$WI = 100 - \sqrt{(100 - L)^2 + a^2 + b^2} \tag{1.9}$$

where L*, a*, and b* are the standard color parameter values and *L*, *a*, and *b* are the color parameter values of the sample.

Thermogravimetric Analysis (TGA)

The thermal stability of the film was examined using thermogravimetry analysis (TGA) with the Shimadzu TGA-50 equipment while under nitrogen gas and a flow rate of 50 mL/min. The temperature range was kept between 20 and 200°C, with a heating rate of 10°C/min. The platinum pan was used to hold 2 milligrams of the film, and the study focused on the thermal characteristics of the reinforced (highest quality) and non-reinforced (plasticized) films [20].

1.2.4.4 Fourier Transform Infrared (FTIR)

The film functional surface groups, as well as chemical bonds (interaction forces), were examined using Fourier transform infrared spectrophotometer (Jasco-FT/IR-6600A). The spectra in terms of percent transmittance were recorded in the range of 4000–400 cm^{-1} with a resolution of 0.4 cm^{-1}.

1.3 Results and Discussion

1.3.1 Characterization of Starch

The pH of the starch solution was measured with a pH meter and found to be 6.2 ± 0.231, which is near to align with the findings of [21] who reported a pH of 5.5.

In food industry, pH testing can be done to detect molds and other unwanted impurities. The swelling power of the starch was determined to be 0.233 ± 0.23%, which is the ability of starch granules to absorb water as the temperature rises. The moisture content was found to be 10.33 ± 0.21% which align with the findings [22], who reported 0.135 ± 0.002% moisture content. The ash content of the starch was found to be 0.41 ± 0.13%, slightly higher than the result obtained by [23] which was around 0.28 ± 0.04%. The amylose content was found to be 19.278%, consistent with the findings of [9]. All data's are available in the appendix.

1.3.2 Evaluation of the Film

Viscosity As depicted in Fig. 1.2, in the first stage, viscosity was slightly increased from 1.491 ± 0.271–1.867 ± 0.213cp in the temperature interval of 55–65°C due to the fact that swelling of starch granules with limited amount of amylose leaching [24, 25]. From the very beginning of gelatinization process, from 55 to 65°C, the viscosity due to the kinetic energy of starch molecules is negligible [26]. In the second phase (65–75°C), the viscosity was sharply raised from 1.867 ± 0.213 to 4.033 ± 0.541cp. This is due to the intermolecular interaction effect of extra granular [27]. The optimum limit of viscosity 3.11 ± 0.31cp was found at a temperature of 75°C. In this temperature interval, the largest percentages of swollen and intact granules are present, and also the starch is considered to be completely pasted [28]. In third phase (75–85°C), the viscosity steadily decreased from 3.11 ± 0.31 to 1.11 ± 0.33cp. Finally, it was also observed that once again the viscosity raised from 2.485 ± 0.132 to 3.368 ± 0.371 while the reduced from

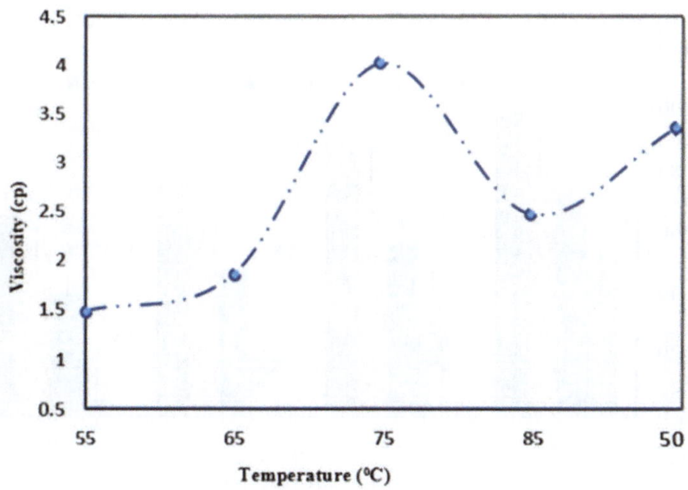

Fig. 1.2 Effect of gelatinization temperature against the viscosity

85 to 50 °C in order to remove the formation of some air bubbles during gelatinization and to create a good network association between the chains. The heated gelatinized solution was cooled down from 85 to 50°C in order to remove air bubbles. It is believed that the initial swelling takes place in the amorphous regions of the amylose where hydrogen bonds are less numerous and the polymers are more susceptible to dissolution. As the structure begins to weaken, the granule absorbs more water and became swelled as a result of the amylose that leach out, and finally, gel formation was observed. Because not all the granules gelatinize simultaneously, different degrees of structural disruption and swelling exist [29, 30].

1.3.2.1 Effects of Parameters on the Product

Effect of Glycerol and Micro-Pottery Clay on the Film's Moisture Uptake
The study's findings, as seen in Fig. 1.3, indicate that an increase in glycerol concentration from 0% to 35% resulted in a rise in the moisture uptake by the films. Conversely, increasing the micro-pottery clay content from 0% to 10% decreased the moisture uptake. The film exhibited its highest moisture content (19.31 ± 0.17) when the glycerol concentration was high (35%) and the micro-pottery clay content was low (0%). This could be attributed to the hygroscopic nature of glycerol, which may have caused it to migrate to the surface of the film and contribute to moisture absorption, leading to an increase in the film's moisture uptake [31]. The study demonstrated that the film's moisture affinity increased with an increase in glycerol concentration. This could be attributed to the hydrophilic nature of glycerol, as the -OH groups in glycerol can form hydrogen bonds with water, resulting in more moisture being absorbed by the film. The research concluded that the starch-glycerol film was effective in absorbing water at all tested

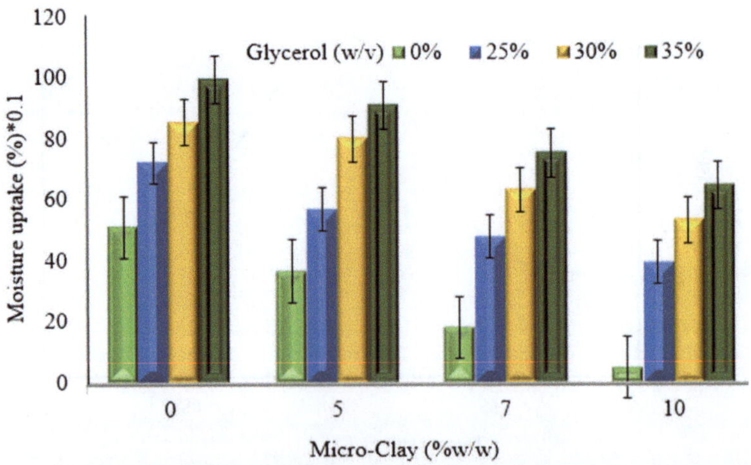

Fig. 1.3 Effect of pottery clay and glycerol on the films' moisture uptake

concentrations. Furthermore, the study found that increasing the micro-pottery clay content reduced the film's moisture uptake, indicating that micro-pottery clay had a hydrophobic effect [32]. Based on the results, it can be concluded that the moisture uptake of starch-based films is influenced by the concentration of glycerol and micro-pottery clay. The introduction of glycerol increases the hydrophilicity of the film by forming hydrogen bonds with water molecules through its -OH groups. On the other hand, the incorporation of micro-pottery clay reduces the moisture uptake, indicating its lower affinity toward water compared to glycerol. As the content of micro-pottery clay increases, the moisture uptake decreases accordingly [33] as compared with reinforced films due to the combined hydrophilicity nature of glycerol and MSK starch. For more information refer to the appendixes.

Effect of Glycerol and Micro-Pottery Clay on the film's Thickness
A digital micrometer was used to measure the thickness of the film. With an increase in the content of micro-pottery clay, the thickness of the films decreases. This is due to the formation of micro-voids within the film matrix caused by the addition of clay particles, which leads to a reduction in the overall thickness of the film. The clay particles also tend to pack together more closely, which further reduces the thickness of the film. Conversely, as the concentration of glycerol increases from 0% to 35%, the thickness of the film increases. The addition of glycerol creates micro-void spaces in the film that are filled by the glycerol molecules, resulting in an increase in film thickness [34]. However, the addition of micro-clay clay decreases the thickness of the film because the clay particles occupy space in the film and reduce the formation of micro-voids. Therefore, the combined effect of glycerol and micro-pottery clay on film thickness is complex and depends on the relative concentration of these components. According to [28] report, as the plasticizer concentration increases, the values of film thickness increased; this is because glycerol causes the films to favor more swelling properties and also the micro-voids are filled by the glycerol as result; the films had become thicker so that interparticle compactness was increased (Fig. 1.4) [32].

The other reason is glycerol, a clear compound, simply dissolves in water, enhances the viscosity of the solution, and easily binds with water. So, the higher the addition of glycerol content, the lower the water evaporation rate would be because some of the water in the film's solution is bounded by glycerol so that it significantly affected the thickness of developed films [34]. The maximum film thickness was obtained at the higher glycerol concentration of 35% w/v which is around 0.43 ± 0.002 and at a lower value of micro-pottery clay content (0%) [35]. The figure clearly had shown that, as the glycerol concentration increases, almost all the thickness of the films was increased [34]. Lower values of film thickness were obtained at 10% micro-pottery clay content; this is due to the formation of microporous media on the film's surface which prevents polymer molecular chains from coming close and compact together. According to [26] the improvement of thickness would be achieved by the addition of plasticizer concentration because glycerol molecules would occupy the micro-voids in the matrix and strongly interact with the film-forming polymer, which caused the distance between

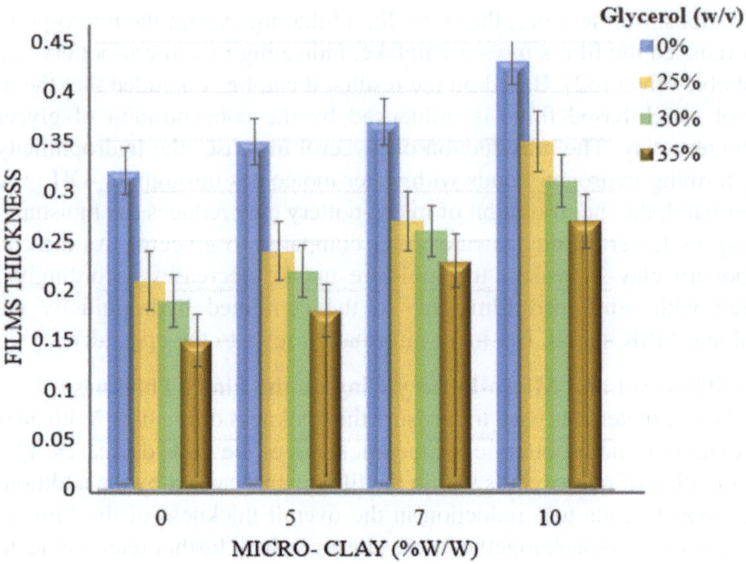

Fig. 1.4 Effect of glycerol and pottery clay on the film thickness

the polymers to increase, thus improving the film thickness. Based on the findings of [21] report, the plasticizer added to the development of packaging film can bind with starch to form starch plasticizer polymer so that the starch-starch bond is substituted by the starch-glycerol-starch bond which led to an improvement of the film thickness.

Effects of Glycerol and Micro-Pottery Clay on the Film's Transparency
Based on Fig. 1.5, the maximum transparency value was obtained at the higher value of micro-pottery clay content (10%) and at a lower value of glycerol concentration which is around 4.93 ± 0.178. As the clay content increases (from 0%, 5%, 7%, to 10%), by keeping the glycerol aspect ratio, the pattern of light passing through the film (%T) was increased from 0.51 ± 0.140, 1.36 ± 0.210, 1.54 ± 0.109, 2.01 ± 0.035, 3.11 ± 0.321, to 4.93 ± 0.178. Based on the observed results, glycerol has an antagonistic effect on the film's response as compared with that of micro-pottery clay. The increased glycerol concentration resulted in a decrease in light transmission. Lower UV light transmission is probably due to the different molecular weight, composition, and size of glycerol concentration that impedes the light transmission properties of those [27]. The experimental results with respect to film transparency ranged between 4.93 ± 0.178 and 0.51 ± 0.140 as presented in Appendix 4.3. The film's transparency value was reduced when its glycerol concentration was increased; this is due to the increase in film thickness. The transparency values of each film produced in this study experienced decreased in the range of 4.93 ± 0.178 to 0.51 ± 0.140 as the glycerol concentration increased from 35%. The higher glycerol concentration caused the color of the developed film to become blurred so that the brightness level passing through the film were decreased; this is

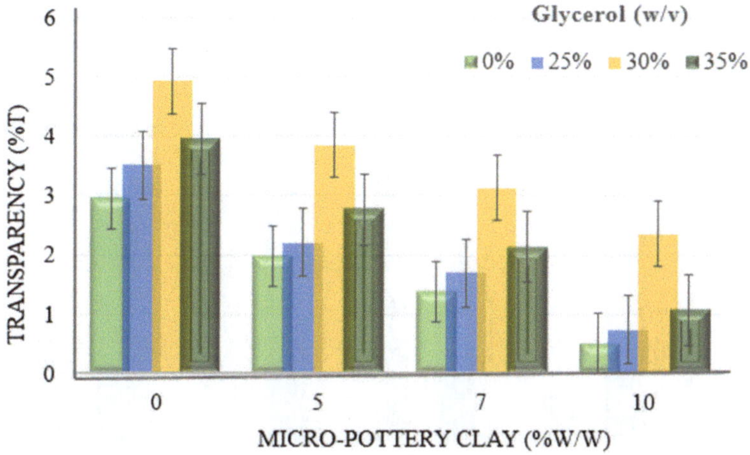

Fig. 1.5 Effects of glycerol and pottery clay on the film transparency

due to the increase in film thickness preventing the light passing through the film [34].

Effect of Glycerol and Micro-Pottery Clay on the Film's Tensile Strength
The mechanical properties of films play a crucial role in determining their durability and effectiveness in enhancing the mechanical integrity of food products. Films with high tensile strength are particularly desirable for food packaging as they provide reliable protection during handling, transportation, and marketing [34]. According to a report referenced as [36] report, it was observed that unplasticized films exhibited significantly higher tensile strength values compared to plasticized films under all relative humidity (RH) conditions. Based on [37] findings, the tensile strength weakening by glycerol is because of the disruption of molecular cohesiveness of starch as well as lowering of intermolecular interaction in the film. As was depicted in Fig. 1.6, initially, it was observed that film tensile was increased as both glycerol concentration (0–25%) and micro-pottery clay content (5%) was increased until it's optimality limit was found. Addition of glycerol concentration beyond 25%, molecules in starch-based films situated between the films bonding chain and be able to interact by forming hydrogen bonds between the film chains, causing interactions between the molecules of films have become less and caused a declined in tensile strength from 5.66 ± 0.012 to 0.95 ± 0.012 as presented in Appendix. The tensile strength gradually decreased from 5.66Mpa to 3.82Mpa as glycerol concentration increases over 25% to 35% w/v due to overplasticization above its critical limit that led to films to become more flexible but less resistant. In fact, film's tensile strength was decreased as glycerol concentration increases after the second four treatments. According to [34] report, the use of plasticizer tends to reduce the tensile strength and increase the percentage elongation of plastics because plasticizers can reduce intermolecular forces and increase the mobility of biopolymer chains.

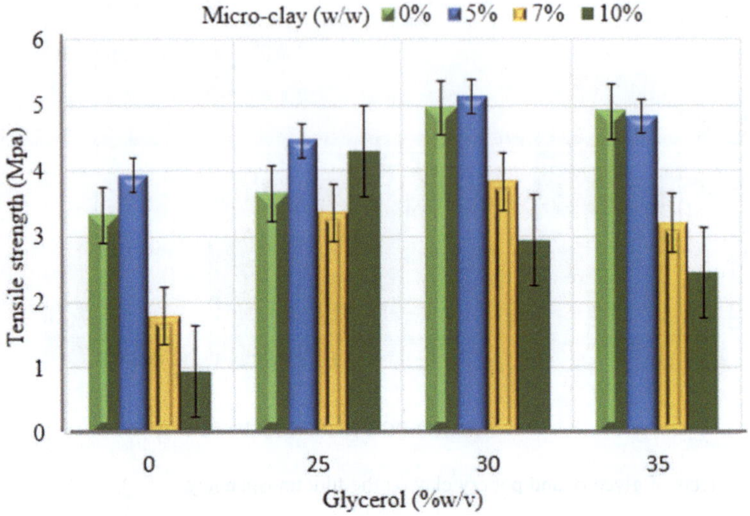

Fig. 1.6 Effect of glycerol and pottery clay on film's tensile strength

Effects of Glycerol and Micro-Pottery Clay on the Film's Biodegradability
According to [9] report, in the soil burial method, mass loss of the biocomposite in normal soil is higher than those found in sterile soil; therefore, the yeasts and molds and total aerobic mesophiles in this active soil probably promote and favor film biodegradation process. The increase in glycerol concentration (0–35%) resulted in an increase in moisture content as pointed out earlier which can in turn influence film biodegradation in various ways due to the essential requirement of water for growth and multiplication of microbes. Hence, film degradation speed is enhanced in the presence of sufficient moisture due to rapid microbial action. In addition, moisture-rich conditions support the process of hydrolysis by generating more chain scission reactions. The starch-based packaging film is favorable for the microbial attack, and hydrolytic enzymes act on the film matrix to reduce their weight [38]. Mass loss was determined by weighing the mass of the film before and after the biodegradation process for 2 weeks. The greater the mass loss means the faster the degradation. As the micro-clay content increases from 0% to 10%, the degradation rate of the film was decreased from 74.38 ± 0.012, 59.93 ± 0.141, 58.37 ± 0.105, 46.62 ± 0.031, 44.76 ± 0.221, 9.91 ± 0.109, etc. as it can be observed in Appendix 4.3. This is due to the micro-clay which resulted in the film's strong compactness within and among polymer chains. However, glycerol has an antagonistic effect on degradation rate as compared with the addition of the pottery clay. The mass loss was increased with increasing concentrations of glycerol because high glycerol concentration caused the film to be hydrophilic with an active hydroxyl-OH group which can initiate the hydrolysis reaction (which leads growth of microorganism), so that the films easily degraded into small pieces [39]. Based on the obtained result in Fig. 1.7, degradability of the films was increased from 9.91 ± 0.109, 15.13 ± 0.122, 16.36 ± 0.208, 58.37 ± 0.105, 59.93 ± 0.141, to 74.38 ± 0.012 as the glycerol concentration

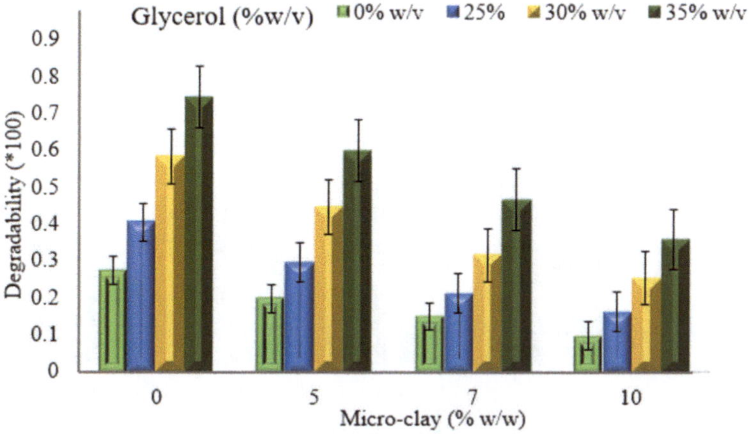

Fig. 1.7 Effects of glycerol and pottery clay addition against on degradability

increased from 0–35% w/v. The maximum weight loss of the film was obtained at a higher percentage of glycerol concentration (35%) which found to be 74.38 ± 0.012.

1.3.2.2 Effects of Glycerol Addition on the Films Colorance

Color attributes are of prime importance because they directly influence consumer's acceptability. The *L*, *a*, *b*, and ΔE* values of tested samples are shown in Appendix 4.3. Since glycerol is a colorless component, this effect of plasticizer was probably related to a dilution effect due to its increasing concentration in the film-forming solution, without any probable association with the plasticizing effect of glycerol (Fig. 1.8).

Color characteristics of the films is one of the physical properties that is very important because it will affect the appearance of the products. Characteristics result of the film color were determined using Eqs. (3.12) and (3.13) as it can be seen in Appendix 4.3. Measurement of color on this research (color detection tool) was based on the measuring value with the hunter system L (0 = black, 100 = white), a (green = −60, + 60 = red), and b (−60 = blue, + 60 = yellow). Brightness level (L) on the addition of 35% w/v glycerol concentrations showed higher brightness than glycerol concentrations of 25% and 30% w/v are likely to produce a low brightness level. This mainly indicated that as the plasticizer content increases, the brightness of the films also increases [40]. As it can be observed in Fig. 1.9, films color brightness were increased with increasing glycerol concentration. Films which made from 35% w/v glycerol concentration and 0% w/w micro-pottery clay had the highest brightness value (89.67%). The first four runs of the films were prepared without glycerol. Films made with no glycerol have had the lowest brightness values as it presented in Appendix 4.3 (31.17%).

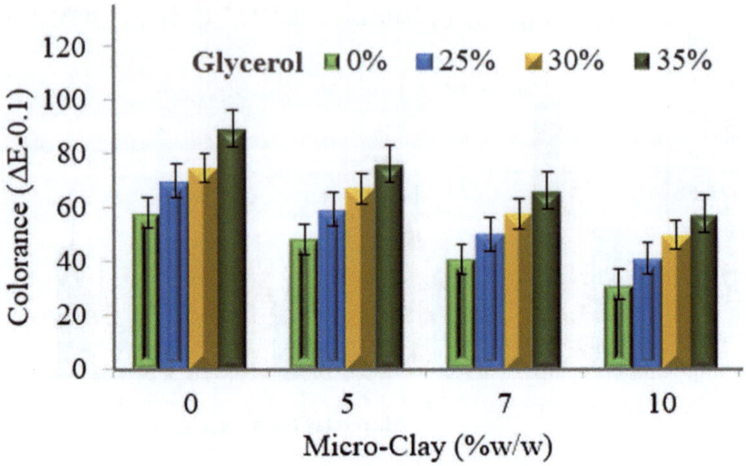

Fig. 1.8 Effects of glycerol concentration on the films colorance

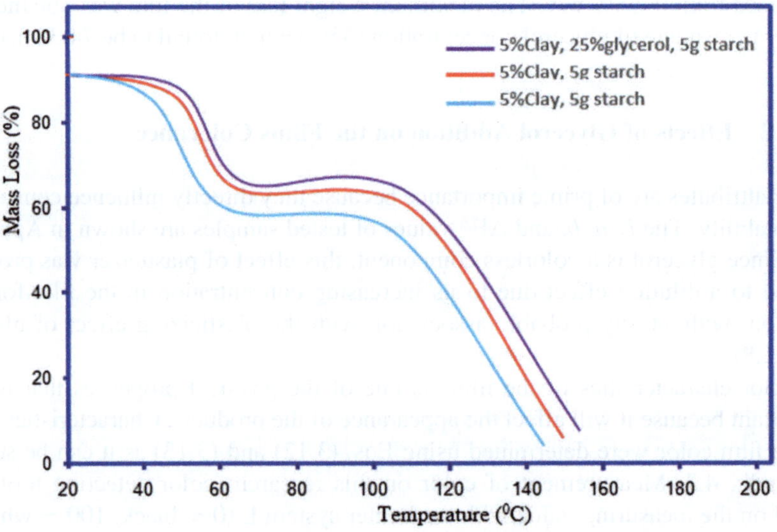

Fig. 1.9 TGA analysis for reinforced clay and glycerol-free micro-composite film

1.3.2.3 TGA Analysis

The biodegradable films were analyzed for thermal degradation and stability using the thermogravimetric analysis technique. The results showed that the initial stage of thermal degradation occurred at a temperature below 90°C. The control films had greater mass loss than the optimized reinforced film at this temperature, indicating that the control film had a higher moisture content. The addition of clay decreased water absorption, resulting in improved thermal stability of the reinforced film beyond a temperature of 135°C. The clay layers acted as barriers to maximize warmth insulation and limit the permeability of volatile degradation products, thus preventing the degradation of starch fragments and enhancing the thermal stability of the biodegradable films. Therefore, the application of pottery clay on starch molecules offers suitable barrier traits [14] due to the fact the water absorption of the starch-based films decreases with the addition of clay (pottery clay immediately influences the crystallinity of starch to expose low water absorption) [15]. Actually, the opposite is true. The addition of micro-pottery clay improved the thermal stability of the starch-based films. The clay acted as a barrier, limiting the permeability of volatile degradation products and preventing the degradation of starch fragments. This was found in a study that investigated the impact of glycerol and micro-pottery clay on the moisture uptake and thickness of starch-based films. While increasing glycerol concentration increased moisture affinity, increasing micro-pottery clay content decreased moisture uptake and thickness. The ideal film formulation balances both concentrations to achieve desired properties [14]. The addition of micro-pottery clay to the starch-based films creates strong polar interactions between the clay, glycerol, and starch matrix. These interactions act as a barrier, limiting the permeability of volatile degradation products and preventing the degradation of starch fragments. As a result, the films exhibit improved thermal stability beyond a temperature of 135°C. Furthermore, the presence of micro-pottery clay decreases the moisture uptake and thickness of the films, while increasing glycerol concentration increases the moisture affinity. The ideal film formulation balances both concentrations to achieve desired properties [16] as evidenced from the results of FTIR, SEM, and 3D Optical Surface Profiler analysis. Also, the improved thermal stability of the starch-pottery clay biodegradable film is attributed to the decrease permeability of oxygen, and the decrease diffusibility of the degradation products, from the starch bulk/matrix because of the exfoliated clay inside the films [17] and clay layers act as barriers to maximize the warmth insulation and to limit the permeability of volatile degradation products via the material [18]. This results in improved thermal stability beyond a temperature of 135°C. Additionally, the clay layers act as barriers to maximize warmth insulation and limit the permeability of volatile degradation products via the material. As a result, the application of pottery clay on starch molecules offers suitable barrier traits to enhance the thermal stability of the biodegradable films and prevent degradation of starch fragments. It's important to balance both glycerol and micro-pottery clay concentrations in order to achieve the desired properties in the film formulation.

1.3.2.4 FTRIR Analysis

The analysis of FTIR is a useful tool for identifying molecular interactions in the constraints of starch-based films. As shown in Fig. 1.10, both adhesive samples with and without filler displayed typical characteristics. The best interactions between filler, plasticizer, and starch were observed at absorption bands around 3100 cm-1

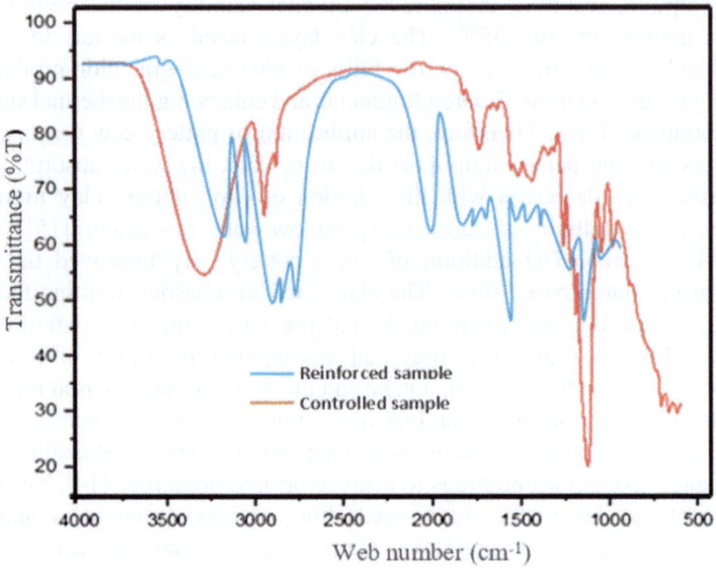

Fig. 1.10 FITR analysis for controlled and reinforced microfilm

and 1250–1100 cm-1. Higher wavenumbers (shorter wavelengths) are associated with higher energy and frequency. The spectra indicated a broad absorption band around 3274 cm-1 for O-H stretching vibrations and a smaller absorption band around 1150 cm-1 for C-H stretching vibrations. These findings demonstrate the potential for FTIR analysis to improve the understanding and development of biodegradable films [41]. The spectra indicated that a broad absorption band around 3274 cm^{-1} for O-H stretching vibrations and a smaller absorption band around 1150 cm^{-1} attributed to C-H stretching vibration. Water molecule is corresponding to a shifting band at a wave number of 1700 cm^{-1} [41]. The addition of micro-pottery clay to starch-based films not only improves the thermal stability of the material but also enhances its adhesive properties. The strong peak observed for -OH groups in the starch molecular chains indicates a moderate amount of moisture uptake, which is important in making good reinforced adhesive. Higher amounts of -OH functional groups provide more accessibility of chains in starch to be hydrolyzed, resulting in an increase in adhesive viscosity. By adding clay materials, the bonding and thickening ability of the adhesive is strengthened, providing a more durable and effective adhesive for various applications [42].

1.4 Conclusion

The addition of micro-pottery clay to starch-based films provides several benefits. The strong polar interactions between the clay, glycerol, and starch matrix act as a barrier to limit the permeability of volatile degradation products, preventing the degradation of starch fragments. The clay also improves the thermal stability of the material and enhances its adhesive properties. FTIR analysis is useful in identifying molecular interactions in starch-based films, with the best interactions between filler, plasticizer, and starch observed at absorption bands around 3100 cm-1 and 1250–1100 cm-1. The -OH functional groups in the starch molecular chains are essential for adhesive properties, and adding clay materials strengthens the bonding and thickening ability of the adhesive. Altering the plasticizer content allows for an increase or reduction of selected physical, chemical, and functional properties of the prepared films.

References

1. Harahap, M. B. (2018). synthesis and characterization of bioplastic from mango seed starch reinforced with microparticle clay using glycerol as plasticizer Utilization of mango seed starch in manufacture of bioplastic reinforced with microparticle clay using glycerol as plastic. 0–7. https://doi.org/10.1088/1757-899X/309/1/012068.
2. Lusiana, S. W., Putri, D., & Nurazizah, I. Z. (2019). Bioplastic properties of sago-PVA starch with glycerol and sorbitol plasticizers. *Journal of Physics Conference Series, 1351*, 012102. https://doi.org/10.1088/1742-6596/1351/1/012102
3. Gadissa, M. (2017). *Production of bioplastic from corn starch and sugarcane bagasse*. Adia Ababa University.
4. Elgindy, A. A. E. (2017). Chemical and technological studies of mango seed kernel. *European Journal of Food Science and Technology, 5*(2), 32–40.
5. Ezeoha, S. L., & Ezenwanne, J. N. (2013). Production of biodegradable plastic packaging film from cassava starch. *IOSR Journal of Engineering, 3*(10), 14–20.
6. Sartika, M., Lubis, M., Harahap, M. B., et al. (2018). Production of bioplastic from avocado seed starch as matrix and microcrystalline cellulose from sugar palm fibers with schweizer's reagent as solvent. *Asian Journal of Chemistry, 30*, 1051–1056. https://doi.org/10.14233/ajchem.2018.21155
7. Maulida, K. T., Harahap, M. B., & Ginting, M. H. S. (2018). Utilization of mango seed starch in manufacture of bioplastic reinforced with microparticle clay using glycerol as plasticizer. In *IOP conference series: Materials science and engineering* (Vol. 309, p. 012068). https://doi.org/10.1088/1757-899X/309/1/012068
8. Alvarez, V. (2018). Surface properties of thermoplastic starch materials reinforced with natural fillers. In *Functional biopolymers* (pp. 131–158).
9. Admase, A. T., Sendekie, Z. B., & Alene, A. N. (2022). Biodegradable film from mango seed kernel starch using pottery clay as filler. *Journal of Polymers and the Environment, 30*, 3431–3446. https://doi.org/10.1007/s10924-022-02449-7
10. Wahyuningtiyas, N. E., & Suryanto, H. (2018). Properties of cassava starch based bioplastic reinforced by Nanoclay. *Environment, 2*, 20–26. https://doi.org/10.17977/um016v2i12018p020
11. Ozsoy, I. (2015). The influence of micro- and nano-filler content on the mechanical properties of epoxy composites. *Strojniski Vestnik/Journal of Mechanical Engineering, 61*, 601–609. https://doi.org/10.5545/sv-jme.2015.2632
12. Fillers and Reinforcements. (2017). Fillers and reinforcements, 4.
13. Chandrappa, D. R. K., & Kamath, M. S. S. (2021). The egg shell as a filler in composite materials – a review. *Journal of Mechanical and Energy Engineering, 4*, 335–340. https://doi.org/10.30464/jmee.2020.4.4.335

14. Sonthalia, M., & Sikdar, D. C. (2015). Production of starch from mango (Mangifera Indica L.) seed kernel and its characterization. *International Journal of Technical Research and Applications, 3*(3), 346–349.
15. NO, A., et al. (2017). Analysis of starch from non-edible root and tubers as sources of raw materials for the synthesis of biodegradable starch plastics. *Journal of Basic and Applied Research International, 3*(1), 27–32.
16. Chakravartula, S. S. N., Soccio, M., Lotti, N., Balestra, F., Dalla Rosa, M., & Siracusa, V. (2019). Characterization of composite edible films based on pectin/alginate/whey protein concentrate. *Materials, 12*(15), 2454.
17. Thakhiew, W., Devahastin, S., & Soponronnarit, S. (2010). Effects of drying methods and plasticizer concentration on some physical and mechanical properties of edible chitosan films. *Journal of Food Engineering, 99*, 216–224. https://doi.org/10.1016/j.jfoodeng.2010.02.025
18. Liu, C., Jiang, S., Zhang, S., et al. (2016). Characterization of edible corn starch nanocomposite films: The effect of self-assembled starch nanoparticles. *Starch-Stärke, 68*, 239–248. https://doi.org/10.1002/star.201500252
19. Vasile, C., Pamfil, D., Râpă, M., et al. (2018). Study of the soil burial degradation of some PLA/CS biocomposites. *Composites. Part B, Engineering, 142*, 251–262. https://doi.org/10.1016/j.compositesb.2018.01.026
20. Yunos, M. Z. B., & Rahman, W. A. W. A. (2011). Effect of glycerol on performance rice straw/starch based polymer. *Journal of Applied Sciences, 11*(13), 2456–2459.
21. Bourtoom, T. (2008). Plasticizer effect on the properties of biodegradable blend film from rice starch-chitosan. Songklanakarin. *Journal of Science & Technology, 30.*
22. Hassan, L. G., Muhammad, A. B., Aliyu, R. U., et al. (2013). Extraction and characterisation of starches from four varieties of Mangifera indica seeds. *IOSR Journal of applied chemistry, 3*, 16–23.
23. Nayak, P., & Rayaguru, K. (2018). Studies on extraction of starch from dried and fresh mango seed kernel. *International Journal of Agriculture Sciences, 10*, 7192–7195.
24. Thomas, D. J., & Atwell, W. A. (1999). Eagan Press Handb Ser 96 (p. 35).
25. Gunorubon, A. J. (2014). Production of cassava starch-based adhesive. *Research Journal of Applied Sciences, 1*(4), 219–214.
26. Mulyati, P. P. (2016). Physical and mechanical properties of agar based edible film with glycerol plasticizer. *International Food Research Journal, 23*, 1669–1675.
27. Nazmi, S. (2017). Effects of plasticizer concentrations on functional properties of chicken skin gelatin films. *International Food Research Journal, 24*, 1910–1918.
28. Jafarzadeh, S., Ariffin, F., Mahmud, S., et al. (2017). Characterization of semolina biopolymer films enriched with zinc oxide nano rods. *Italian Journal of Food Science, 29*, 195–208. https://doi.org/10.14674/1120-1770/ijfs.v516
29. Alcázar-alay, S. C., Angela, M., & Meireles, A. (2015). Physicochemical properties , modifications and applications of starches from different botanical sources. *Food Science and Technology, 35*, 215–236.
30. Alcázar-alay, S. C., Angela, M., & Meireles, A. (2015). Physicochemical properties, modifications and applications of starches from different botanical sources. *Food Science and Technology, 35*, 215–236.
31. Nor, M. H. M., Nazmi, N. N. M., & Sarbon, N. M. (2017). Effects of plasticizer concentrations on functional properties of chicken skin gelatin films. *International Food Research Journal, 24*, 1910–1918.
32. Farahnaky, A., Saberi, B., & Majzoobi, M. (2013). Effect of glycerol on physical and mechanical properties of wheat starch edible films. *Journal of Texture Studies, 44*(3), 176–186. https://doi.org/10.1111/jtxs.12007
33. Zeng, Q. H., Yu, A. B., Lu, G. Q., & Paul, D. R. (2005). Clay-based polymer nanocomposites: Research and commercial development. *Journal of Nanoscience and Nanotechnology, 5*(10), 1574–1592. https://doi.org/10.1166/jnn.2005.411

34. Khairunnisa, S., & Rostini, I. (2018). The effect of glycerol concentration as a plasticizer on edible films made from alginate towards its physical characteristic. *World Scientific News, 112*, 130–141.
35. López, O., Castillo, L., Zaritzky, N., et al. (2015). Talc nanoparticles influence on thermoplastic corn starch film properties. *Procedia Materials Science, 8*, 338–345. https://doi.org/10.1016/j.mspro.2015.04.082
36. Nogueira, G. F., Fakhouri, F. M., & de Oliveira, R. A. (2018). Extraction and characterization of arrowroot (Maranta arundinaceae L.) starch and its application in edible films. *Carbohydrate Polymers, 186*, 64–72. https://doi.org/10.1016/j.carbpol.2018.01.024
37. Lutfi, M., Sumarlan, S. H., Susilo, B., Zenata, R., & Perdana, L. P. R. (2017). The glycerol effect on mechanical behaviour of biodegradable plastic from the walur (Amorphophallus paeoniifolius Var. sylvestris). *Nature Environment and Pollution Technology, 16*(4), 1121–1124.
38. Pathak, V. M. (2017). Review on the current status of polymer degradation: A microbial approach. *Bioresources and Bioprocessing*. https://doi.org/10.1186/s40643-017-0145-9
39. Manshor, N. M., Rezali, M. I., Jai, J., et al. (2018). Synthesis of biodegradable plastic from tapioca with N-Isopropylacrylamide and chitosan using glycerol as plasticizer Synthesis of biodegradable plastic from tapioca with N- Isopropylacrylamide and chitosan using glycerol as plasticizer. In IOP *conference series: Materials science and engineering* (Vol. 345, No. 1, p. 012049). IOP Publishing. https://doi.org/10.1088/1757-899X/345/1/012049.
40. Indrianingsih, A.W., & Rosyida, V. T. (2018). Characteristics of water solubility and color on edible film from bioselulosa nata nira siwalan with the additional of glycerol. In *Journal of physics: Conference series* (Vol. 983, p. 012191). IOP Publishing.
41. Güler, P., & Doğan, İ. (2022). Kahramanmaras Sutcu Imam University. *Journal of Engineering Science, 22*, 135–152.
42. Baini, R., Demong, M., Hui, J. L. C., et al. (2021). Characterization of bio-adhesive derived from Sarawak native sago starch. *IOP Conference Series: Materials Science and Engineering, 1101*, 012041. https://doi.org/10.1088/1757-899x/1101/1/012041

Chapter 2
Assessment of Residual Elastic Properties in Damaged Composite Materials

Destayehu Addisu Durressa, Ermias Gebrekidan Koricho, Addisu Negash Ali, and Getnet Ayele Kebede

2.1 Introduction

2.1.1 Background of the Research

Nowadays, composite materials are the most widely used materials for developing various industry fields. Because of its excellent noise reduction qualities, mechanical strength, stiffness, resistance to corrosion, lightweight nature, and energy absorption capabilities, composite materials are frequently utilized in structural engineering. However, its widespread use is limited by many factors, such as high raw material costs or high production volume requirements. Among other issues, the complexity of the failure development process plays an important role in limiting their applications, particularly for structural component applications. Composite materials typically exhibit many interaction failure modes. and the gradual and rapid reduction of mechanical properties is a common challenge [1]. In this context, methods to assess damage levels and predict residual structural strength in composites are starting to gain more significance. Numerous methods have been developed to evaluate the structural integrity of composite components without causing damage [2]. Nondestructive damage detection was applied to map the remaining elastic properties on the damaged plate. Therefore, the damage detection method allows a local assessment of the composite material's remaining elastic characteristics after damage. A thorough description of the mechanical characteristics of the composite sheet can be

D. A. Durressa (✉) · A. N. Ali · G. A. Kebede
Faculty of Mechanical and Industrial Engineering, Bahir Dar Institute of Technology, Bahir Dar University, Bahir Dar, Ethiopia
e-mail: destayehu2123@asu.edu.et

E. G. Koricho
Addis Ababa Science and Technology University, Addis Ababa, Ethiopia

K. Mequanint et al. (eds.), *Sustainable Development Research in Materials and Energy*, Green Energy and Technology,
https://doi.org/10.1007/978-3-031-69860-6_2

achieved by mapping the component's elastic properties and taking the values determined by the finite element model into account.

Given the time and cost of physical testing, finite element (FE) engineering offers the possibility of developing numerical models capable of accurately describing effects and predicting internal failure mechanisms complex in a relatively short time. The complex behavior of composite materials can be simulated based on developed models provided by several analytical studies [2–7]. It is crucial to follow the experimental circumstances as precisely as possible when building and meshing the model. In certain instances, it becomes imperative to establish specific assumptions in order to streamline the model and expedite the analysis. To maximize accuracy and minimize computing time, a suitable trade-off must be made between variables including sludge size, form, and complexity of the model. Therefore, the FE method is a practical and effective tool for engineering research when used with a solid knowledge of finite element theory [3]. The finite element method (FEM) is a numerical technique for analyzing any physical object. Comprehensive understanding and quantification of physical phenomena such as structure and fluid behavior, heat transport, and wave propagation require the use of mathematics. Most of these processes are described using partial differential equations (PDEs), for example, a dynamic problem:

$$[\mathrm{M}]\{\ddot{X}\} + [\mathrm{C}]\{\mathrm{X}\} = \{F^{\mathrm{Ext}}\} \tag{2.1}$$

To replicate events that occur naturally or intentionally, explicit FEM is utilized. This numerical method serves as the basis for simulation software so that engineers including aerospace engineers, civil engineers, and as a whole mechanical engineer can assess their designs for failure and damage in various loading conditions, before prototyping or implementation stages. Prior to doing actual research, it is crucial to evaluate the mechanical properties of laminated composite materials using a theoretical framework. Fiber-reinforced composites are anisotropic (orthotropic) and microscopically nonhomogeneous. Consequently, compared to traditional materials, the mechanics of fiber-reinforced composites [8] are more intricate. A software application is designed to estimate the mechanical properties of laminated composites with unidirectional reinforcement using the classical theory of laminate-based algorithms.

In this paper, the objectives are to use finite element method (FEM) modeling approaches for different loading conditions, to develop a stepwise damage model of laminated structures by projectile impact using ABAQUS, and to develop the FEM procedures of damage modeling of fiber-reinforced polymer composite (FRPC) laminates in ABAQUS. To forecast the kind and degree of damage over the laminate thickness, a user-defined 3D damage model with solid components was created and integrated into the finite element program Abaqus/Explicit. The Abaqus general contact algorithm was used to simulate contact between the impactor and the surfaces of the composite panel, as well as contact between layers, by defining suitable contact pair properties. The numerical results from the simulation were evaluated and demonstrated based on the proposed damage modeling approach to predict the type and extent of impact-induced damages.

2.2 Materials and Methods

In this section, we first describe in detail the materials considered. Then numerical methods are presented. The composite material consists of reinforced by six layers of glass fabric and a matrix made of epoxy resin. The GFRP laminate is treated as symmetric in the numerical analysis, with six layers of 0.5 mm thickness that are orientated in accordance with the stacking sequence to yield a total thickness of 3 mm. The model contains a rigid spherical impactor of diminution of 8 mm diameter and a target of 250 mm × 25 mm × 3 mm deformable laminated plate, and both are modeled using Abaqus. The laminate is stacked in [(0/90)3] S, [(30/−60)3] S and [(45/−45)3] S sequence. It is observed from the result that GFRP with [(0°/90°)3] S fiber orientation yields high residual elastic when compare to other degrees of orientations for the same load, size, and shape in addition. This spherical-shaped projectile is fired at the center of the GFRP-laminated composite plate. The model has formed from six woven layers [(0/90)3] s fiber direction as ASTM standard. In the current investigation, the element is employed based on Hashin's damage criteria. Hashin's damage criterion confirms that the continuum shell element type was used to mesh every element in the current model. A mesh refinement procedure was performed so that the size of the elements did not affect the results. A mesh of the composite sample model is shown in Fig. 2.4. In the present work, Hashin's failure criteria are used to predict the translaminar progressive fracture and get the failure load of the present composite specimen according to ASTM standard. Hashin's failure criteria is commonly used to predict anisotropic damage in elastic-brittle materials. The effect of damage extent was studied by changing the impact energy value on the fracture behavior of GFRP composite material. The model contains a rectangular thin laminated plate and is impacted by a rigid tool dropped with mass 40, 50, 90, and 150 J and a diameter of 9 mm.

2.2.1 *Elastic Properties of Orthotropic Lamina*

Anisotropic and orthotropic materials require 21 and 9, respectively, in independent elastic constraints to be characterized. The nine independent elastic constants for an orthotropic material are E11, E22, E33, G12, G13, G23, υ12, υ13, and υ23. Orthotropic materials fall into a specific class known as cross-ply-oriented fiber composites. The elastic characteristics are equal in the 2–3 direction if the fiber is in the 1–2 plane, which means that E22 = E33, υ12 = υ13, and G12 = G13. As a result, for a cross-ply-oriented fiber [9] composite, there are only five independent elastic constants left: E11, E22, G12, υ12, and υ23. It is common to refer to these composites as transversely isotropic. The elasticity characteristics of a cross-ply continuous fiber lamina can be computed using the subsequent formulas.

Longitudinal modulus referring to figures

$$E_{11} = E_{\mathrm{f}} V_{\mathrm{f}} + E_{\mathrm{m}} V_{\mathrm{m}} \tag{2.2}$$

And major Poisson's ratio

$$\upsilon_{12} = V_{\mathrm{f}} \upsilon_{\mathrm{f}} + V_{\mathrm{m}} \upsilon_{\mathrm{m}} \tag{2.3}$$

The transverse modulus is

$$E_{22} = (E_{\mathrm{f}}/E_{\mathrm{m}})/(E_{\mathrm{f}} V_{\mathrm{m}} + E_{\mathrm{m}} V_{\mathrm{f}}) \tag{2.4}$$

And minor Poisson's ratio

$$\upsilon_{21} = E_{22}/E_{11} \upsilon_{12} \tag{2.5}$$

Shear modulus referring to Figs. 2.1 and 2.2

$$G_{12} = (G_{\mathrm{f}} V_{\mathrm{m}} + G_{\mathrm{m}} V_{\mathrm{f}}) \tag{2.6}$$

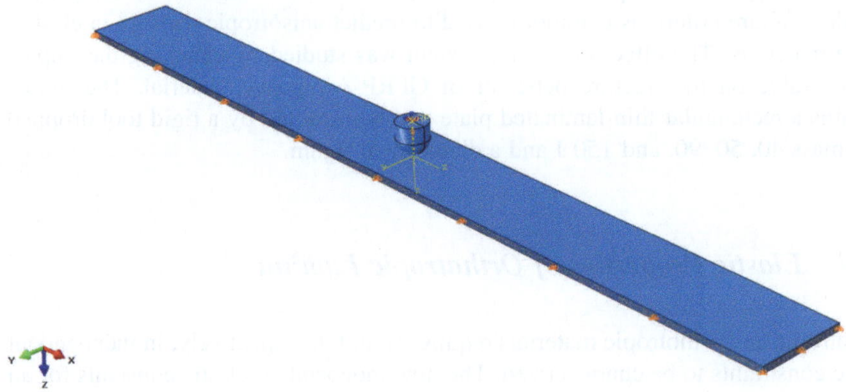

Fig. 2.1 Boundary condition

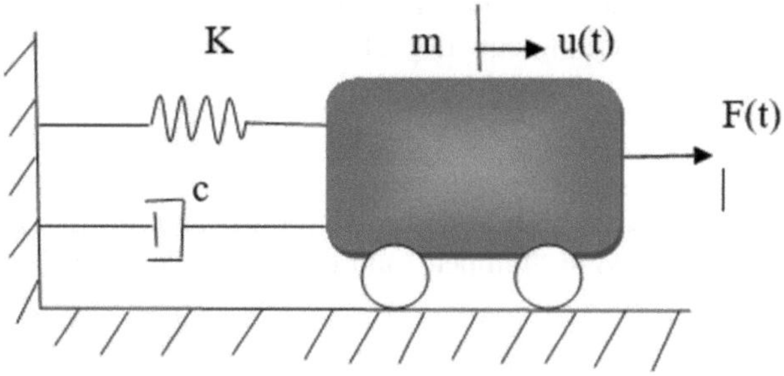

Fig. 2.2 Schematic of dynamic system

Table 2.1 Elastic and damage properties of cross-ply GFRP composite laminas

Elasticity properties								
E_{11}	E_{22}	E_{33}	ν_{12}	ν_{13}	ν_{23}	G_{12}	G_{13}	G_{23}
32.9 GPa	32.9 GPa	6.7 GPa	0.29	0.49	0.49	2.46 GPa	2.25 GPa	2.25 GPa

Table 2.2 Hashin model constants

Strength properties					
X^t	X^c	Y^t	Y^c	S_L	S_T
991.9 MPa	502.7 MPa	548.3 MPa	53 MPa	65 MPa	26.5 MPa

2.2.2 *Elastic Properties of Orthotropic Lamina*

The cantilever plate size was 250 mm × 25 mm × 3 mm. The boundary condition of the clamping area is U1 = U2 = U3 = UR1 = UR2 = UR3 = 0. In this simulation, the first three output of natural frequency and mode shape were investigated. The fixed plate size was 250 mm × 25 mm × 3 mm. The boundary condition of the clamping area is U1 = U2 = U3 = UR1 = UR2 = UR3 = 0. At this point, the maximum frequency of interest or the number of requested eigenvalues must be defined when there is a fixed-free boundary requirement.

2.2.3 *Numerical Results*

The initial data for this investigation was gathered from the manufacturer's handbook regarding epoxy and E-glass fiber (Tables 2.1 and 2.2).

2.2.3.1 Explicit Dynamic Analysis Scheme

To explain the dynamic explicit analyses scheme in the Abaqus FEM tool. How to model this kind means how to model the energy loss called damping.

Schematic of the Dynamic System
The governing equation of the system is

$$m\ddot{u} + c\dot{u} + \mathrm{ku} = \{F^{\mathrm{Ext}}\} \tag{2.7}$$

Explicit analyses aim to solve for acceleration $\{\ddot{u}\}$. In most cases, the mass matrix is considered as "lumped" and thus a diagonal matrix.

$$\ddot{u}(\mathrm{t}) = M^{-1}\left[\mathrm{F}(t)^{\mathrm{ext}} - \mathrm{F}(t)^{\mathrm{int}}\right]$$

The explicit method should be used when the strain rates/velocity is higher.

$$U,\ \dot{u},\ \ddot{u}$$

$$\dot{u}_{\mathrm{i}-1/2} = \frac{u_i - u_{i-1}}{\Delta t} \tag{2.8}$$

$$\dot{u}_{\mathrm{i}+1/2} = \frac{u_{i+1} - u_i}{\Delta t} \tag{2.9}$$

The equilibrium equation at *i* is

$$\dot{u}_{\mathrm{i}} = \frac{u_{i+1} - u_{i-1}}{2\Delta t} \tag{2.10}$$

$$\ddot{u}_{\mathrm{i}} = \frac{\dot{u}_{i+1/2} + u_{i-1/2}}{\Delta t} \tag{2.11}$$

$$\ddot{u}_{\mathrm{i}} = \frac{\frac{u_{i+1} - u_i}{\Delta t} + \frac{u_i - u_{i-1}}{\Delta t}}{\Delta t} \tag{2.12}$$

$$\therefore [\mathrm{M}]\left\{\frac{u_{i+1} - 2u_i + u_{i-1}}{(\Delta t)^2}\right\} + [\mathrm{C}]\left\{\frac{u_{i+1} - u_{i-1}}{\Delta t}\right\} + [\mathrm{K}]\{u_i\} = \{F^{\mathrm{Ext}}\} \tag{2.13}$$

Unknowns u_i, u_{i-1}, u_{i+1}?

$$\mathrm{U}_{i+1}\left(\frac{[M\}}{\Delta t^2} + \frac{[C]}{2\Delta t}\right) + \mathrm{u}_i\left(\frac{2[M]}{\Delta t^2} + [\mathrm{K}]\right) + \mathrm{u}_{i-1}\left(\frac{[M\}}{\Delta t^2} + \frac{[C]}{2\Delta t}\right) = \{F^{\mathrm{Ext}}\}$$

$$P_i = U_{i+1} K_i \tag{2.14}$$

$$\therefore Ui + 1Ki = F^{\mathrm{Ext}} – Fi \tag{2.15}$$

The equilibrium equation at *i* is

$$\mathrm{M}\ddot{u}_{\mathrm{i}} + \mathrm{C}\dot{u}_{\mathrm{i}} + \mathrm{K}u_i = Fi \tag{2.16}$$

Since

$$K_i = \frac{M}{\Delta t^2} + \frac{C}{2\Delta t} \tag{2.17}$$

$$\dot{u}_i = \frac{u_{i+1} - u_{i-1}}{2\Delta t} \;\; ; \;\; \ddot{u}_i = \frac{u_{i+1} - 2u_i + u_{i-1}}{(\Delta t)^2} \tag{2.18}$$

$$\dot{u}_0 = \frac{u_1 - u_{-1}}{2\Delta t}; \;\; \ddot{u}_o = \frac{u_1 - 2u_0 + u_{-1}}{(\Delta t)^2} \tag{2.19}$$

$$u_{-1} = u_0 - u_0 \Delta t + \frac{1}{2} u_0 \Delta t^2 \tag{2.20}$$

$$\ddot{u}_i = \frac{p_0 - C\dot{u}_0 - ku_0}{m} \tag{2.21}$$

2.3 Result Discussion

Evaluate the residual elastic behavior of glass-epoxy laminate composite specimens by varying impact energy. To determine damage characteristics and to evaluate the effect of impact energy on GFRP composite laminate by Abaqus software that evaluate the damage modulus of an orientation. From impact energy response envelope curves obtained damage parameters by using the logarithmic decrement method. This is described by the alignment of fibers in the lamina which contributes to the stiffness of the specimens (Figs. 2.3, 2.4, and 2.5).

Figure 2.8 shows that the absolute maximum force increases with increasing impact energy. This is because the time required to completely perforate the GFRP laminate (i.e., contact force drops to 0) decreases as the impact energy increases. As

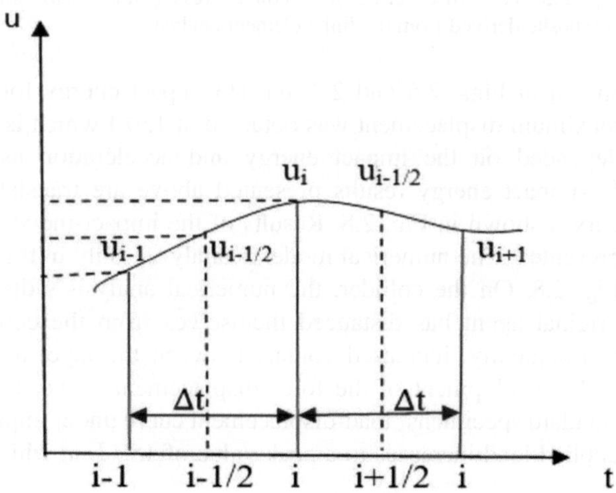

Fig. 2.3 Displacement time graph

Fig. 2.4 Meshing of element

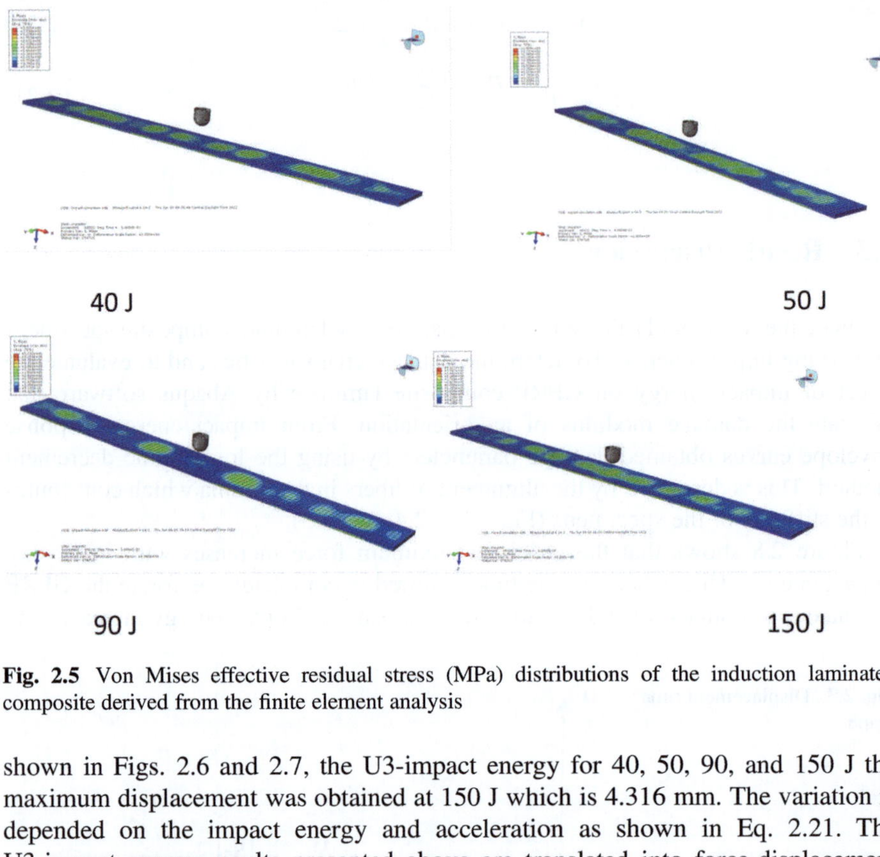

Fig. 2.5 Von Mises effective residual stress (MPa) distributions of the induction laminated composite derived from the finite element analysis

shown in Figs. 2.6 and 2.7, the U3-impact energy for 40, 50, 90, and 150 J the maximum displacement was obtained at 150 J which is 4.316 mm. The variation is depended on the impact energy and acceleration as shown in Eq. 2.21. The U3-impact energy results presented above are translated into force-displacement curves shown in Fig. 2.8. Results of the impact-induced damage analysis are here presented. The numerical model is analyzed only in the elastic domain as shown in Fig. 2.8. On the collider, the numerical analysis's displacement is assessed. The original agent has distanced themselves from the composite sheet, nevertheless. Consequently, increased contact between the layer and the impactor defined the early development of the force-displacement curve. Figure 2.8 shows the ASTM standard specimens' load-displacement curve under impact load. It is evident that the applied load increases to a peak value of 150 J, at which point the matrix is totally

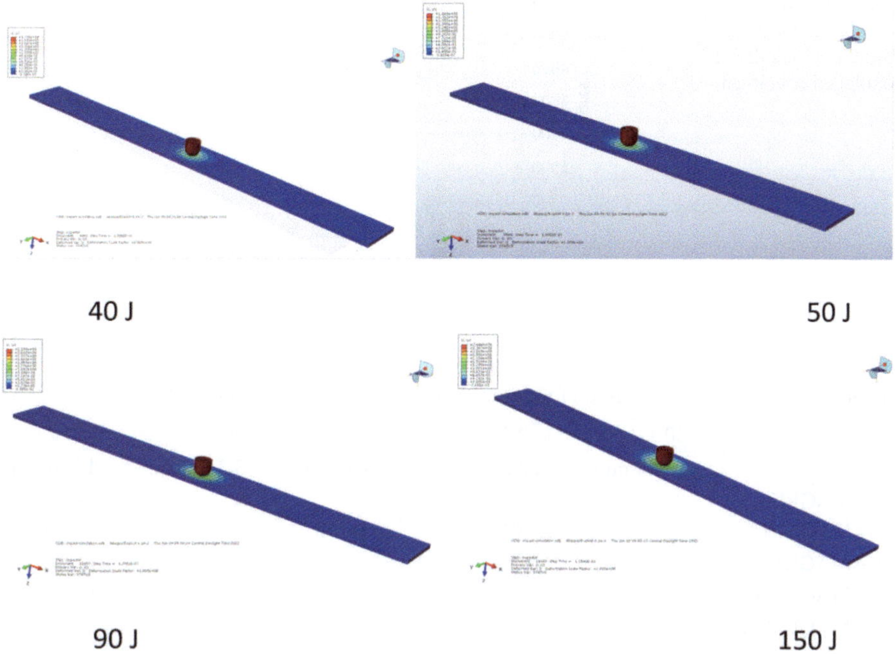

Fig. 2.6 Deflection in Z-direction (mm) distributions of the induction laminated composite derived from the finite element analysis

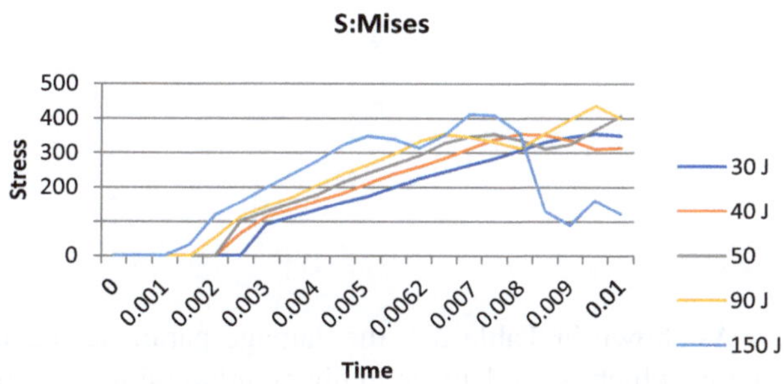

Fig. 2.7 Von Mises effective residual stress (MPa) distributions of the induction laminated composite derived from the finite element analysis

destroyed, and then abruptly decreases. The 150 J specimen's applied load peak value is higher than other impact energy levels, as would be expected. By adjusting the impact energy value, the influence of damage extent on the fracture behavior of GFRP composite materials was investigated.

As shown in Table 2.3, the variation of impact energy is affected by the mode of failure. The pop-in events can be expressed as the damage growth in load, and then the fibers sustain the load again before the matrix is completely damaged.

The main conclusions from this study can be summarized as follows (Fig. 2.9):

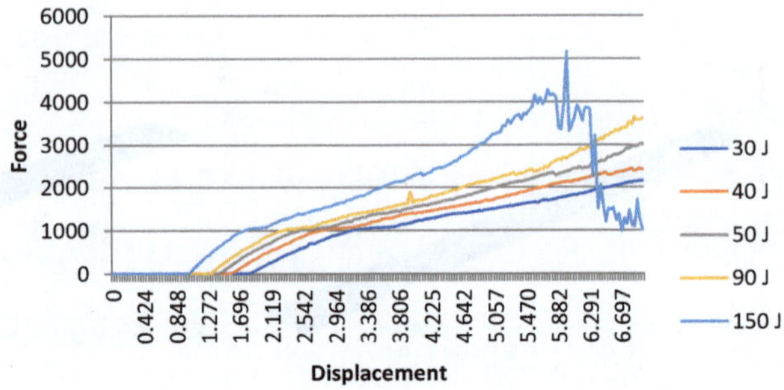

Fig. 2.8 Force-displacement graph of numerical results of damaged specimen

Table 2.3 Damage characteristic of the composite samples summarized as follows

Parament	30 J	40 J	50 J	90 J	150 J
U3	1.934 mm	2.232 mm	2.496 mm	3.344 mm	4.316 mm
DAMAGEFC	0	0	0	0	0.02254
DAMAGEFT	0	0.01904	0.00127	0.3155	0.1719
DAMAGEMC	0.9252	0.9986	1	1	1
DAMAGEMT	0.9852	1	1	1	1
DAMAGESHR	0.9852	1	1	1	1

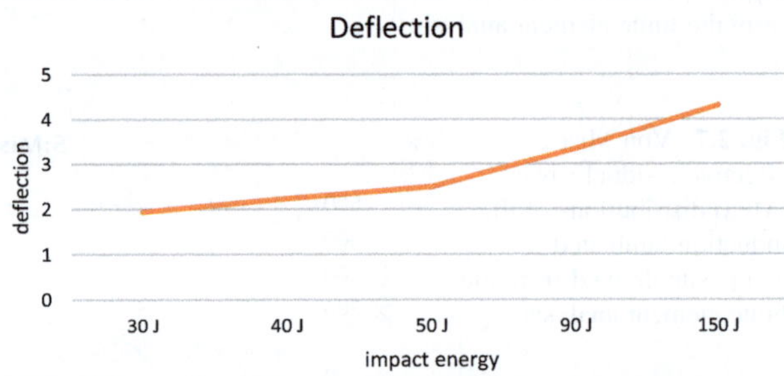

Fig. 2.9 Deflection versus impact energy

As shown in Table 2.3, the damage parameter result obtained different impact energy which is for damage in fiber compression. For all impact energies, the values are 0 for DAMAGEFC. For the damage in matrix tension (DAMAGEMT) and damage in shear (DAMAGESHR), damage parameters affect the elastic properties of the damaged composites independent of load and displacement values for impactor Z-direction and [0/90]3 s. As reported in Fig. 2.1, the elastic properties have been estimated at the center locations of the composite plate in four different impact energy. Table 2.4 reports the impact energy values and the corresponding Young's moduli.

The dynamic method involved calculating Young's modulus based on the univocal relationship between it and the dimensions, weight, and natural frequencies of the beam for a given support configuration. The natural frequency of the system

Table 2.4 Results of Young's modulus

	Young's modulus E [GPa]		
Specimens	Mode 1	Mode 2	Mode 3
30 J	15.3101	15.2663	15.2085
40 J	14.9011	14.8591	14.8039
50 J	14.6639	14.6225	14.5664
90 J	14.5416	14.5006	14.4470
150 J	14.2366	14.1963	14.1416

increases as the stiffness of the system increases. It is also shown in Table 2.4. In correspondence with the external noses, whereas the impact energy is 30, 40, 50, 90, and 150 J and no damage should be present, the elastic modulus is different that is small which is significantly different from the original value of 32.879 GPa, 32.529 GPa, and 31.39 GPa which is for 0 J impact energy of three modes, respectively. As the result shows their application constraints are largely due to the intricacy of the damage evolution, particularly for applications involving structural components. The value is less than the original Young's modulus.

2.4 Conclusions

In this paper the impact response and impact-induced damage for a simple cross-ply [0/90] 3 s composite laminate under four different impact energy levels was numerically investigated. Until complete failure that occurred, an impact-induced damage analysis was performed on a six-layer fabric composite laminate. The method was implemented on the commercially available Abaqus/Explicit.

To map the residual elastic properties on the damaged GFRP composite plate, the numerical analysis was adopted. Four impact loads have been considered along the middle point of the GFRP composite laminate. In the finite element model, a continuous curve was then considered to account for the variable elastic modulus.

The results showed that the elastic properties of damaged composites basically depend on different damage parameters such as fiber compression (FC), fiber tension (FT), matrix compression (MC), matrix tension (MT), and shear (SHR). The applied impact loads give different force-displacement relations leading to different damage parameters and elastic modulus. However, the MT and SHR damage parameters affect the elastic properties of the damaged composites independent of load and displacement values. The variation of the elastic modulus in the finite element model has been taken into consideration by using a continuous force-displacement curve.

Accordingly, the numerical method allows a local assessment of the remaining elastic properties of the composite materials destroyed by GFRP. An accurate representation of the mechanical behavior of the composite sheet can be obtained in a finite element model by mapping the elastic properties of the component and examining the evaluated values. The findings indicate that their applicability is severely limited, particularly in the case of structural component applications due to the intricacy of damage evolution. The value is lower than the original Young's modulus.

References

1. Wesolowski, M., Barkanov, E., Rucevskis, S., Chate, A., & La Delfa, G. (2009, July). Characterisation of elastic properties of laminated composites by non-destructive techniques. In *Proceedings of the 17th international conference on composite materials* (p. F5: 12), Edinburgh, United Kingdom.
2. Davies, G., & Olsson, R. J. T. A. J. (2004). Impact on composite structures. *The Aeronautical Journal, 108*(1089), 541–563.
3. Matthews, F. L., Davies, G., Hitchings, D., & Soutis, C. (2000). *Finite element modelling of composite materials and structures*. Elsevier.
4. Donadon, M., Iannucci, L., Falzon, B. G., Hodgkinson, J., & de Almeida, S. F. J. C. (2008). A progressive failure model for composite laminates subjected to low velocity impact damage. *Computers & Structures, 86*(11–12), 1232–1252.
5. Faggiani, A., & Falzon, B. G. (2010). Predicting low-velocity impact damage on a stiffened composite panel. *Composites Part A: Applied Science and Manufacturing, 41*(6), 737–749.
6. Iannucci, L., & Ankersen, J. J. C. S. (2006). An energy based damage model for thin laminated composites. *Composites Science and Technology, 66*(7–8), 934–951.
7. Yokoyama, N., Donadon, M., & De Almeida, S. J. C. S. (2010). A numerical study on the impact resistance of composite shells using an energy based failure model. *Composite Structures, 93*(1), 142–152.
8. Gieske, J., & Allred, R. J. E. M. (1974). Elastic constants of B-A1 composites by ultrasonic velocity measurements. *Experimental Mechanics, 14*(4), 158–165.
9. Rokhlin, S., & Wang, W. (1989). Ultrasonic evaluation of in-plane and out-of-plane elastic properties of composite materials. In *Review of Progress in quantitative nondestructive evaluation* (pp. 1489–1496). Springer.
10. Vishnuvardhan, J., Krishnamurthy, C. V., & Balasubramaniam, K. (2008). Determination of material symmetries from ultrasonic velocity measurements: A genetic algorithm based blind inversion method. *Composites Science and Technology, 68*(3–4), 862–871.
11. Rahimian Koloor, S. S., Karimzadeh, A., Yidris, N., Petrů, M., Ayatollahi, M. R., & Tamin, M. N. J. P. (2020). An energy-based concept for yielding of multidirectional FRP composite structures using a mesoscale lamina damage model. *Polymers, 12*(1), 157.
12. Steenhuis, H.-J., & de Bruijn, E. J. (2006). Building theories from case study research: The progressive case study. In *OM in the new world uncertainties. Proceedings (CD-ROM) of the 17th annual conference of POMS* (pp. 546–558)*, 28 April-1 May 2006, Boston, USA*. Production and Operations Management Society (POMS).
13. Matthews, F. (1999). Damage in fibre-reinforced plastics; its nature, consequences and detection. In *Key engineering materials* (Vol. 167, pp. 1–16). Trans Tech Publications.
14. Montalvao, D., Maia, N. M. M., & Ribeiro, A. M. R. J. S. (2006). A review of vibration-based structural health monitoring with special emphasis on composite materials. *Shock and Vibration Digest, 38*(4), 295–324.
15. Lifshitz, J. M., & Rotem, A. (1969). Determination of reinforcement unbonding of composites by a vibration technique. *Journal of Composite Materials, 3*(3), 412–423.
16. Fällström, K. E., & Jonsson, M. J. P. C. (1991). A nondestructive method to determine material properties in anisotropic plates. *Polymer Composites, 12*(5), 293–305.
17. Hu, H., & Wang, J. J. E. S. (2009). Damage detection of a woven fabric composite laminate using a modal strain energy method. *Engineering Structures, 31*(5), 1042–1055.

18. Perez, M. A., Gil, L., & Oller, S. J. C. S. (2014). Impact damage identification in composite laminates using vibration testing. *Composite Structures, 108*, 267–276.
19. Paolino, D. S., Geng, H., Scattina, A., Tridello, A., Cavatorta, M. P., & Belingardi, G. J. C. P. B. E. (2017). Damaged composite laminates: Assessment of residual young's modulus through the impulse excitation technique. *Composites Part B: Engineering, 128*, 76–82.
20. Boursier Niutta, C. J. A. S. (2020). Enhancement of a new methodology based on the impulse excitation technique for the nondestructive determination of local material properties in composite laminates. *Applied Sciences, 11*(1), 101.
21. Abrate, S. (1998). *Impact on composite structures*. Cambridge University Press.
22. Belingardi, G., Cavatorta, M. P., & Paolino, D. S. (2009). On the rate of growth and extent of the steady damage accumulation phase in repeated impact tests. *Composites Science and Technology, 69*(11–12), 1693–1698.

Chapter 3
Analysis on Thermal Behavior of Bamboo Leaf Ash-Blended Cement Paste

Gashaw Abebaw Adanu, Bahiru Bewket Mitikie, Shumet Getahun Reda, and Kassahun Admassu Abegaz

3.1 Introduction

Ordinarily in terms of volume, Portland cement is one of the most significant binding materials. It uses a lot of energy since it is made at extremely high temperatures. Along with massive levels of energy usage, it generates toxic gasses that contaminate the atmosphere [1]. Besides energy consumption and the generation of hazardous gasses, calcium hydroxide, a byproduct of cement hydration, is a nuisance for the construction industry. This has an impact on the durability of Portland cement pastes, mortars, and concretes [1]. Natural pozzolans are vitreous cementitious materials with little or no cementing value on their own, but when finely crushed in the presence of moisture, they will chemically react with calcium hydroxide at room temperature to generate hydrated phases with cementing capabilities [2]. It is generally known that hydrated phases generated during pozzolanic reaction typically increase the performance of concrete.

G. A. Adanu (✉)
Department of Civil Engineering, Debre Tabor University, Debre Tabor, Ethiopia
e-mail: gashabe@dtu.edu.et

B. B. Mitikie
Department of Civil Engineering, Adama Science & Technology University, Adama, Ethiopia

S. G. Reda
Faculty of Civil and Water resource, Bahir Dar Institute of Technology, Bahir Dar University, Bahir Dar, Ethiopia

K. A. Abegaz
Adjunct at the Department of Construction Engineering and Management, Institute of Technology, University of Gondar, Gondar, Ethiopia
e-mail: kassahun.admassu@uog.edu.et

K. Mequanint et al. (eds.), *Sustainable Development Research in Materials and Energy*, Green Energy and Technology,
https://doi.org/10.1007/978-3-031-69860-6_3

Concrete has a better thermal behavior in fire than steel because it has a lower thermal conductivity. Despite the fact that concrete has a better thermal behavior in the event of a fire, the loss of human life and material is immeasurable [3]. According to various studies on the thermal characteristics of concrete, no changes are seen at temperatures below 80°C. At 100°C, free and physically bonded water evaporates, producing drying shrinkage and the onset of hydrothermal reactions. At 105°C, hydrothermal reactions begin. At temperatures over 120°C, the degradation of calcium silicate hydrate (CSH) is a continuous process; as porosity increases, microcracks form, and compressive strength decreases. At temperatures between 400 and 600°C, portlandite ($Ca(OH)_2$) dissociates, and fissures appear on the top of the hydrated cement paste, whereas temperatures over 700°C result in decarbonization of calcite ($CaCO_3$), i.e., CaO + CO_2 [4]. The escape of chemically bonded water, on the other hand, occurs about 800°C, at which stage all concrete crumbles [4].

Thermodynamic properties of concrete are a fascinating and difficult subject to study. To determine the extent of structural member deterioration and to determine how long and how serviceable they are, several innovative techniques are constantly being developed. Commonly utilized methods include X-ray diffraction (XRD), thermal analysis (TA), scanning electron microscopy (SEM), and spectroscopy analysis, which includes Raman spectroscopy (RS) and infrared spectroscopy (IR) [5]. The reduced amount of material required for research is one of the key advantages of spectroscopic analysis. OPC specimens were exposed to temperatures ranging from 27 to 1000°C and evaluated using XRD and TG/DTA. The level of damage to the specimens' surface and core changed for the portlandite phase, but not for the calcite phase [5]. At high temperatures, hydrated paste undergoes irreversible breakdown, resulting in changes in mineralogical composition. At temperatures above 400°C, hydrated samples lose weight and their capillary pores expand. PC with changing specific surface area (SSA) and compressive strength was analyzed by TA and XRD after 1–6 years of hydration. Up to 450°C, XRD examination revealed a continual decline in portlandite peak. DSC and XRD were used to examine limestone and fly ash (FA) specimens that had been heated to temperatures ranging from 250°C to 850°C [6]. At 250°C, there was an increase in compressive strength because the deposition of CSH caused non-hydrous grains to produce more hydration. A significant absorbance peak of portlandite, belite, and alite phases appeared in the 500 and 900 cm ranges. The stretching of bonds between chemical oxides results from a shift toward the lower wavenumber due to the lowering of polymerization [5].

According to literature sources, FTIR is employed as a quantification tool to predict the quantities of different phases contained in a cement sample. For determining the changes occurring in the different phases contained in the hydrated PC at various temperatures, FTIR and thermal analyses (TGA) were extensively employed. However, no literature was discovered that linked the physical damage seen in hydrated bamboo leaf ash mixed cement paste with OPC-43 specimens to

TG/DTA and FTIR measurements. A pozzolan material has a content of siliceous or siliceous and aluminous materials; it holds little or no property of cement. In the presence of water/moisture, cement responds chemically with calcium hydrate at room temperature to create compounds having cementing properties. Calcium hydroxide is released into the environment during the hydration of cement, troubling the construction industry [7]. This affects the durability of Portland cement concrete, paste, and mortar. But when certain pozzolanic substances containing amorphous silica are added during the insertion of the Portland cement, it responds with lime providing an additional amount of calcium silicate hydrate (C-S-H), the main cementing constituent [8]. Hence the property of pozzolanic material condenses total $Ca(OH)_2$ content and raises the amount of the C-S-H bond. Thus, if high-quality pozzolanic material at reasonable amounts is added during the hydration of Portland cement, cement bonding quality increases.

Investigations are ongoing globally on the suitability of using locally available bamboo leaf ash (BLA) for various purposes in construction. Ethiopia is blessed with deposits of bamboo, although with varied purity. Ethiopia contains approximately 1.5 million hectares of bamboo forest, accounting for 67% of African bamboo forest cover and 7% of global bamboo forest cover, according to data [9]. However, the potentials of these bamboo have not been widely studied due to the lack of scientific knowledge about their properties. Recently, researchers have been experimenting with Ethiopian bamboo leaf ash as cement replacing materials up to 10% of the weight of cement [10]. In light to the above, this study examines the thermal properties of BLA-blended concrete. An experimental investigation will be carried out to examine the impact of adding BLA to thermal properties of blended concrete through FTIR and TGA analysis. Due to a lack of laboratory equipment, thermal analysis is restricted to FTIR and TGA analysis. These experiments are available at the Bahir Dar Institute of Technology's Chemical Engineering Laboratory in Ethiopia.

To evaluate the thermal behavior of lowland Ethiopian bamboo leaf ash deposits in West Amhara, Ethiopia, the paper employed several characterization techniques. Due to a dearth of knowledge about their properties, which are rarely discussed in literature, these Ethiopian bamboo leaf ashes have been underused. The characteristics of this ash are still being studied scientifically, which will increase its usefulness. Thermogravimetric analysis in conjunction with Fourier transform infrared (FTIR) spectroscopy and differential thermal analysis (TGA/DTA) were used to the bamboo leaf ash states to investigate their chemical composition, thermal reactions, mineralogical phases, and structural functional groups. Examining Ethiopian lowland bamboo leaf ash-mixed concrete's thermal characteristics is the aim of this study. Using TGA and FTIR analyses, thermal characteristics were examined. This work looks into how BLA concentration affects the performance of thermal characteristics in concrete using TGA and FTIR. First and foremost, FTIR and TGA will be used to examine the thermal characteristics of BLA in concrete.

3.2 Methods

3.2.1 Materials

Cement In this study, ordinary Portland cement (OPC) from a Dangote Cement Factory was available in markets and tested according to ASTM standards. The physical properties of the cement are depicted in Table 3.1.

Bamboo Leaf Ash For this study, lowland Ethiopian bamboo leaf ash (BLA) that met ASTM C618-2015 standards was used. The obtained BLA is class N according to ASTM C618-2015 specifications [10]. The chemical and physical properties of Portland cement and BLA ash used in this study are shown in Table 3.2.

Fine Aggregate Local river sand was used as fine aggregate in this study. The following tests and methods are carried out, as shown in Table 3.3.

Table 3.1 Cement properties [10]

Properties	Result	Test standard
Soundness (mm)	0.85	ASTM C-151
Initial setting time (min)	121	ASTM C-191
Final setting time (min)	235	ASTM C-191
Consistency (%)	25	ASTM C-187

Table 3.2 Chemical compositions of bamboo leaf ash [10]

Oxide	BLA composition
SiO_2	65.66
Al_2O_3	6.41
Fe_2O_3	4.28
CaO	15.22
MgO	2.48
Na_2O	2.76
K_2O	4.84
MnO	0.24
TiO_2	0.33
H_2O	1.03
LoI	9.65

Table 3.3 Physical properties of fine aggregate

Properties	Result	Test standard
Fineness modulus	2.56	ASTM-C136
Specific gravity	2.75	ASTM-C128
Absorption	2.1%	ASTM-C128
Moisture content	1.03%	ASTM-C566

Table 3.4 Mix proportion summary

Mix designation	BLA (%)	Cement	Water	Fine aggregate	W/C ratio
OPC C	0	365	196.74	778	0.49
BLA5	5	346.5	196.74	778	0.49
BLA10	10	328.5	196.74	778	0.49
BLA15	15	310.25	196.74	778	0.49
BLA20	20	292	196.74	778	0.49

3.2.2 Mix Proportion

In this study, five mix proportions were designed: a control mix (0 percent BLA), a 5% BLA, a 10% BLA, a 15% BLA, and a 20% BLA. Table 3.4 contains information on the mix proportions.

3.2.3 Specimen Preparation

Cubes of 150 × 150 × 150 mm^3 are cast for the 0% BLA, 5% BLA, 10% BLA,15% BLA, and 20% BLA. Specimens were cast and cured in an open tank at the 28th and 56th days, with three cube specimens for each duration and mix type. This equates to a total of 24 cubes being formed. Collecting a concrete sample from broken past specimens and immersing it in ethanol with a 90% alcohol content for 3 days, followed by drying the sample in an oven at 150°C for 24 h and grinding it to make powder for thermal experiments using TGA and FTIR. For thermal loading, a group of three specimens is placed in the muffle furnace. Once the specimens have attained the correct temperature, they were returned to the furnace to cool to room temperature. The specimens taken out from the furnace at the different temperatures are evaluated under a compressive testing equipment to determine their compressive strength.

After the samples' compressive strengths at the appropriate temperatures have been determined, the test specimen cores are gathered for TGA and FTIR analysis. A planetary ball mill is used to crush the gathered samples to a uniform fineness. Sample-to-weight ratio of 1:10 is achieved by dispersing 30 g of sample in wiles using toyline. The process takes 30 min to complete, with a 5-min rest period in between each 5-min round of the planetary ball mill set to run at 300 rpm. In order to create a dry powder for additional research, the sample is then gathered and dried at room temperature. The pulverized samples that were taken out of the mill at various temperatures are examined using FTIR and thermal analyses. An FTIR specimen is created by combining 250 mg of potassium bromide (KBr) with 3 mg of cement powder, and then compressing the mixture into a pellet.

TGA operates in accordance with national and international standards such as ASTM E 914, ASTM E 1131, and ASTM E 1868, since the TG analysis was performed to assess the relative quantity of CH in the cement paste, percent mass loss, and thermal stability of BLA-blended cement paste.

3.2.4 *Tests*

Compressive Strength

The compressive strength test was carried out in accordance to ASTM standard using 150-mm concrete cubic specimens on the 28th day using a compressive strength testing machine, which has a maximum capacity of 2000 KN at the in Debre Tabor University, Civil Engineering Laboratory. Table 3.4 shows the mix proportions in detail. The samples' mixing design and curing conditions remained consistent. Dry ingredients (aggregate, cement, and pozzolan) were mixed for at least 10 min in the initial step of mixing concrete components. After that, water was added to the batch, and it was mixed for 4 min. Finally, the mold was filled with concrete. After 24 h of casting, the samples were demolded and cured in a water tank for 28 days. To minimize the threat of concrete spalling during the heating process, the samples were relocated to the laboratory and cured for again for 28 days, at which point they were tested at high temperatures [11]. Ambient temperature (28°C), 100°C, 200°C, 300°C, 350°C, 400°C, 450°C, 500°C, 650°C, and 800°C were the temperatures studied. Three and two replicas, respectively, were put through mechanical and durability tests. In this inquiry, the average of the test outcomes is shown. The mechanical specification tests were performed using a 2000 KN testing machine, with the heat source being an electric furnace with a 1200°C heat capacity. The concrete samples were kept in the furnace as long as necessary to achieve a steady-state thermal condition. The samples were heated at a rate of 1.66–3.61°C/min until the desired temperature was reached to avoid spalling. The compressive strength of the specimens would checked immediately after they were removed from the oven [11, 12].

Thermal Analysis

The use of HTC BEJEING HENVEN (as shown in Fig. 3.1) was used to perform simultaneous thermal behaviors of bamboo leaf ash (BLA)-blended cement paste

Fig. 3.1 HTC thermogravimetric analysis equipment

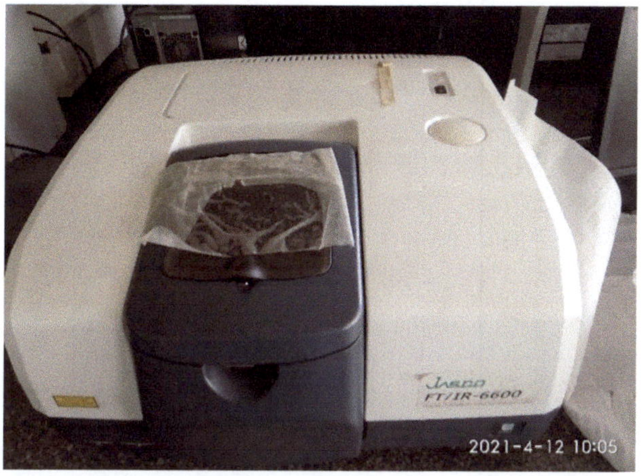

Fig. 3.2 Infrared spectroscopy equipment, FTIR- 6600- Jasco brand

samples with 0% BLA, 5% BLA, 10% , 15% BLA, and 20% BLA replacement. The samples were heated at a rate of 10°C/min in an argon environment with a starting temperature of 27°C and temperatures rising up to 1000°C to determine their reaction (i.e., the percentage of mass loss with respect to the original weight). The researcher prioritizes TGA analysis since the data obtained from it is important for determining the drying and ignition temperatures, compound stability temperatures, and material purity and composition.

Fourier Transform Infrared Analysis Structural functional groups of the raw and thermally treated bamboo leaf ash (BLA)-blended concrete were identified by FTIR spectrometer (FTIR- 6600-Jasco shown in Fig. 3.2) set to scan frequencies across a range of 4000–400 cm^{-1} with a resolution of 4 cm^{-1}. Sample preparation was done by KBr pellet technique, which entails placing a solid sample combined with a transparent alkali halide (KBr) in a mold that is clamped to produce a clear pill that can be inserted into the analytical apparatus. The researcher prioritizes FTIR analysis because it helps users understand materials and related matters. FTIR can perform quantitative and qualitative analyses on organic and inorganic materials. Fourier transform infrared spectroscopy (FTIR) creates an infrared absorption spectrum to discover chemical connections in a molecule.

3.3 Results and Discussion

3.3.1 Compressive Strength

Using 150-mm concrete cubic specimens, the compressive strength test was performed in accordance with ASTM C39 The variation in compressive strength as temperature rises is shown in Fig. 3.3. Control mix (0% BLA) has a compressive strength of 38.56MPa at room temperature. At room temperature, the compressive

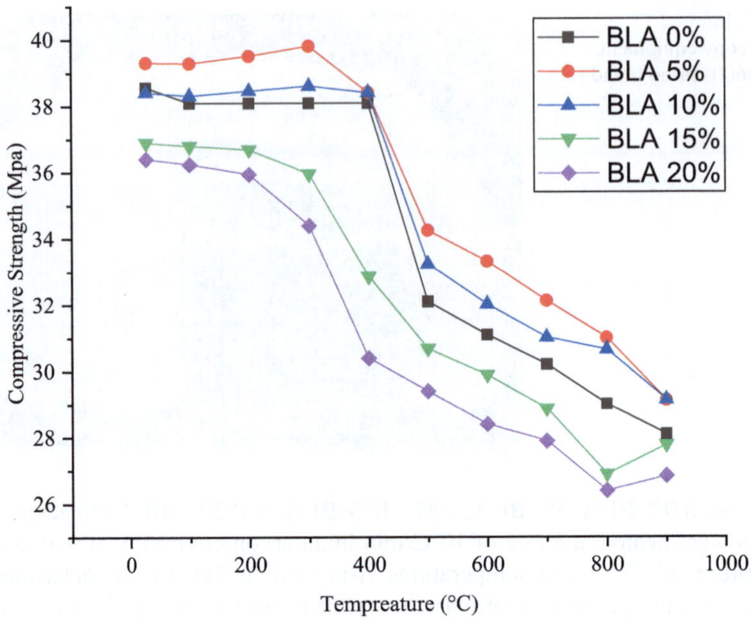

Fig. 3.3 The variation of compressive strength of concrete with temperature

strength was highest, and as the temperature rose, it decreased. Normal and pozzolanic concrete's compressive strength varies similarly as temperature rises. The compressive strength of the material substantially reduced when the temperature rose to 100°C. The `reduction of resistance occurs at this temperature because of vapor pressure within the concrete pores. Concrete regained its strength as the temperature climbed higher [11, 12].

Bamboo leaf ash (BLA) replacement of 15% and 20% of cement weight lowered compressive strength at normal temperature by 4.33% and 5.60%, respectively. Compressive strength for 15% and 20% is 36.89 and 36.40Mpa, respectively. In contrast, the compressive strength increases by 1.09% after 5% replacement when compared to the control mix. The highest compressive strength was at room temperature and decreased as the temperature goes up. The compressive strength fluctuations of regular and pozzolanic concrete as temperature increases varies. The compressive strength for a 5% replacement varies with temperature. The compressive strength first dropped significantly as the temperature increased to 400°C. The reduction of resistance in this temperature range is caused by vapor pressure within the concrete pores.

Initially for the temperature range 100°C, the compressive strength had increased at 5% and decreased for 10% BLA, 15% BLA, and 20% BLA replacement as compared to control mix. The compressive strength at 100°C for control mix (0% BLA replacement), 5% BLA, 10% BLA, 15% BLA, and 20% BLA replacement are 38.1, 39.29, 38.31, 36.79, and, 36.24Mpa, respectively. As per the researcher's findings, 5% BLA and 10% BLA replacement had good thermal resistance as compared to control mix. Similarly, the compressive strength at 400°C had some loss in strength by 0.389% as compared to the control mix for 5% and 10% replacement. The compressive strength in this temperature range for 5% BLA and 10% BLA are 38.41Mpa, while the compressive strength for 15% BLA and 20% BLA replacement decreased by 14.76% and 21.16%, respectively. Earlier studies showed that the increase in resistance at these temperatures attributed to the stiffening of the cement gel due to cement layers contraction and an increase of surface forces between cement gel layers resulting from the removal of water from concrete [11–13].

3.3.2 *Thermogravimetric Analysis (TGA)*

In a controlled setting, thermogravimetric analysis (TGA) determines the absolute quantity and rate of change in a sample's weight as a function of time or temperature. TGA may be used to evaluate volatile content, moisture content, oxidative stability, thermal stability, and atmospheric impacts under various conditions. Figures 3.4 and 3.5 depict the graphs derived from the TGA analysis of the BLA-blended cement paste specimens. To obtain qualitative and quantitative details on the effects of various heat treatments on materials of all kinds, for example, the hydration of

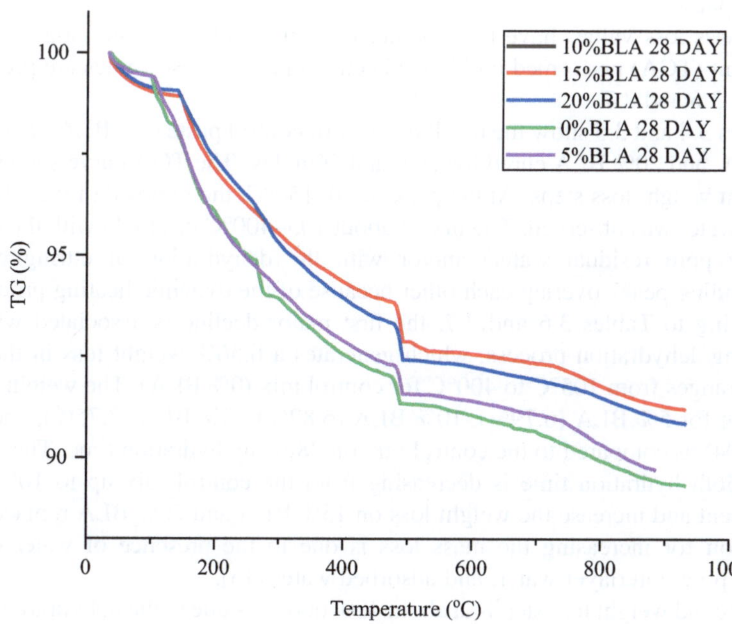

Fig. 3.4 TGA curves of BLA-blended cement paste on 28th day hydration time

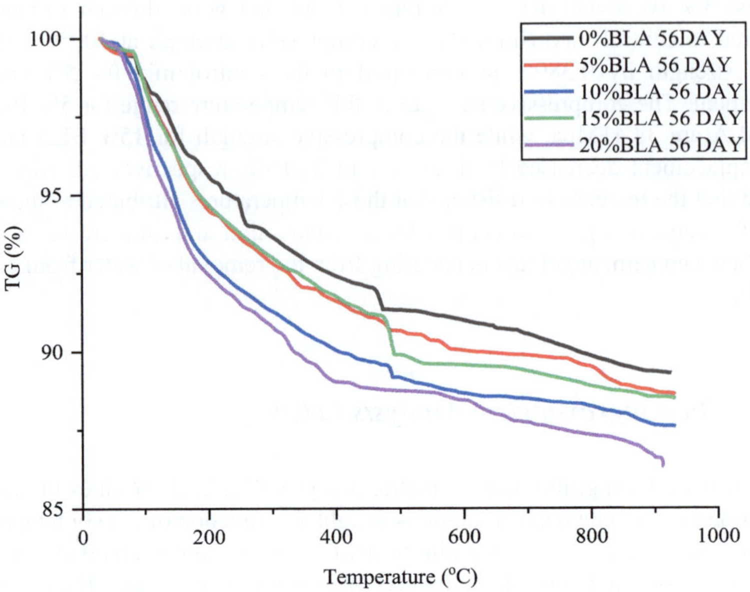

Fig. 3.5 TGA curves of BLA-blended cement paste on 56th day hydration time

concrete one can uses thermal analysis methods such as differential scanning calorimetry and thermogravimetry (TG). Table 3.5 gives a temperature range for different phases.

The mass loss values have been obtained from the classical thermogravimetric experiment (TGA) performed on blended cement paste. These values are presented in Tables 3.6 and 3.7.

Figures 3.4 and 3.5 show the result of TGA of control paste (0% BLA), 5% BLA, 10% BLA, and 20% BLA cured for 28th and 56th day. The TGA curve shows three significant weight loss steps. At the peak of 50–150°C, the removal of the adsorbed and free water was observed. The first at about 105–400°C has to do with the drying (capillary pore residual water) and/or with the dehydration of ettringite. The corresponding peaks overlap each other because of the dynamic heating process.

Referring to Tables 3.6 and 3.7, the first major decline is associated with the dewatering/dehydration process, which generates a 6.56% weight loss in the temperature ranges from 105°C to 400°C for control mix (0% BLA). The weight loss is increasing for 5% BLA (6.79%), 10% BLA (6.82%) 15% BLA (7.75%), and 20% BLA (8.04) as compared to the control mix on 28th day hydration time. The weight loss on 56th hydration time is decreasing from the control mix up to 10% BLA replacement and increase the weight loss on 15% BLA and 20% BLA replacement. The reason for increasing the mass loss is due to the presence of water on the capillary pore, interlayer water, and adsorbed water [15].

The second weight loss step is at about 400–600°C is due to the dehydration of Ca $(OH)_2$. It can be seen from Tables 3.6 and 3.7 that there is a decrease in % weight

Table 3.5 Temperature ranges of different phases [14]

Region	Temperature range (°C)
Dehydration of water	105–400
Dihydroxylation of CH	400–600
Decarbonation of $CaCO_3$	600–1000

Table 3.6 Mass loss values during TGA tests on 28th day hydration time

	Mass loos at each range of temperature on 28th day hydration time			
Mix code	Dehydration 105–400°C	Dihydroxylation 400–600°C	Decarbonization 600–1000°C	Total mass loss
0% BLA	7.083	1.074	1.816	9.973
5% BLA	7.002	1.09	1.728	9.820
10% BLA	6.98	1.08	1.84	9.900
15% BLA	7.57	1.43	2.02	11.020
20% BLA	7.85	1.25	2.35	11.450

Table 3.7 Mass loss values during TGA tests on 56th day hydration time

	Mass loos at each range of temperature on 56th day hydration time			
Mix code	Dehydration 105–400°C	Dihydroxylation 400–600°C	Decarbonization 600–1000°C	Total mass loss
0% BLA	6.56	1.066	1.757	9.383
5% BLA	6.79	1.143	1.37	9.303
10% BLA	6.82	1.36	1.15	9.330
15% BLA	7.75	1.75	1.084	10.584
20% BLA	8.04	1.33	2.07	11.440

loss with an increase in bamboo leaf ash content. Referring to Figs. 3.4 and 3.5, the temperature gradient of mass loss (TGA) exhibits a high peak at about 450–500°C, which represents the dehydration of $Ca(OH)_2$. The dehydration of $Ca(OH)_2$ can be determined by the edge of that peak. Based on this the 20% BLA and 15% BLA replacement have dehydrated the CH earlier at a temperature of 475.5°C and 480°C, respectively, while the 0% BLA, 5% BLA, and 10% BLA replacement dehydrate the $Ca(OH)_2$ at 484.5°C, 483°C, and 483.85°C, respectively. The third weight loss step is at about 600–1000°C can be attributed to the decarbonation of $CaCO_3$. The total $CaCO_3$ content was observed from the weight loss recorded in the TGA curve between 600°C and 800°C, whose intensity increases with the addition BLA; the loss of weight caused by the CO_2 effect is due to the decarbonization of $CaCO_3$. The effect of BLA on the decarbonization process can be seen in Tables 3.6 and 3.7 for 28th and 56th day hydration time, respectively.

Referring to the Fig. 3.6 graph, the mass loss (TGA) has a large peak at around 450–500°C, which stands for the dehydration of $Ca(OH)_2$. The dehydration of

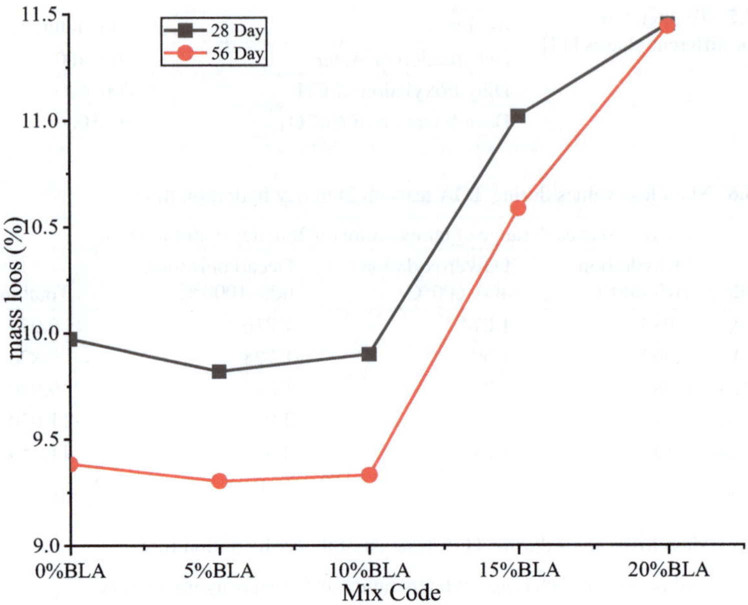

Fig. 3.6 Mass loss of BLA-blended cement

$Ca(OH)_2$ can be determined by the edge of that peak. Based on this the 20% BLA dehydrate the CH earlier on a temperature of 400.01°C, while the 0% BLA, 5% BLA, 10% BLA, and 15% BLA replacement dehydrate the $Ca(OH)_2$ at 470.0°C, 475.5°C, 480°C, and 486°C, respectively. $Ca(OH)_2$ is created and cement begins to harden. With respect to this, 20% BLA and 15% BLA replacement is dehydrating or decomposing the $Ca(OH)_2$ at the latter temperature, so those mixes are not thermal stable as compared to others [16].

Figure 3.6 illustrates that the % weight loss of BLA-blended cement paste is decreases as the curing age increase, while the weight loss decreases for 5% BLA and 10% BLA replacement for both 28th and 56th curing ages. The % weight loss increase is above 10% BLA replacement. The TGA analysis on BLA-blended concrete has indicated that the 5% BLA and 10% BLA replacement were more thermal stable than the control mix (0% BLA), while the 15% BLA and 20% BLA replacement were not thermal stable as compared to the control mix. Based on this analysis, partial replacement of cement improves the thermal stability of concrete. The increased reactivity of BLA contributed to better mechanical behavior and durability of BLA-containing cementitious matrices. An earlier study has shown that thermogravimetric analysis (TG) of the BLA/$Ca(OH)_2$ system revealed a significant weight loss between 100°C and 300 °C of temperature, the primary zone of dihydroxylation of the main hydrated phases formed during the pozzolanic reaction such as silicates, aluminates, and aluminum silicate hydrates [17].

3.3.3 Fourier Transform Infrared Analysis

The result of FTIR on cement paste replaced by bamboo leaf ash on 28th and 56th days of hydration time is given in Figs. 3.7 and 3.8, respectively. Table 3.8 depicts the position of infrared bands and functional group in Fourier transform infrared spectroscopy (FTIR) spectra of cement paste. In this research, silicate and aluminate phases have been used to identify infrared vibration bands of a previous study with the bamboo leaf ash-blended cement paste. Table 3.8 highlights the wavenumber and the associated functional group.

Referring to Table 3.8 and Fig. 3.7, the bending vibration of Si–O–Si (Si–O–Al) is evident in the 600–1200 cm^{-1} range. Moreover, compared with the control, after the addition of the bamboo leaf ash (BLA), the absorption peak of Si–O–Si (Si–O–Al) shifts toward a higher wavenumber. The peak stretching vibration of the control mix (0% BLA) on the 28th day hydration time is 1090.54 cm^{-1}, whereas the 5% BLA, 10% BLA, 15% BLA, and 20% BLA are 1105.018 cm^{-1}, 1099.22 cm^{-1}, 1098.262 cm^{-1}, and 1106.94 cm^{-1}, respectively. This indicates that the aluminum

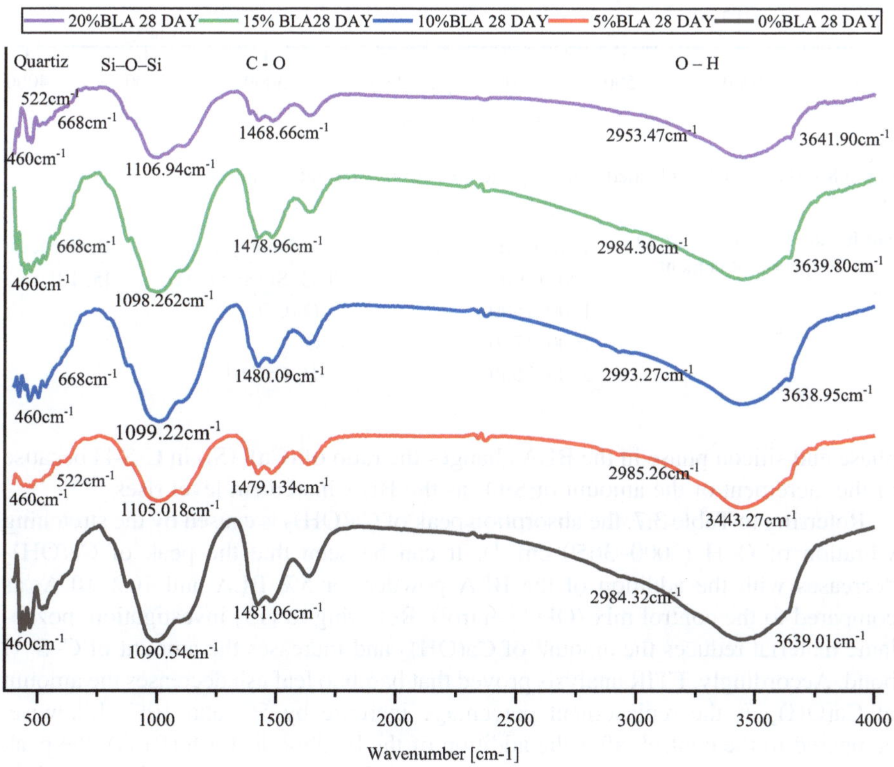

Fig. 3.7 FTIR of BLA-blended cement paste at 28th day hydration time

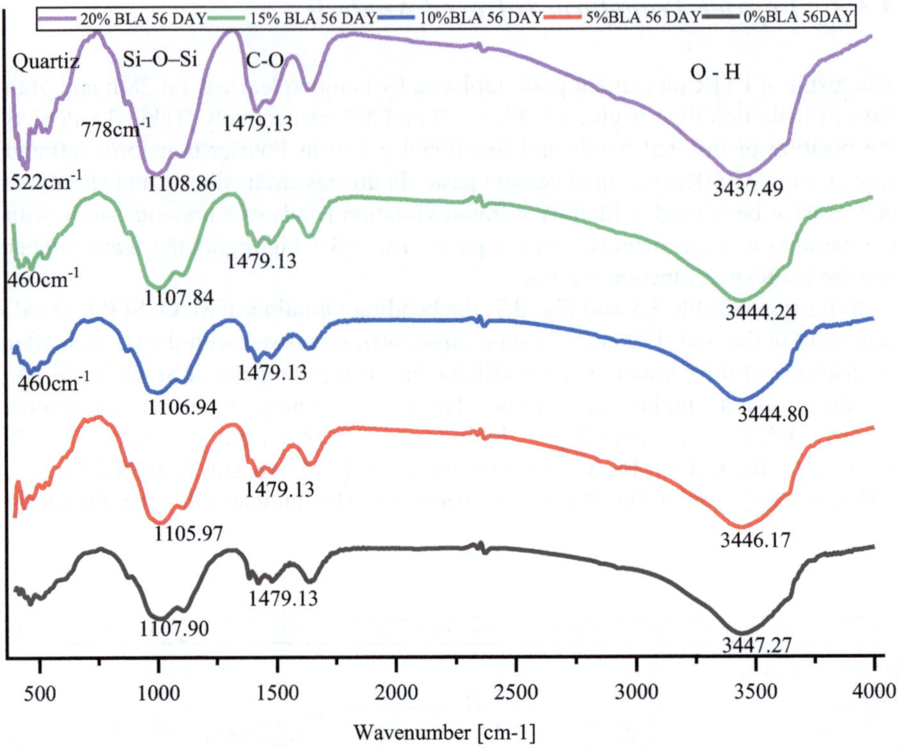

Fig. 3.8 FTIR of BLA-blended cement paste on 56th day hydration time

Table 3.8 Functional group in FTIR spectra of cement paste

Wavenumber (cm^{-1})	Functional groups	References
600–1200	Si–O–Si (Si–O–Al),	[5, 18]
1400–1500	C–O (CO_3^{2-})	
1600–1700	H_2O	
3000–3650	O–H, $Ca(OH)_2$	

phase and silicon phase in the BLA changes the ratio of (Ca)/ (Si) in C-S-H because of the increment of the amount of SiO_2 as the BLA increment level rises.

Referring to Table 3.7, the absorption peak of $Ca(OH)_2$ is caused by the stretching vibration of O–H (3000–3650 cm^{-1}). It can be seen that the peak of $Ca(OH)_2$ decreases with the addition of the BLA powder for 5% BLA and 10% BLA, as compared to the control mix (OPC control). Referring to [19] investigation, pozzolanic material reduces the amount of $Ca(OH)_2$ and increases the amount of C–S–H bond. Accordingly, FTIR analysis proved that bamboo leaf ash decreases the amount of $Ca(OH)_2$ as the replacement percentage increase by 5% and 10%. Likewise, compared to the control, after the addition of the bamboo leaf ash (BLA), the peak of $Ca(OH)_2$ shifts toward a lower wavenumber. The peak stretching vibration of the control mix on the 28th day hydration time is 3639.02 cm^{-1}, while the 5% BLA and

10% BLA are 3443.27 cm^{-1} and 3638.95 cm^{-1}, respectively. But the $Ca(OH)_2$ peak is increasing by 15% and 20% as compared to the control mix. The wavenumber for 15% and 20% BLA is 3641.90 cm^{-1} and 3639.80 cm^{-1} individually.

The FTIR analysis witnessed that; the addition of pozzolanic material reduces the amount of $Ca(OH)_2$ and increases the amount of C–S–H bond [19]. The FTIR analysis on concrete after the 28th day hydration time proved that the 5% BLA and 10% BLA replacement improve the C–S–H gel formation. This is the reason that for the 28th day compressive strength attains the target mean strength for 5% and 10% BLA replacement. The bands in the 1400–1500 cm^{-1} range correspond to the asymmetric stretching of $CO_3{}^{2-}$.

Referring to [20] investigations, undissolved quartz particles are traced at 1145 cm^{-1}, 1084 cm^{-1}, 778 cm^{-1}, 697 cm^{-1}, 668 cm^{-1}, 522 cm^{-1}, and 460 cm^{-1}. Figure 3.7 shows the infrared spectra of BLA-blended cement paste on the 28th day hydration time. Referring to the figure, the concentration of the undissolved peak of the BLA-blended cement paste was higher than that of the control mix (0% OPC), due to the bamboo leaf ash silica oxide concentration which is good for the formation of C_2S; hence ,C_2S reacts slowly. Likewise, for all mixes, the concentration of the undissolved peak increase noticeably with increasing BLA content, due to the higher SiO_2 content of the BLA leading to the slow hydration reaction.

Referring to Table 3.8 and Fig. 3.8, the bending vibration of Si–O–Si (Si–O–Al) is evident in the 600–1200 cm^{-1} range. Moreover, compared with the control, after the addition of the bamboo leaf ash (BLA), the absorption peak of Si–O–Si (Si–O–Al) shifts toward a lower wavenumber up to 15% BLA replacement. The peak stretching vibration of the control mix (0% BLA) on 56th day hydration time is 1107.90 cm^{-1}, whereas the 5% BLA, 10% BLA, 15% BLA, and 20% BLA are1105.97 cm^{-1}, 1106.94 cm^{-1}, 1107.84 cm^{-1}, and 1108.86 cm^{-1}, respectively. This is the reason that more SiO_2 react in latter hydration time as BLA increment level increases and forms more C–S–H gels in latter hydration time.

Referring to Fig. 3.8, the absorption peak of $Ca(OH)_2$ is caused by the stretching vibration of O–H (3000–3650 cm^{-1}). It can be seen that the peak of $Ca(OH)_2$ decrease with the addition of the BLA powder for all mix, compared to the control mix (OPC control). Referring to [19] investigation, pozzolanic material reduces the amount of $Ca(OH)_2$ and increases the amount of C–S–H bond. FTIR analysis proved that bamboo leaf ash decreases the amount of $Ca(OH)_2$. Similarly, compared to control, after the addition of the bamboo leaf ash (BLA), the peak of $Ca(OH)_2$ shifts toward a lower wavenumber.

The FTIR analysis witnessed that; the addition of pozzolanic material reduces the amount of $Ca(OH)_2$ and increase the amount of C-S-H bond [19]. The FTIR analysis on concrete after the 56th day hydration time proved that the BLA replacement improves the C–S–H gel formation at later hydration time. On 28th day hydration time the C–S–H gel formation is limited to 5% BLA and 10% BLA replacement level. The bands in the 1400–1500 cm^{-1} range correspond to the asymmetric stretching of $CO_3{}^{2-}$.

Referring to [20] investigation, undissolved quartz particles are traced at 1145 cm^{-1}, 1084 cm^{-1}, 778 cm^{-1}, 697 cm^{-1}, 668 cm^{-1}, 522 cm^{-1}, and 460 cm^{-1}. Figure 3.8 shows infrared spectra of BLA-blended cement paste on 56th day hydration time. Referring to the figure, there is no concentration of the undissolved peak of the BLA-blended cement paste of the control mix (0% OPC) and 5% BLA. The concentration peak of undissolved solid was similar for 10% BLA and 15% BLA replacement; there is more undissolved peak on 20% BLA replacement.

Researchers investigated the thermal properties of BLA-blended concrete by SEM and XRD microstructure analysis methods [21, 22], while referring to [17] and [23], they try to investigate the FTIR analysis of bamboo leaf ash. This research finding is mainly concerned on FTIR analysis of BLA-blended cement paste rather than bamboo leaf ash, but as compared to this analysis, the peak turning point is similar, but the peak intensity and concertation are different. These differences would be related to the different species of bamboo, harvesting soil, weather condition, and stage of development.

3.4 Conclusions

The purpose of this study was to ascertain whether BLA replacement has superior thermal performance when subjected to heat loads between 27°C and 1000°C. Compared to the 0% BLA replacement, the 5% and 10% BLA replacement specimens showed noticeably more physical changes. To gain a deeper understanding of the phenomena, TGA and FTIR are also being utilized. The research yielded the following conclusions:

- The mass loss taking place in BLA-blended cement paste specimens are almost similar observed from the TGA. The 5% BLA and 10% BLA replacements were found to be more thermostable than the control mix (0% BLA) in the TGA study of BLA-blended concrete; however, the 15% BLA and 20% BLA replacements were not.
- The percentage weight loss of BLA-blended cement paste reduces as the curing age increases, whereas the weight loss for 5% BLA and 10% BLA replacement decreases for both the 28th and 56th curing ages. The percentage weight loss increment is greater than 10% BLA replacement. According to the investigation, partial replacement of cement with BLA improves the concrete's thermal stability.
- The FTIR spectroscopy of concrete after the 56th day of hydration showed that the BLA substitution promotes C–S–H gel formation at later hydration times. On the 28th day of hydration, the C–S–H gel formation is restricted at 5% BLA and 10% BLA replacement levels. According to FTIR analysis, the concentration of $Ca(OH)_2$ reduces throughout the mixture as the BLA replacement increase. This finding indicated that Bamboo Leaf ash has good pozzolanic properties.

Acknowledgment We would like to thank Debre Tabor and Bahir Dar Institute of Technology University Civil Engineering Laboratory Technicians. All laboratory experiments for analytical chemistry, chemical engineering, and construction materials were examined at Bahir Dar Institute of Technology and Debre Tabor University, respectively.

Availability of Data and Materials Some or all data, models, or code generated or used during the study are available from the corresponding author by request.

Competing Interest The authors state that the publishing of this work does not include any conflicts of interest.

Funding No funding was obtained for this study.

Author Contribution Gashaw Abebaw performed all experimental analyses, data collection and interpretation, manuscript writing, editing, and submission. Kassahun Admassu, Bahiru Beweket, and Shumet Getahun comment, evaluate, and advice the research work.

References

1. Ba-Shammakh, M., Caruso, H., Elkamel, A., Croiset, E., & Douglas, P. L. (2008). Analysis and optimization of carbon dioxide emission mitigation options in the cement industry. *American Journal of Environmental Sciences, 4*(5), 482–490. https://doi.org/10.3844/ajessp.2008.482.490
2. Somia, B., & Djamal, Z. (2018). Effect of natural pozzolan on the fresh and hardened cement slurry properties for cementing oil well. *World Journal of Engineering, 15*(4), 513–519. https://doi.org/10.1108/WJE-10-2017-0337
3. Chen, C., Habert, G., Bouzidi, Y., & Jullien, A. (2010). Environmental impact of cement production: Detail of the different processes and cement plant variability evaluation. *Journal of Cleaner Production, 18*(5), 478–485. https://doi.org/10.1016/j.jclepro.2009.12.014
4. Trník, A., Scheinherrová, L., Kulovaná, T., & Černý, R. (2016). Simultaneous differential scanning calorimetry and Thermogravimetric analysis of Portland cement as a function of age. *International Journal of Thermophysics, 37*(1), 1–9. https://doi.org/10.1007/s10765-015-2028-7
5. Fernández-Carrasco, L., Torrens-Martín, D., Morales, L. M., & Martínez-Ramírez, S. (2012). Infrared spectroscopy in the analysis of building and construction materials. In *Infrared spectroscopy: Materials science, engineering and technology*. https://doi.org/10.5772/36186
6. Criado, M., Fernández-Jiménez, A., & Palomo, A. (2007). Alkali activation of fly ash: Effect of the SiO2/Na2O ratio. Part I: FTIR study. *Microporous and Mesoporous Materials, 106*(1–3), 180–191. https://doi.org/10.1016/j.micromeso.2007.02.055
7. Newman, J., & Choo, B. S. (2003). *Advanced concrete technology constituent materials*. Elsevier Ltd.
8. Mccarthy, M. J., & Dyer, T. D. (2019). *Pozzolanas and Pozzolanic materials* (5th ed.). Elsevier Ltd..
9. B. D. Programme, U. N. Environment, U. Nations, and U. N. Industrial. (2019). Perspectives Sustainable Development and Bamboo Value Chains: Ethiopia's Green Growth Opportunities Within the 'Sino-Dutch-East Africa Bamboo Development Programme 2016–19.'
10. Abebaw, G., Bewket, B., & Getahun, S. (2021). Experimental investigation on effect of partial replacement of cement with bamboo leaf ash on concrete property. *Advances in Civil Engineering, 2021*, 9.

11. Abdi Moghadam, M., & Izadifard, R. A. (2019). Experimental investigation on the effect of silica fume and zeolite on mechanical and durability properties of concrete at high temperatures. *SN Applied Sciences, 1*(7), 1–11. https://doi.org/10.1007/s42452-019-0739-2
12. Nari, V., Praneeth, P. H., & Manchana, V. (2020). A comparative study on the thermal behaviour of PPC and OPC cement. *Materials Today Proceedings, 39*(xxxx), 1588–1593. https://doi.org/10.1016/j.matpr.2020.05.708
13. Gernay, T. (2019). Fire resistance and burnout resistance of reinforced concrete columns. *Fire Safety Journal, 104*, 67–78. https://doi.org/10.1016/j.firesaf.2019.01.007
14. Deboucha, W., Leklou, N., Khelidj, A., & Oudjit, M. N. (2017). Hydration development of mineral additives blended cement using thermogravimetric analysis (TGA): Methodology of calculating the degree of hydration. *Construction and Building Materials, 146*, 687–701. https://doi.org/10.1016/j.conbuildmat.2017.04.132
15. Dweck, J., Melchert, M. B. M., Viana, M. M., Cartledge, F. K., & Bu, P. M. (2013). Importance of quantitative thermogravimetry on initial cement mass basis to evaluate the hydration of cement pastes and mortars. *Journal of Thermal Analysis and Calorimetry, 113*, 1481–1490. https://doi.org/10.1007/s10973-013-3112-5
16. Ramachandran, V. S., & Beaudoin, J. J. (2001). *Handbook of analytical techniques in concrete*. William Andrew Publishing, LLC Norwich.
17. Frías, M., Savastano, H., Villar, E., Sánchez, M. I., Rojas, D., & Santos, S. (2012). Cement & concrete composites characterization and properties of blended cement matrices containing activated bamboo leaf wastes. *Cement and Concrete Composites, 34*, 1019–1023. https://doi.org/10.1016/j.cemconcomp.2012.05.005
18. Bruice, P. Y. (2011). Beauchamp spectroscopy tables 1. *Organic Chemistry, 2620*, A16–A17.
19. Dwivedi, V. N., Singh, N. P., Das, S. S., & Singh, N. B. (2006). A new pozzolanic material for cement industry: Bamboo leaf ash. *International Journal of Physical Sciences, 1*(3), 106–111.
20. Rosas-Casarez, C. A., et al. (2014). Experimental study of XRD, FTIR and TGA techniques in geopolymeric materials. *International Journal of Advanced Computer Science and Applications, 4*(4), 25–30.
21. Ikumapayi, C. M. (2017). Crystal and microstructure analysis of Pozzolanic properties of bamboo leaf ash and locust beans pod ash blended cement concrete. *Journal of Applied Sciences and Environmental Management, 20*(4), 943. https://doi.org/10.4314/jasem.v20i4.6
22. Singh, N. B., Das, S. S., Singh, N. P., & Dwivedi, V. N. (2007). Hydration of bamboo leaf ash blended Portland cement. *Indian Journal of Engineering and Materials, 14*(February), 69–76.
23. Moraes, M. J. B., et al. (2019). Production of bamboo leaf ash by auto-combustion for pozzolanic and sustainable use in cementitious matrices. *Construction and Building Materials, 208*, 369–380. https://doi.org/10.1016/j.conbuildmat.2019.03.007

Chapter 4
Parametric Analysis on Heavy Quadricycle Vehicle Front Crash for Safety Optimization

Seid Belay Gesese and Haileleoul Sahle Habte

4.1 Introduction

Nowadays, frequently occurring vehicle collision accidents have become a serious threat to the safety of both passengers and drivers. The crashworthiness performance of the vehicle is of great importance in vehicle safety design. That is because this performance of the vehicle support to absorb energy and to prevent occupant injuries in the event of an accident [1–3]. Conventionally, crashworthiness always depends upon the strength of body structure which has a direct relation with the structure design, material thickness, and material strength grade [3, 4]. The front-end structure of a vehicle, especially the bumper and crash box, has the purpose of protecting passengers and the vehicle at low and medium vehicle impact speed. As for the bumper, the minor deflection is acceptable during a low-speed collision because it also undertakes a protection role. To enhance the crashworthiness of these components, distributing more material or using more suitable material was adopted as the general method [4–6]. However, increasing thickness or unscientific material randomly may lead to low weight efficiency of the car body.

According to the report from the International Aluminum Institute, for an automobile weight loss of 10%, the fuel consumption and emissions will be decreased by 8% and 4%, respectively. With the fuel energy shortage, the energy-saving and lightweight design can make vehicles more economical have become quite a signif-

S. B. Gesese · H. S. Habte (✉)
School of Mechanical and Industrial Engineering, Addis Ababa Institute of Technology, Addis Ababa University, Addis Ababa, Ethiopia
e-mail: haileleoul.sahle@aait.edu.et

K. Mequanint et al. (eds.), *Sustainable Development Research in Materials and Energy*, Green Energy and Technology,
https://doi.org/10.1007/978-3-031-69860-6_4

icant tendency. And most researchers approach to reduce the vehicle body weight by optimizing the design of materials, replacing parent material with low-density material (steel with aluminum foam) or by using high-strength steel materials with lesser thickness [2–4].

One of the main structural parts of heavy quadricycle vehicle, bumper and splash guard, will subjected to high deformation in crash accident [7, 8]. This part is better to withstand crashes by plastically deform to reduce possible damage to be encountered in impact condition. That is reasonable because this class of vehicle is currently seen mixed with ordinary vehicle in cities in provision for personal and commercial transportation services.

David et al. [9] examine road safety for Urban Lightweight Vehicles (ULV) to find critical crash scenarios from which future crash testing methods for urban vehicles can be derived. The term ULV is specific engine-powered three- and four-wheeled vehicles categorized by the European Commission. Other attributes than the wheel geometry are engine power and the mass of the vehicle and according to the standard for light quadricycle four-wheeler with a maximum mass of 350–400 kg with an internal combustion engine capacity of 50 cc or less, internal combustion or electric engine with maximum power of 4 kW, and/or the maximum impact speed must exceed 45 km/h.

Dange et al.'s [8] study focused on a simulation of bumper beam that was done under low-velocity impact as per the standards of automotive stated in European commission for ECE United Nations Agreement, Regulation No. 42, 1994. The research uses the bumper beam material, shape, and impact condition to improve the crashworthiness. The study results showed that an M220 material (martensitic steel) minimizes the bumper beam deflection.

Marzbanrad et al. [10] simulate low-speed frontal impact of original bumper under ECE United Nations Agreement, Regulation No. 42, 1994, using LS-DYNA. Weight and impact behavior parameter were compared for composite and aluminum material resulted that modified SMC bumper beam can replace the use of glass mat thermoplastic (GMT) and aluminum.

Among many quadricycle vehicle types in local market of developing countries, we can find Bajaj Qute, a four-wheeler Bajaj. This vehicle got the attention for this study because this heavy quadricycle vehicle has been used as passenger and commercial vehicle in Far East countries and few African countries, where minibus and automobiles are highly demanded for taxi service provision.

This study aims in identifying the crash performance of an existing front bumper and splash guard of a Bajaj Qute using a full- and partial-front impact. The dynamic/explicit finite element analysis was done using commercially available software, Abaqus. Further, a parametric analysis was done to identify an enhanced crash safety using new bumper and splash guard designs.

4.2 Material and Method

4.2.1 Data

The 2012 model Bajaj Qute which was observed in most urban cities of developed countries was chosen for this study. The overall geometrical dimensions of the front-side bumper and splash guard was recorded using different measuring tools accordingly.

To easy the geometrical modeling process, the 3D CAD model was developed using SolidWorks 2020, and then exported to finite element analysis software ABAQUS in .sat file format. The 3D CAD models are depicted in Fig. 4.1. The thickness of the bumper and splash guard is 2 mm, and model was adopted as a shell element taking into consideration the thickness which is compared to other dimension of the model.

4.2.2 Material

This study used polypropylene (PP) as a reference material for the bumper and splash guard with the consideration of famous material deployed to develop vehicle bumpers [11, 12]. The mechanical properties related to mass and stiffness was adopted from material handbook: density 9.94e-10 ton/mm^3 and elastic modulus 915 MPa.

While the plastic property of the PP considering Johnson-Cook elastic-plastic parametric related to impact behavior were adopted, see Table 4.1. The

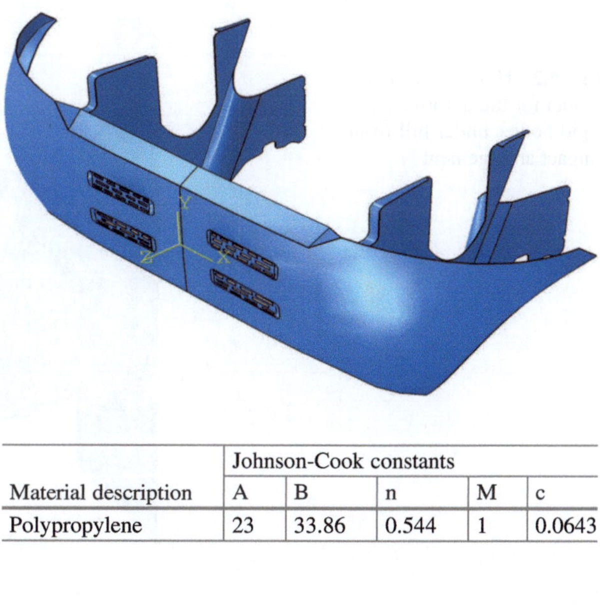

Fig. 4.1 Three-dimensional model for the front side of Bajaj Qute bumper and splash guard

Table 4.1 Johnson-Cook elastic-plastic material constant for polypropylene [12]

Material description	Johnson-Cook constants				
	A	B	n	M	c
Polypropylene	23	33.86	0.544	1	0.0643

Johnson-Cook parameter was selected for this study considering its advantage of processing strain hardening and strain rate effects related to material behavioral change due to impact load on components [4]. Thermal effect is neglected in this study.

4.3 Finite Element Analysis

The finite element analysis was done by adopting dynamic/explicit non-linear behavior with impact time elapsed to 0.05 sec. The time step was defined considering the approximate time that undertakes the impact to elapse a total deformable body deformation where the impact speed is 50 km/h.

The dynamic/explicit finite element analysis that mimics an impact load, which leads to a nonlinear behavior, was simulated using Abaqus/Explicit. The 3D CAD model that represents the physical model of Bajaj Qute bumper and splash guard was imported to Abaqus/CAD, and after other material, contact and boundary conditions were defined.

In the simulation environment, the imported 3D CAD model of the bumper and splash guard were modeled as flexible bodies, while a simple rectangular plane was modeled to represent the rigid body. Hence, the study considers the bumper and splash guard as deformable body which was assembled within the rigid body (impacted body), and the vehicle frame has been considered as non-deformable. These models were discretized using shell elements that match the part models. As it is shown in Fig. 4.2, the deformable body was meshed with S4R element with fine element length of 8 mm (see Fig. 4.3), while the rigid body meshes with course CR4 element with equal length of 40 mm.

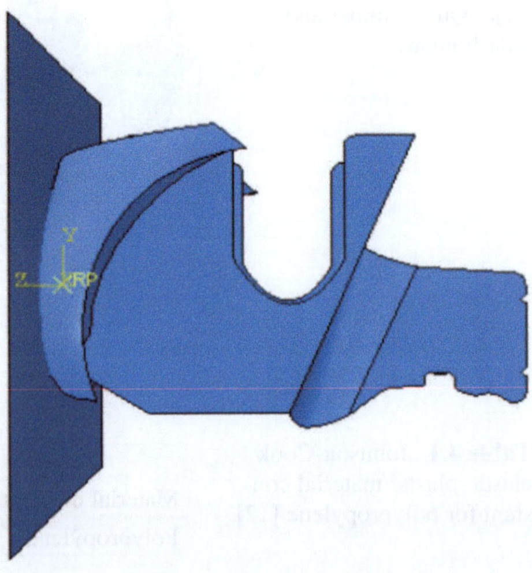

Fig. 4.2 Three-dimensional model for the deformed and rigid bodies under full-front impact arrangement

Fig. 4.3 Meshed deformable body

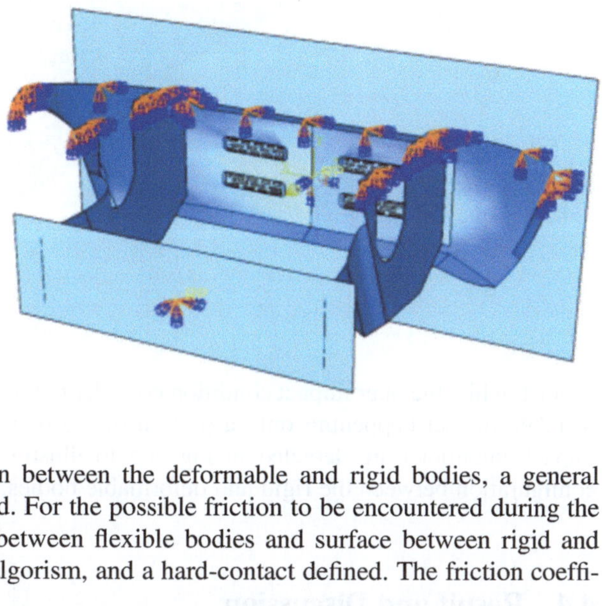

Fig. 4.4 Constrained conditions for deformable body

To define the interaction between the deformable and rigid bodies, a general contact module was selected. For the possible friction to be encountered during the surface to surface contact between flexible bodies and surface between rigid and flexible, a penalty friction algorism, and a hard-contact defined. The friction coefficient of 0.15 was used.

Further, in the finite element analysis, the boundary conditions were set. The constraint surfaces were set mimicking the connection parts of the bumper and slipping guard with other body of the vehicle. The front end of the bumper and splash guard end was constrained for all degree of freedom and the top edge constrained to linear constrain along vertical direction and with all angular rotation. Figure 4.4 illustrates these constraint conditions totally.

This study considered two of the impact conditions, which are suggested in impact test of a vehicle. These are the full-front and partial-front impact. The former one considered the impact between the deformable and the rigid one totally in

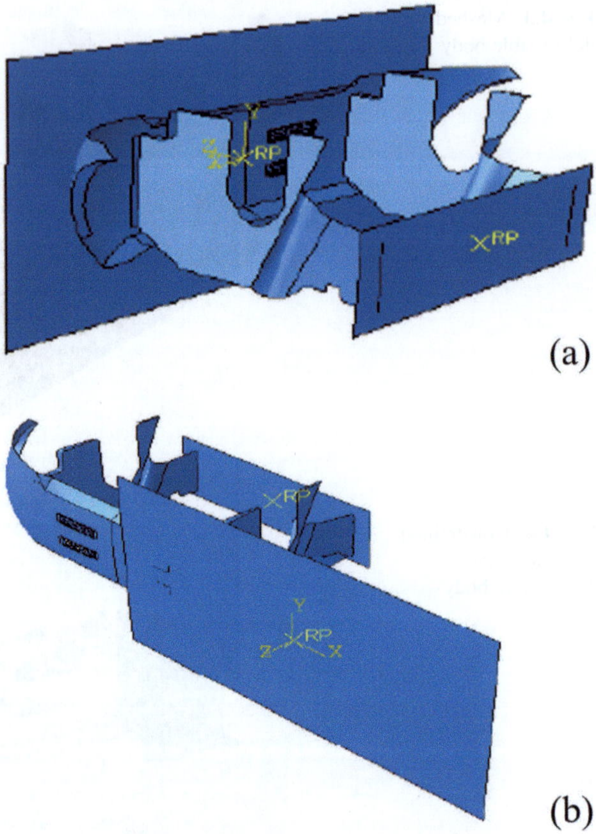

Fig. 4.5 (**a**) Full-front and (**b**) partial-front impact models

contact, while the later impact condition consider the partial-front (50%) impact that simulate impact happening only a portion of the deformable front surface. These impact conditions are depicted in Fig. 4.5 to illustrate the full- and partial-front configuration between the rigid and deformable bodies.

4.4 Result and Discussion

4.4.1 Deformation

The undeformed and deformed shape of the deformable body is illustrated in Fig. 4.6. From the figure, the color contour supports to visualize the degree of deformation possessed in different section of the deformable body. The beam section of the bumper and splash guard is observed to be displaced and deformed to the rear end of the support, where the deformation process of the deformable beam support

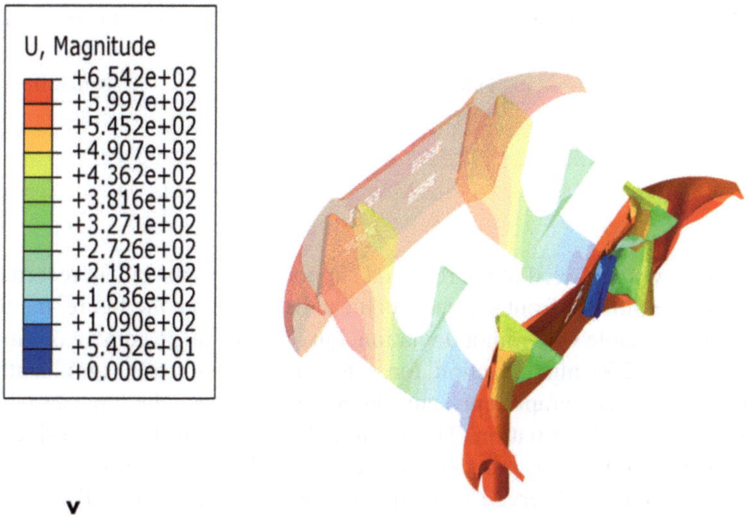

Fig. 4.6 Deformed and undeformed model of the full-front impact of the deformable body

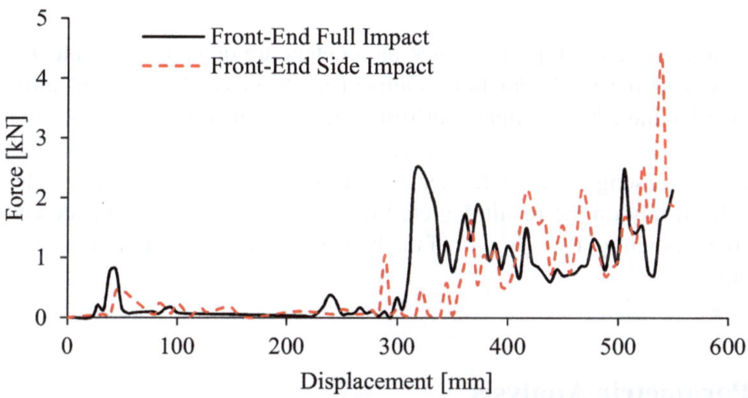

Fig. 4.7 Force versus displacement graph for full-front and partial-front impact for reference material

shows sever compression and bending. In addition, Fig. 4.6 shows the deformable body displaced a maximum of 6.54e2 mm in full-front impact condition.

4.4.2 *Force-Displacement*

Figure 4.7 shows the force versus displacement diagram of bumper and splash guard under full-front (front-end full) and partial-front (front-end side) impact. The

Table 4.2 Crash performance parameter results for full-front and partial-front impact using reference material

	Full-front impact	Partial-front impact
F_{peak}(kN)	2.48	4.43
F_{mean}(kN)	0.74	0.58
EA(MJ)	0.41	0.32

reaction force was retrieved at the rear-end support and displacement recorded from the center of the bumper front surface. As it can be seen from the figure, the value of reaction force is stable with minor deviation, but once the displacement of the energy absorbed passes 200 mm, reaction force is stated to increase significantly. The non-smooth but zigzag graph represents the relation between the force and displacement of the plastic deformation. The identified relation can be synthesized as the complex internal forces are process compression and bending. From the graph, the highest peak force is observed near the end, which justifies that the deformable bodies gaining a rigidity behavior can have excessive load yields for fracturing the body. The energy absorption (EA) is approximated using the area under the force and displacement graph for the maximum deformable body displacement of 550 mm.

The result of the crash performance parameters are depicted in Table 4.2.

The result of the nonlinear finite element analysis yields the front bumper and splash guard to have better impact performance when the impact orientation is full as compared to side impact. Indirectly this suggests that the severity of impact on the driver and/or passenger will be higher if the Bajaj Qute face a front-end side impact. Besides, been rely on the result depicted in Table 4.2, it can be considered that the crash force efficiency (F_{mean}/F_{peak}) of the bumper with splash guard will be very low, below 50%.

4.5 Parametric Analysis

4.5.1 Effect of Thickness

In different research findings [13–15], it is mentioned that the thickness of a bumper is one of the factors that can be considered for maximizing the efficiency of the energy absorption with tradeoff-specific energy absorption. The existing Bajaj Qute bumper and splash guard developed with 2 mm thickness. With this in reference, this study considered a case of increasing thickness of the deformable body to 3 mm and 4 mm to study the crash performance for full-front and partial-front impact condition at lower speed.

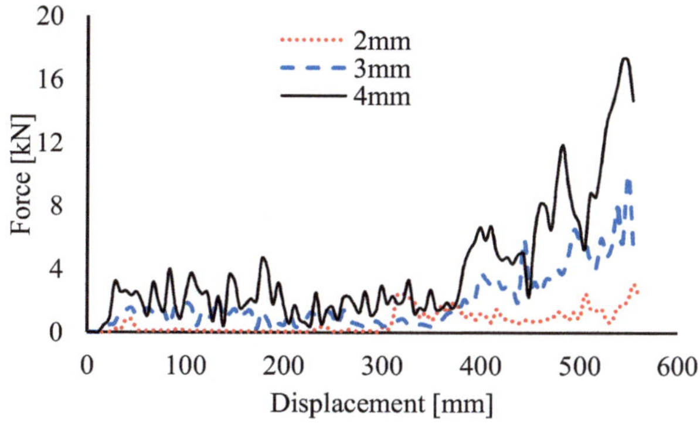

Fig. 4.8 Force versus displacement diagram for full-front impact of reference material with different thickness

Table 4.3 Crash performance parameter results for full-front impact for reference material with different thickness

Description	Thickness of the material		
	2 mm	3 mm	4 mm
F_{peak} [kN]	2.48	10.05	17.33
F_{mean} [kN]	0.74	2.01	4.07
EA [MJ]	0.41	1.05	1,57

4.5.1.1 Front-End Full Crash Test with Different Thickness

The full-front impact on bumper and splash guard made of polypropylene (reference material) with different thickness is shown in Fig. 4.8. The relation shows a zigzag pattern for all models with different thicknesses. The graph also supported to observe that when the thickness of the model increased, the total area covered by force versus displacement region increases. This indicates the energy absorption by model with 4 mm is greater as it followed the model with 3 mm and 2 mm thicknesses, respectively.

Further to the result of force versus displacement, this study identifies the specific energy absorption and crash force efficiency results; see Table 4.3. Result in Table 4.3 shows that energy absorption is enhanced as the thickness increased with tradeoff mass increment. Moreover, the specific energy absorption and crash force efficiency are also affected by increased deformed body thicknesses.

For 4 mm thickness, the mass of the bumper and splash guard increased two times the reference one which is 100% increment. To select between 2 mm, 3 mm, and 4 mm, the Pareto frontier analysis [16] was done and result illustrated in Fig. 4.9.

The utopic point on the Pareto graph is 0.22 at 4 mm thickness. Therefore, selecting 4 mm thickness will be the wise choice considering both specific energy absorption (SEA) and crash force efficiency (CFE) in the case of full-front crash for polypropylene.

Fig. 4.9 Pareto frontier result for full-front impact for reference material at different thickness

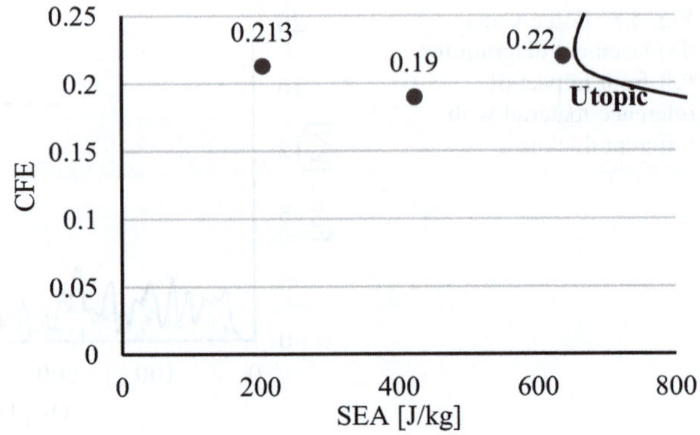

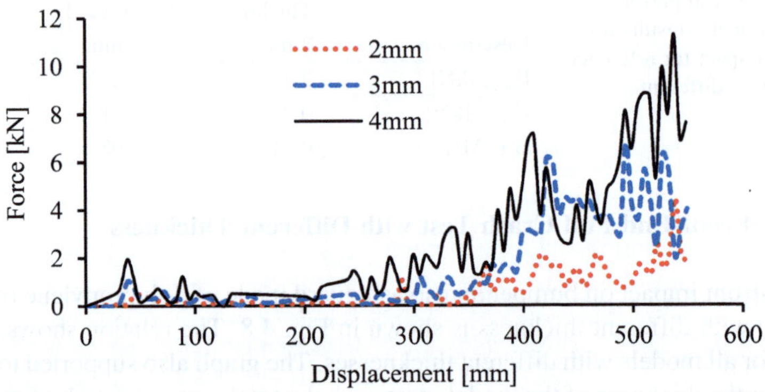

Fig. 4.10 Force versus displacement diagram for partial-front impact of reference material with different thickness

Table 4.4 Crash performance parameter results for partial-front impact for reference material with different thickness

Description	2 mm	3 mm	4 mm
F_{Peak} [kN]	4.43	6.73	11.27
F_{Mean} [kN]	0.72	1.80	2.84
EA [MJ]	0.32	0.99	1.56
CFE	0.16	0.26	0.25

4.5.1.2 Partial-Front Impact Test with Different Thickness

This follows the same procedures as the previous one. The effect of thickness in the case of partial-front impact on front bumper with splash guard was simulated adopting the same finite element procedure. The force versus displacement graph that illustrates the result is shown in Fig. 4.10.

While using the force versus displacement results, crash performance criteria were analyzed and depicted in Table 4.4.

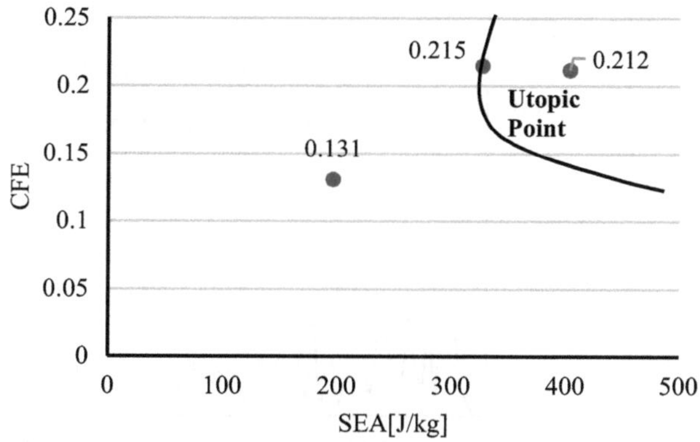

Fig. 4.11 Pareto frontier result for partial-front impact for reference material at different thickness

Table 4.5 Elastic behavior for mild steel and aluminum 6061 materials [17, 18]

Material	E(GPa)	ρ(Kg/mm^3)
Mild steel	200	7850
Aluminum 6061	69.5	2700

Table 4.6 Johnson-Cook constants for Mild steel and Aluminum 6061 materials [17, 18]

Material	Johnson-Cook constants				
	A	B	n	m	c
Mild steel	217	233	0.643	1	0.0756
Aluminum 6061	324	114	0.42	1.34	0.002

To choose the best thickness, the Pareto set graph is used as follows:

Again, the utopic point for front-side crash is 0.212 at 4 mm thickness. Therefore, 4 mm is the best thickness during the side impact for the existing model (Fig. 4.11).

4.5.2 *Material Effect*

To investigate the effect of the material used to develop the bumper and splash guard, this study use aluminum 6061 and mild steel metal as potential candidate to check their potential related with crash safety. The nonlinear finite element analysis was done following the same procedure done for the reference material (PP). Besides, the material modeled using the candidate material behavior is depicted in Tables 4.5 and 4.6. The elastic-plastic behavior of the candidate materials were modeled using Johnson-Cook.

4.5.2.1 Energy Absorption Capacity of Model Made of Aluminum 6061

For this analysis, the nonlinear finite element analysis procedure was adopted using aluminum material substituting the reference material.

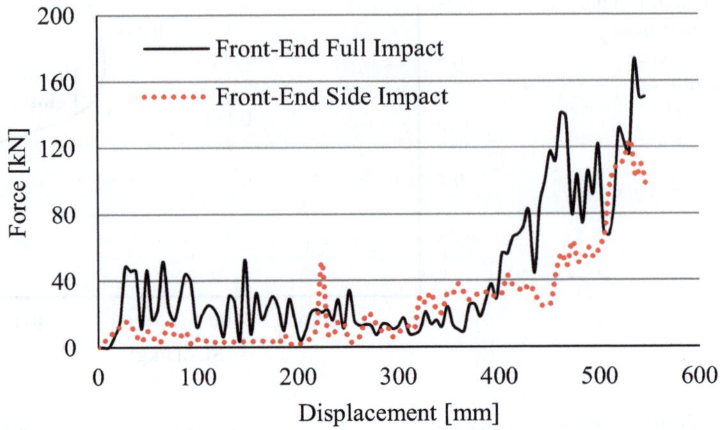

Fig. 4.12 Force versus displacement graph for full-front and partial-front impact for aluminum 6061 material

Table 4.7 Crash performance parameter results for full-front and partial-front impact for aluminum 6061 material

Description	Full-front impact	Partial-front Impact
F_{peak} [kN]	172.60	123.74
F_{mean} [kN]	41.98	28.37
EA [MJ]	22.92	15.52

The force versus displacement results for bumper and splash guard modeled with aluminum 6061 were depicted in Fig. 4.12. In this figure, full-front (front-end full) and partial-front (front-end side) impact results can be observed, where the full-front impact was observed to process bending and stretching throughout the impact period. While partial-front impact showed a minimal bending and stretching for the first 220 mm of model displacement, and then proceed with enhanced deformation for the remaining impact length.

The area under the force-displacement graph gives the energy absorption capacities of the front bumper and splash guard with different impact orientations. As it is depicted in Table 4.7, results show energy absorption of front bumper and splash guard modeled with aluminum 6061 under full-end impact which yields higher EA compared with the reference material (Fig. 4.7).

4.5.2.2 Energy Absorption Capacity of Model Made of Mild Steel

Mild steel is considered as low carbon steel as it is defined as having a low carbon content of 0.05–0.25%. It is one of the popular material with good ductility and formability and can be been used for designing and developing different automobile body parts, like crash box and bumper. To investigate the energy absorption capacity

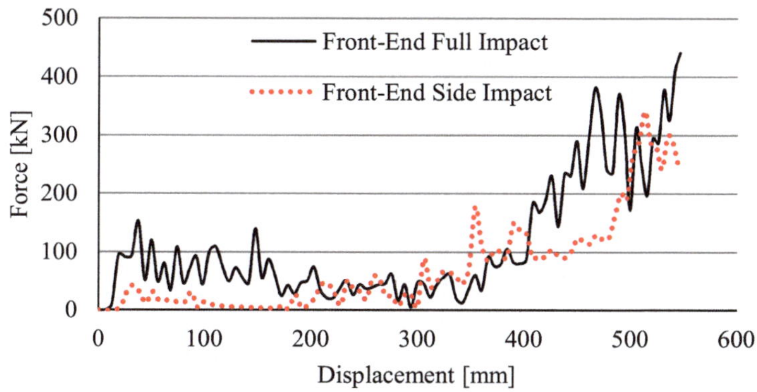

Fig. 4.13 Force versus displacement graph for full-front and partial-front impact for mild steel material

Table 4.8 Crash performance parameter results for full-front and partial-front impact for mild steel material

Description	Full-front impact	Partial-front Impact
F_{peak} [kN]	441.78	341.55
F_{mean}[kN]	117.56	85.37
EA [MJ]	64.28	44.75

of this material on this study substituting the reference material, first the force versus displacement relation were observed, which is depicted in Fig. 4.13.

The results on full-front (front-end full) and partial-front (front-end side) impact for bumper and splash guard modeled with mild steel, Fig. 4.13, show a zigzag relation that supports to consider the candidate material exercise stretching and bending in the process of their deformations. Both results show that the peak force near the end are the highest.

Considering the area under the force-displacement graph of Fig. 4.13, the energy absorption capacities of the front-end bumper and splash guard with different impact orientations are identified. As it is depicted in Table 4.8, results show that energy absorption of front bumper modeled with mild steel under full-end impact yields higher EA, while the specific energy absorption by rear-end bumper with full-impact shows the better.

4.5.2.3 Material Effect on the Energy Absorption Capacity

Mild steel and aluminum 6061 T6 were analyzed and compared with the reference material polypropylene. Both materials showed best performance in terms of energy absorption capacity and specific energy absorption. The only problem on these materials is their weight compared to reference material which is very small. It is difficult to identify the best material from mild steel and aluminum for the existing model. That is because bumper and splash guard model with mild steel has higher

Fig. 4.14 Pareto set for (a) full-front and (b) partial-front impact using aluminum and mild steel

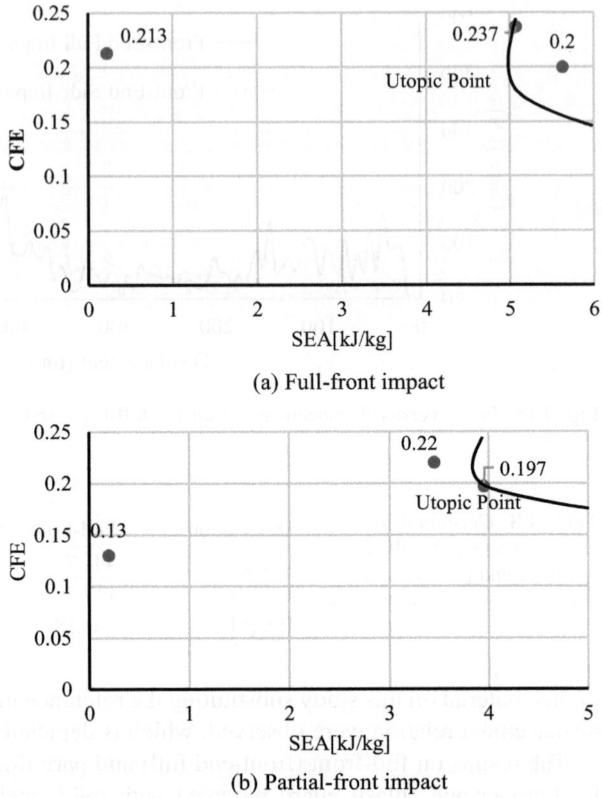

energy absorption capacity; besides the model developed with aluminum has better specific energy absorption. To choose the best one, the Pareto analysis was used, and results are shown in Fig. 4.14.

4.6 Conclusion

The study targeted to analyze the crash performance of Bajaj Qute front bumper and splash guard, and optimize without affecting the vehicle frame structure. To achieve the goal, different parameters are considered, and similar preconditions are applied. Most previous research finding indicate, changing parameters like thickness, material and geometrical configuration could potential affect the crash force efficiency. Keeping those things in mind, the present work analyzed the effect of thickness and material on crash safety.

The existing structure's response, which was modeled using a 2-mm-thick polypropylene material, to full-front and partial-front impact crash test was analyzed employing explicit dynamic finite element analysis. The results of the existing model impact test has shown low crash performance compared with models having higher thickness and models with better plastic deformation behavior.

Using Pareto frontier method, designing bumper and splash guard with Aluminum material was found to be the best solution than Mild steel and Polypropylene in relation to crash safety performance. On the other hand, mild steel has shown the highest energy absorption capacity and the strongest of all, but due to its mass, it placed second in the Pareto frontier graph.

References

1. Davies, H. C., & Bastien, C. (2021). An approach for the crash safety assessment of smaller and lightweight vehicles. *Transport Policy, 105*, 12–21.
2. Natarajan, N., Joshi, P., & Tyagi, R. K. (2021). Design improvements of vehicle bumper for low speed impact. *Materials Today: Proceedings, 38*(1), 456–465.
3. Sheshadri, A. (2006). *Design and analysis of a composite beam for side-impact protection of occupants in a sedan*. M.S. Thesis. Wichita State University, College of Engineering, Dept. of Mechanical Engineering.
4. Li, Z., Qiang, Y., Zhao, X., Man, Y., Shi, P., & Yan, C. (2017). Crashworthiness and lightweight optimization to applied multiple materials and foam-filled front end structure of auto-body. *Advances in Mechanical Engineering, 9*(8), 168781401770280. https://doi.org/10.1177/1687814017702806
5. Habte, H. S., & Abdul-Latif, A. (2020). Finite element modeling of non-conventional energy dissipating systems using metallic like-composite tubular structures. *International Journal of Impact Engineering, 139*, 103511. https://doi.org/10.1016/j.ijimpeng.2020.103511
6. Sonawane, C. R., & Shelar, A. L. (2018). Strength enhancement of car front bumper for slow speed impact by FEA method as per IIHS regulation. *Journal of International Engineering (India): Series C, 99*(5), 599–606.
7. Phanmueang, K., Jongpradist, P., & Ruangjirakit, K. (2021). Analysis of aluminum front bumper beam for heavy quadricycles under impact loading. *IOP Conference Series: Materials Science and Engineering, 1137*, 012013.
8. Egertz, D. (2011). *Novel safety requirements and crash test standards for light-weight urban vehicles*. M.S. Thesis. School of Engineering Sciences (SCI), Aeronautical and Vehicle Engineering, KTH.
9. Dange, M. V., Buktar, D., Rajesh, B., Raykar, D., & Nilesh, R. (2015). Design and analysis of an automotive front bumper beam for low-speed impact. *IOSR Journal of Mechanical and Civil Engineering., 12*, 17–27.
10. Marzbanrad, J., Alijanpour, M., & Kiasat, M. S. (2009). Design and analysis of an automotive bumper beam in low speed frontal crashes. *Thin-Walled Structured, 47*(8–9), 902–911.
11. Shokrieh, M. M., & Joneidi, V. A. (2015). Characterization and simulation of impact behavior of graphene/polypropylene nanocomposites using a novel strain rate–dependent micromechanics model. *Journal of Composite Materials, 49*, 2317–2328.
12. Muhamad Fitri, S., & Mahzanb, F. A. (2020). The mechanical properties requirement for polymer composite automotive parts – a review. *International Journal of Advanced Technology in Mechanical, Mechatronics and Materials, 01*, 125–133.

13. Hosseinzadeh, R., Shokrieh, M. M., & Lessard, L. B. (2005). Parametric study of automotive composite bumper beams subjected to low-velocity impacts. *Composite Structures, 68*(4), 419–427.
14. Zhang, Z., Liu, S., & Tang, Z. (2009). Design optimization of cross-sectional configuration of rib-reinforced thin-walled beam. *Thin-Walled Structures, 47*, 868–878.
15. Hambali, A., Rabiatul, A., Rahim, A., & Taufik, S. (2012). Design optimization of automotive bumper beam through energy absorption analysis. *International Journal Engineering of Science, 3*(1), 34–39.
16. Kasprzak, E. M., & Lewis, K. E. (2001). Pareto analysis in multiobjective optimization using the collinearity theorem and scaling method. *Structural and Multidisciplinary Optimization, 22*(3), 208–218.
17. Johnson, G. R., & Cook, W. H. (1985). A constitutive model and data for metals subjected to large strains, high strain rates, and high temperatures. *Engineering Fracture Mechanics, 21*, 31–48.
18. Boyer, H., & Gall, T. (1985). *Metals handbook.* Desk edition. American Society for Metals.

Chapter 5
Experimental Investigation of the Impact Strength of Sunflower Oil-Treated Ethiopian Highland Bamboo Fiber-Reinforced Polyester Composites

Berihun Abebaw Mebratie, Addisu Alamirew Meku, Ermias Wubete Fenta, and Samuel Getachew Chekol

5.1 Introduction

Composite materials are the combination of various kinds of materials to form a special combination of mechanical, chemical, or other properties. Composite materials are being formed to acquire important qualities and enhance performance characteristics that are not feasible with any component material alone. They can be categorized as metal, polymer, or ceramic matrix composites, based on the matrix of the material. Since the manufacturing process of polymer matrix-based composites is simple compared to the other types, they get more attention for material development. Fibers are significant in providing polymer composite strength, stiffness, thermal stability, and insulation. The reinforcing polymer binds the fibers, provides the composite's shape and structure, and protects the reinforcing fibers against mechanical damage and chemical attack [1–4].

Mechanical properties of laminated fibers are affected by fiber loading, fiber length, and fiber orientation. Researchers indicate that 35%/65% fiber/matrix weight ratio has relatively good mechanical properties. Among those mechanical properties, the impact strength of the composite is better at optimum fiber loading (30–40%) by weight [5–7].

Fiber-reinforced plastic polymers have comparable strength to traditional metals. They have low density and good strength-to-weight ratio which are markedly superior to materials of metals. They are also characterized by their high internal

B. A. Mebratie (✉) · A. A. Meku · E. W. Fenta · S. G. Chekol
Faculty of Mechanical and Industrial Engineering, Bahir Dar Institute of Technology, Bahir Dar University, Bahir Dar, Ethiopia

K. Mequanint et al. (eds.), *Sustainable Development Research in Materials and Energy*, Green Energy and Technology,
https://doi.org/10.1007/978-3-031-69860-6_5

damping and good modulus, which enables the material to absorb vibrational energy more effectively. Contrarily, reinforced composites of natural fibers have a high moisture-absorbing capacity [1, 8–11].

There are many composite materials manufacturing methods. Among those, lay-up (manual lay-up, spray lay-up, and filament winding) and quick-cure (vacuum bagging and resin transfer molding (RTM)) are the most commonly simple manufacturing techniques. Hand lay-up technique is the most straightforward, widely available, and affordable type of composite lamination. When using the hand lay-up method, resins are manually infused into fibers to create woven, stitched, knitted, or bonded fabrics [11]. The extraction of fiber can also be done chemically, mechanically, or by combining mechanical and chemical techniques. Alkaline treatments combined with the roller-milled method (RMT) have been investigated and used for fiber extraction.

5.1.1 Bamboo Fiber-Reinforced Polymer Composites

Fibers of plants are the most prevalent natural fibers which are available in the forms of grass, leaves, and wood. Bamboo belongs to the grass family of Bambusoideae that is composed of cellulose and lignin [12–14]. They are low-cost, quickly growing plants with good mechanical properties. In Latin America, Asia, and parts of Africa 1200 known species of bamboo cover an estimated 36 million hectares of growing space [15, 16]. Bamboo has approximately 60.8% cellulose, 32.2% lignin (which makes the fiber brittle), and 7% other constituents. But these constituent quantities vary depending on the kind of bamboo. The chemical makeup of those many species determines how strong they are [5].

The microstructure of the bamboo culm comprises various vascular bundles sandwiched in parenchyma tissue that are widely dispersed close to the outside of the bamboo culm wall and are surrounded by fibrils [14]. Smaller components like sieve tubes, sclerenchyma cells, fiber strands, and vessels with companion cells make up each vascular bundle. Elementary fibers are the building blocks of vascular bundles with pentagonal and hexagonal forms of nano-fibrils bonded and arranged together with hemicellulose and lignin [17–20].

Bamboo may mature in 3 years, although most species need between 4 and 6 years to reach full maturity [21, 22]. However, height growth of bamboo stops around 6–7 months after the shoot emerges [17, 18].

Many experiments were done on bamboo to develop new composites because bamboo attains better mechanical properties as other natural fiber plants. Various studies have shown that bamboo fiber-reinforced composites have specific strength and stiffness that are comparable to those of glass-reinforced composites. However, bamboo culm is not easy to use in the form of structural materials, plates, or panels [22]. Through various chemical treatments, the mechanical characteristics of composites can be improved [23–26].

The researchers found that the mechanical properties of bamboo fiber/polyester composites, such as bending, tensile, hardness, and impact, were improved compared to pure bamboo [5, 7]. The composite has the best impact strength as fiber loads increase up to a certain level [27]. The oil treatment of bamboo is better than the Plato wood treatment process in the early wood property improvement technique. After the treatment, the surplus oil on the surface can be removed by extracting it using 100% ethanol, which improves its properties [21]. Unlike other natural fibers, extraction of long bamboo fiber is relatively difficult. In contrast to conventional extraction techniques, roller milled technique (RMT) have been used as fiber extraction technique to get excellent mechanical strength [16, 23, 24, 28].

In addition to chemical treatments, water absorption has its own effect on different properties of the laminate. In a composite, water absorption occurs due to moisture being diffused into the micro- or macro-holes of the laminates. The main determinant of how much moisture or water is retained by the composite depends on how many fibers are sandwiched in a matrix [26, 27]. The quantity of fiber used as reinforcement has an impact on water absorption characteristics' thus, it's essential to balance these two factors while making fiber-reinforced polymers. Consequently, in the field of composite science, the optimal fiber loading always maintains a compromise between enhanced mechanical load withstanding capabilities and diminished impacts of water absorption [28–33].

5.1.2 Application of Natural Fiber-Reinforced Polymer Composites

Fiber-reinforced composites have various applications, including the aerospace, automobile, and marine manufacturing industries. Epoxy and polyester (both saturated and unsaturated) are the most widely used thermoset polymers in structural applications and have several uses in the construction industry [34]. Laminates of natural fiber composites are easier to handle and have good acoustic, thermal, and insulation properties. Due to their remarkable qualities, natural fiber-reinforced polymer composites are used in various applications, as shown in Table 5.1 [1, 11, 18, 19, 35, 36].

5.2 Methods

In order to measure the mechanical properties of bamboo fiber/polyester composites, experimental approaches were followed. Mixture rule was used to calculate the fraction of component materials of the laminate.

Table 5.1 Applications of natural fiber/polymer-laminated composites [1, 11, 18, 19, 35, 36]

General application areas	Specific applications
Home infrastructures	Panels, decorating products, chairs, tables, bath units, safety cases
Transportation applications	Tricycle bodies, door trim, dashboard, roof lining, safety cover of the spare tire, car trunk, boat hull, seat chairs, railway coach interior, passenger seat frames
Everyday applications	Helmets, lampshades, suitcases
Storage devices	Post boxes, grain storage silos, biogas containers
Energy-converting unit	Wind turbine blades, impeller blades

5.2.1 Fiber Extraction Techniques of Bamboo

This investigation used a 2-year-old Ethiopian highland bamboo, which is found in an ecological zone 2560 meters above sea level. The species of bamboo that grows in this area is called *Yushania alpina* (K. Schum.) W.C.L., and it typically reaches elevations of 2200–3500 [37].

The node and bark sections of the culm were removed out during extraction, and the remaining culm was used to extract the fiber. The culm was peeled longitudinally to make strips that were 10 mm wide and 0.35 to 1.5 mm thick. Then, for 3 days, these strips were soaked in water to soften them. To eliminate the wax and other undesirable components and provide a rough surface for effective adhesion, the dry strips were then treated with 3% sodium hydroxide for 5 h. After being thoroughly cleaned with distilled water, the strips were rolled with a cylindrical roller load on a smooth, flat surface to weaken the lignin component and combed with a knife. In this investigation, mechanical fiber extraction was used to separate 160 mm x 0.7 mm x 0.7 mm-sized fibers of bamboo.

To lower the moisture content, the fibers were solar-dried for 7 days [15]. Then, for 7 days, half of the dried fiber was immersed in sunflower oil. After the fibers had been treated with oil, excess oil on their surface was removed using ethanol that has a 97% alcohol content. The fiber was then exposed to sunlight for 7 days on black plastic to minimize the amount of moisture on its surfaces [26].

The fibers were arranged in the same direction as weft on a simple conventional weaving machine, and a very thin yarns were used as warp in the transverse direction. Long, continuous fibers were used to make laminates with any necessary orientations and regular arrangements very easily. By measuring and cutting the made-up mat to the necessary sample size and fiber orientation, any orientation can be obtained. As seen in Fig. 5.1, a wide mat was used to made unidirectional, cross-ply, and angled ply mats.

The mat is then captured with a camera and microscope, and each fiber is measured with a Java-based image processing application (ImageJ), which developed at the National Institutes of Health's Laboratory for Optical and Computational Instrumentation. It was created with an open-access architecture that enables extensions through Java plugins and recordable macros. Using the distribution curve of the origin plot, the size of the fibers in the mat was examined using the data received from the ImageJ software (Fig. 5.2) [27].

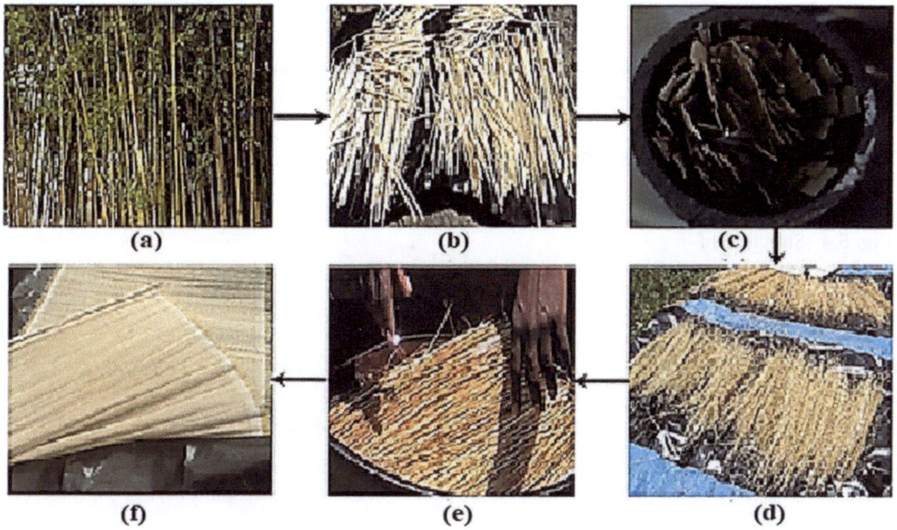

Fig. 5.1 Preparation of material: (**a**) raw bamboo, (**b**) bamboo strip, (**c**) treatments of the bamboo strip by NaOH, (**d**) bamboo fiber, (**e**) oil treatments of bamboo fiber, and (**f**) unidirectionally aligned bamboo fiber

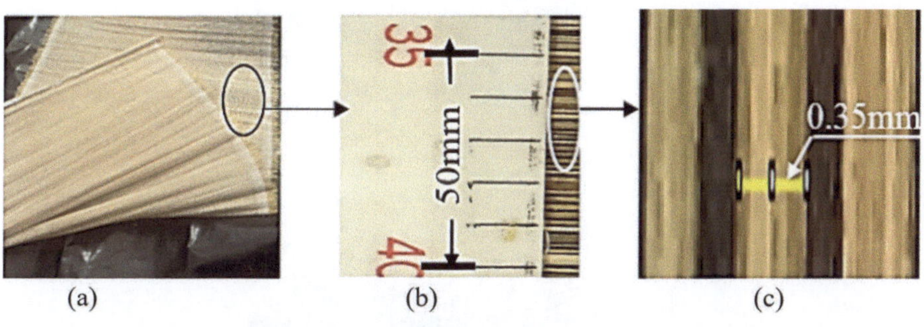

Fig. 5.2 Setup for fiber size measurement using ImageJ. (**a**) Mat of bamboo fiber. (**b**) An initial distance photograph for a 176-mega-pixel image. (**c**) Measurement of fiber size

5.2.2 Geometry and Laminate Structure of Experimental Specimens

The structure of the laminate is intended to have good impact strength properties. Depending on their nature, natural fibers could have a certain lamination structure. Based on ASTM D256 (for the impact strength test), the samples' size (in millimeter), geometry (Fig. 5.3), and the laminate structure of the compound (Fig. 5.4a–c) were selected [18, 32, 38, 39].

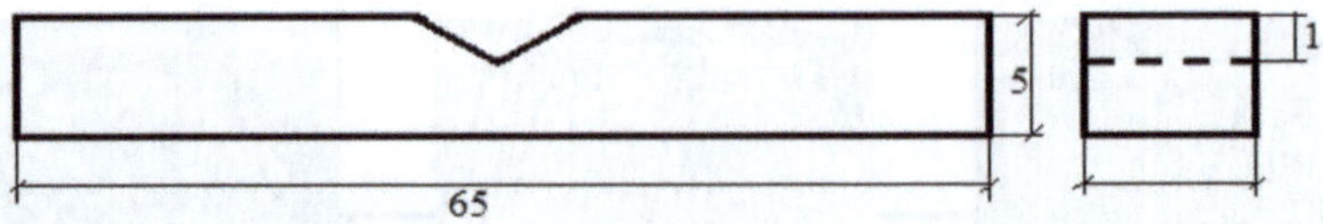

Fig. 5.3 Geometry of specimen for impact strength test

Fig. 5.4 Structure of laminate for impact strength test: (**a**) $[0/\pm 45/0]_S$, (**b**) $[0_7]$, and (**c**) $[(0/90)_3/0]$

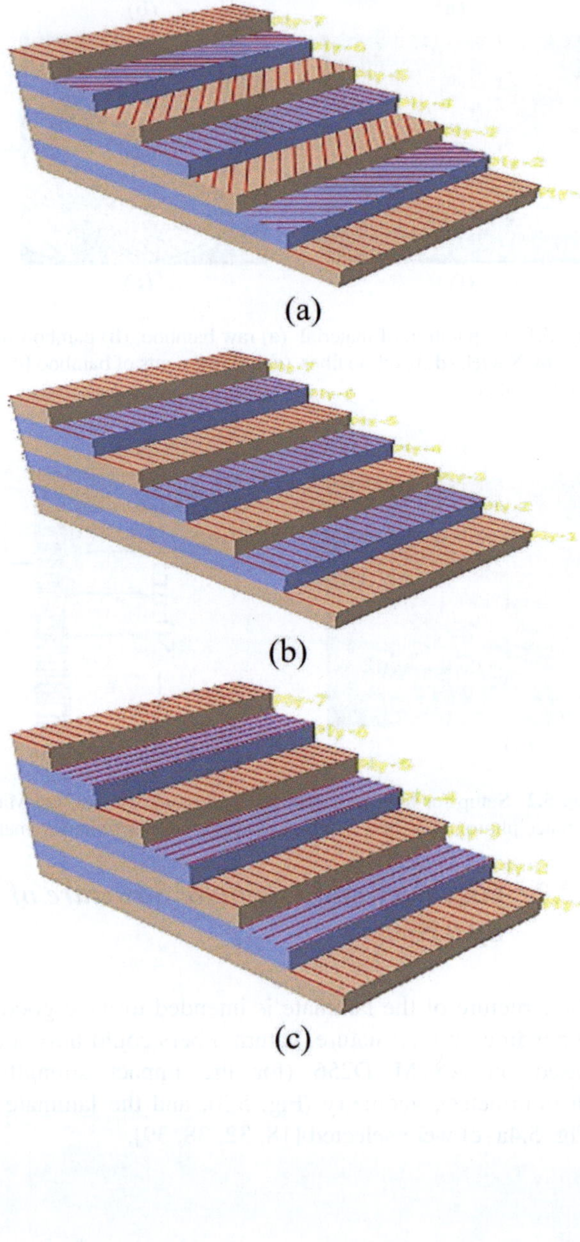

5.2.3 Composite Manufacturing Methods

5.2.3.1 Resin and Fiber Fractions of the Composite

Based on mixture rule of the composite, mass fraction of the fiber and polymer resin can be obtained as follows.

$$m_f = \frac{M_f}{M_f + M_r}, m_r = \frac{M_r}{M_f + M_r} \tag{5.1}$$

$$\begin{aligned} M_f + M_r &= M_r \\ m_f + m_r &= 1 \end{aligned} \tag{5.2}$$

$$v_f = \frac{V_f}{V_f + V_r}, v_r = \frac{V_r}{V_f + V_r} \tag{5.3}$$

$$\begin{aligned} V_f + V_r &= V_c \\ v_f + v_r &= 1 \end{aligned} \tag{5.4}$$

The densities of the fiber(ρ_f), resin(ρ_r), and composite(ρ_c) can be obtained as given in (5.5). The density in gram/cubic centimeter (g/cc) of component materials is computed by the mass-to-volume ratio.

$$\rho_f = \frac{M_f}{V_f}, \rho_r = \frac{M_r}{V_r}, \rho_c = \frac{M_c}{V_c} \tag{5.5}$$

where

(M_f), (m_f)	Mass of fiber and its fraction
(M_r), (m_r)	Mass of resin and its fraction
(V_f),(v_f)	Volume of fiber and its fraction
(V_r),(v_r)	Volume of resin and its fraction
(ρ_f)	Density of fiber
(ρ_r)	Density of resin

Using the mixture rule formulated above, the fraction of fiber and resin are shown in Table 5.2. The mass percent of polyester employed in the laminates varied as well, with values of 0.673 and 0.686, respectively, due to the differences in densities of oil-treated and untreated fibers.

Table 5.2 Mass fraction and volume fraction of fiber and resin in the composite

Materials	Density (g/cc)	Mass fraction, (m_f)	Volume fraction, (v_f)
Untreated BF	1.061	0.314	0.35
Oil-treated BF	1.126	0.327	0.35
Polyester	1.25	–	0.65

5.2.3.2 Lamination Process of the Composites

Testing samples of 300 × 300 × 3.5 mm^3 were prepared based on the standards of the specimen. At atmospheric pressure and temperature, a general-purpose polyester was combined with a solidifier (MEKP) in a 100:1.5 weight ratio. A 24 kN/m^3 unit-weight concrete block was used to provide pressure over the mold area. The total masses of the loads were 115 kg for the 300 × 300 mm mold.

The pressure load (force per unit area) was 12.5 kPa. The composite was taken out of the mold after 24 h and exposed to air to dry naturally. Aluminum sheets were used to cover the composites to avoid moisture absorption till the testing was done (Fig. 5.5).

The amount of bamboo polyester (P) and fiber (BF) utilized in this lamination (mold size 300 × 300 × 3.5 mm^3) is based on the 35%/65% ratio of fiber/polyester mixture by volume as shown in Table 5.3.

The density of the laminate based on mixture rule made from untreated fiber was 1.18 g/cc, whereas the measured or actual density of the laminated product was 1.15 g/cc. The difference in density between the measured and calculated densities was 2.61%.

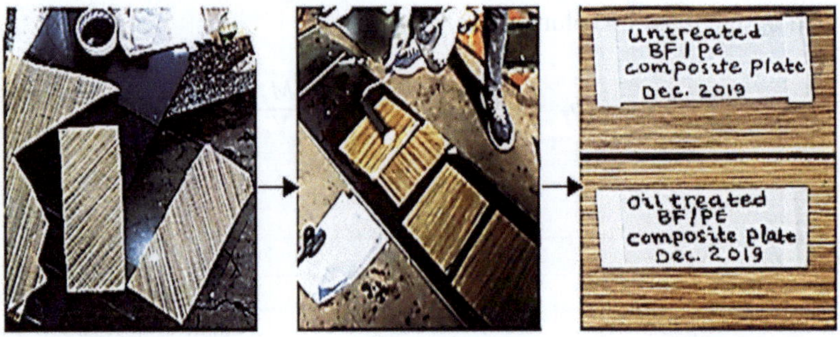

(a) (b) (c)

Fig. 5.5 (**a**) Bamboo mat, (**b**); Lamination process, and (**c**) OBFP/UBFP composite

Table 5.3 Quantity of materials used in the composite

	For samples based on oil-treated fiber		For samples based on untreated fiber	
(i) Materials	Mass (g)	Volume (cc)	Mass (g)	Volume (cc)
Polyester	685.4	548.3	681.5	545.2
Bamboo fiber	332	295	313	295
Laminate	1017.4	843.3	994.5	545.2

Table 5.4 Experimental design of the study

Materials	Tests	Samples	Fiber orientation (degree)	Test repetition
Laminated composite	Water absorption	UBFP	$[0]_7$	5
			[0/±45/0]s	5
			$[(0/90)_3/0]$	5
		OBFP	$[0]_7$	5
			[0/±45/0]s	5
			$[(0/90)_3/0]$	5
Laminated composite before and after water absorption	Impact strength	UBFP	$[0]_7$	3
			[0/±45/ 0]s	3
			$[(0/90)_3/0]$	3
		OBFP	$[0]_7$	3
			[0/±45/ 0]s	3
			$[(0/90)_3/0]$	3

5.2.4 Experimental Design

The experimental design details of the study for an optimum fiber loading are summarized in Table 5.4. The composites' properties were examined both before and after water absorption. Complying with internationally acceptable standardized geometries, a total of 48 specimens in both before and after water absorption were used [32].

5.2.4.1 Experimental Setups for Impact Strength Testing of Composites

In this study, the impact energy absorbed by a material during fracture was determined by the Charpy V-notch test using the JBS-500B impact testing machine. The maximum energy capacity of the machine is 250 or 500 joules with an impact speed of 5.4 m/s, a pre-rising of the pendulum of 1500 mm, and a specimen bearer distance of 40 mm. The composite specimens (Fig. 5.6) for the Charpy impact test for both before and after water absorption of the composite were based on the ASTM D256 standard [32, 39].

When the energy of the applied load exceeds the material's capacity, the material experiences damage. To quantify this damage, the Charpy impact tests can be performed [34]. In order to understand this relationship, this study investigated the influence of oil treatment and water absorption on the impact strength (toughness) of the bamboo fiber-reinforced polymer composites.

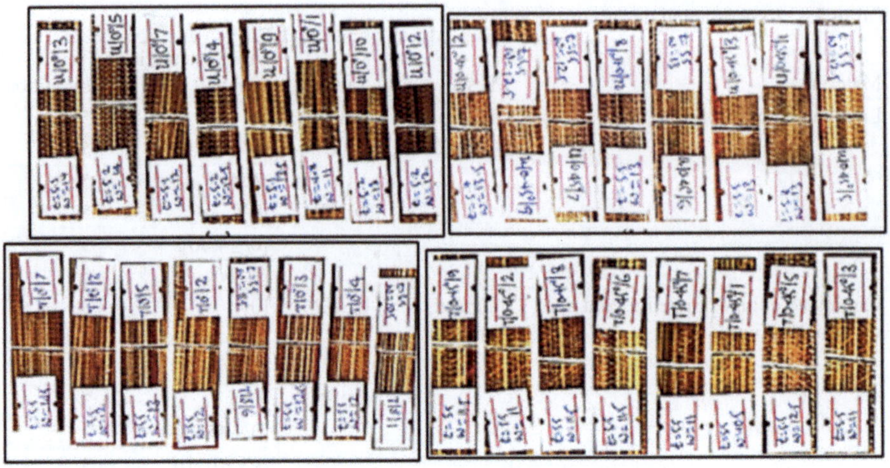

Fig. 5.6 Specimens for impact strength test

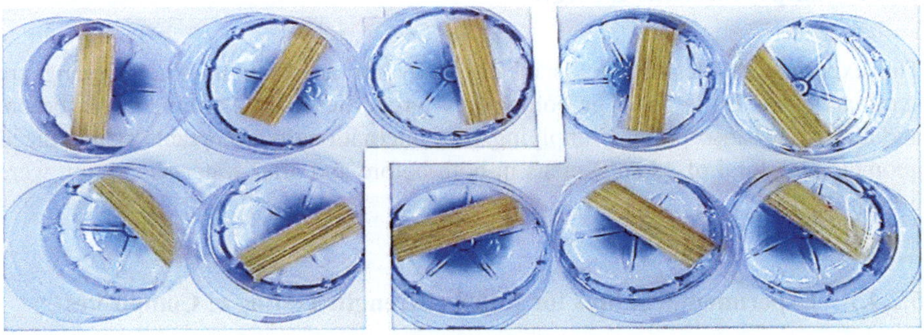

Fig. 5.7 Specimens (OBFP and UBFP) of water absorption test

5.2.4.2 Experimental Setups for Water Absorption Tests of the Composite

Water molecules' capacity to permeate inside a sample structure is demonstrated by a composite's water absorption behavior. As shown in Fig. 5.7, the dimensions of the specimens were 76.2 × 25.4 × 3.5 mm^3-sized samples designed to test impact properties. The samples were submerged in distilled water for 480 h, with each specimen's mass being recorded every 48 h. The nature of water-absorbing in a composite could be computed by the ratio of the change in mass to the dry mass of the specimen, given as (5.6).

$$\%M = \frac{M_f - M_i}{M_i} \times 100 \tag{5.6}$$

5.3 Results and Discussions

Since the investigation focuses on the impact strength of laminated fiber-reinforced polyester, discussions were held depending on the results obtained from each specific objective.

5.3.1 *Basic Characteristics of Bamboo Fiber*

5.3.1.1 Densities of Bamboo Fiber

The water displaced in a 60-mL pycnometer at a temperature of 28°C was used to calculate the volume of fiber. The volume of the fiber is equal to the difference between the pycnometer's capacity and how much water has been added to it over the fibers [40].

In this study, a pycnometer at room temperature was used to measure the average density of bamboo fibers. The densities of oil-treated and untreated bamboo fibers are shown in Fig. 5.8, where the densities of solidifier (MEKP) and polyester are taken from the respective standard data. Following the oil treatment, bamboo fiber's average density rose from 1.062 g/cc to 1.126 g/cc. The mass of the fiber is increased by 64 g for each cubic meter of treated fiber. This occurred as a result of the fiber bundle's noncellulosic components absorbing oil during treatment.

The size of the fibers plays a vital role in producing more surface area to adsorb the matrix, which influence the mechanical properties of the laminated composite. The composite would likely have a more homogeneous character similar to a laminate as the fiber size decreased. The size variety of the fibers are ranged from

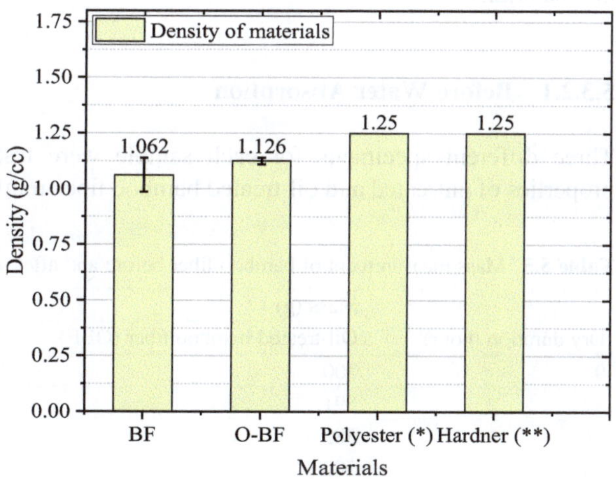

Fig. 5.8 Density of materials [11, 41]

78.6 μm to 1341.0 μm. Predominantly, fiber sizes were extracted in the range of 375–800 μm, with an average fiber size of 531.2 μm. As a result, the fiber extraction technique used in this study was effective as most of the fiber sizes were admissible [5].

5.3.1.2 Mass of Bamboo Fiber

The mass balance in this investigation is used to measure the sum of the masses of bamboo fiber and distilled water. Based on the relationship between volume, mass, and density, the density of distilled water at 28°C is 0.996232 g/mL, and its volume in the pycnometer can be calculated accordingly. The mass change of the bamboo fiber was obtained by the ratio of the change in mass before and after oil treatment to the original mass of the fiber before oil treatment. Using the relation explicated above, the oil treatment of bamboo fiber increased its mass by 14.8%. Table 5.5 shows the mass measurement of bamboo fiber before and after oil treatment [22].

5.3.2 Impact Properties of Bamboo Fiber-Reinforced Polyester Composite

The impact test aims to determine the material's ability to absorb energy during a collision. From this experiment, two properties (impact energy and fracture toughness) were obtained. The experiments were done via Charpy impact testing methods (before and after water absorption) with a drop weight of 250 J. The collision between the specimen and its weight generally destroys the specimen, but the transfer of energy between the two is used to determine the fracture mechanics of the material.

5.3.2.1 Before Water Absorption

Three different specimens for each sample were tested to observe the impact properties of untreated and oil-treated bamboo fiber-reinforced polyester composites

Table 5.5 Mass measurement of bamboo fiber before and after oil treatment

	Mass (g)	
Dry duration (hour)	Oil-treated bamboo fiber (OBF)	Untreated bamboo fiber (UBF)
9	700	508.5
27	591	507.5
36	585	507.0
44	582	506.0
60	581	506.0

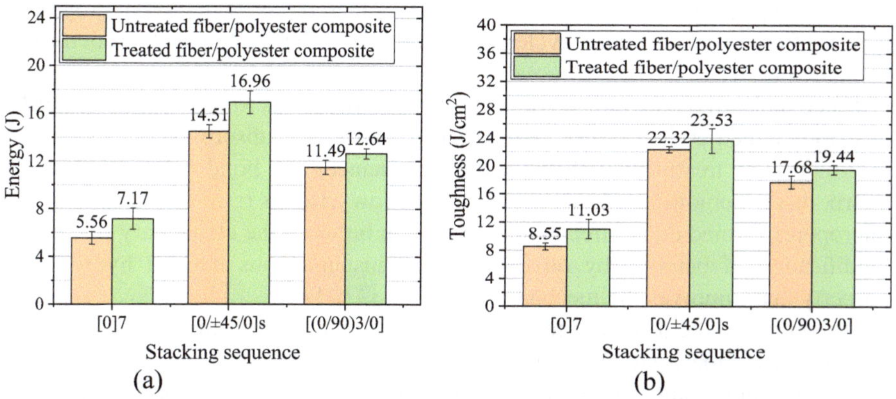

Fig. 5.9 Average impact strength of laminated bamboo at different stacking sequence before water absorption: (**a**) impact energy and (**b**) toughness

before moisture absorption. Among fiber orientations, a unidirectional fiber orientation has the lowest impact load resistance. The result shows that impact resistance also increased as the number of fiber orientations increased.

The impact strength and toughness of the untreated and oil-treated bamboo fiber-reinforced composites before moisture absorption with three different samples and different fiber orientations are shown in Fig. 5.9a and b. Fiber orientations were selected based on the findings of different researchers on various natural fiber composites. Bamboo fiber-laminated composites with $[0/\pm45/0]_S$ and $[0]_7$ attained maximum and minimum impact energies, respectively.

The experimental results showed that the orientation of fibers significantly influences the composite's impact properties. As the number of fiber orientations in a laminate increase, its impact properties also increase due to the greater probabilities of fiber locking in all directions of the composite plane. Because of their lower fracture resistance across the fiber orientation of the composite, unidirectional fiber orientations result in small impact energy. Laminates based on oil-treated bamboo fiber increase their impact energy by 10–28.9% for different fiber orientations. Oil makes the bamboo fiber more flexible and softens the lignin constituent. Therefore, treating bamboo fibers with oil increases the energy-absorbing capacity of bamboo fiber laminates.

5.3.2.2 After Water Absorption

Water absorption of a composite begins at the outer layers and steadily diminishes as it reaches the matrix's core. High water absorption by composite materials typically causes an increase in the weight of wet profiles, a significant deflection and swelling, a potential reduction in their strength, and placing strain on neighboring structures.

These can result in warping, buckling, and an increased risk of microbial growth, as well as freeze and unfreeze cycles that deteriorate the mechanical properties of laminate [41, 42].

Since moisture and water absorption were the problems to fully utilizing NFRP, this study aimed at the extent to which OBFP and UBFP absorb water as shown in Fig. 5.10. The treatment of bamboo fibers enhances the bond between fiber and matrix that is demonstrated by the fact that the low-viscous resin can already fill the micropores formed due to treatment and flow in between the elementary fibers. The modification of bamboo by oil develops an instantaneous internal hydrophobic property that improves the mechanical strength [43, 44].

According to the results of many research, laminates of oil-treated bamboo fiber exhibited less capacity to absorb water. The number of fibers sandwiched in a matrix is the most important determinant factor of water/moisture absorbed by the composite. When moisture diffuses into the micro- or macro-holes in the laminates, the composite becomes susceptible to water absorption [21, 45].

Bamboo's hydrophobicity and dimensional stability of the fiber increased after the oil treatment, and the oil-treated bamboo fiber become an anti-moisture absorbent [43]. Hence, the laminates of oil-treated bamboo fibers tended to absorb less moisture than untreated bamboo fibers (Fig. 5.11). Quantitatively, those two samples' minimum and maximum saturation water absorption rates are 6.9 and 9.5%, respectively.

Taking it into account, the results (Fig. 5.12a and b) of bamboo fiber-laminated composite after water absorption with $[0/\pm45/0]_S$ and $[0]_7$ attained maximum and minimum impact energy, respectively. Impact load resistance of untreated bamboo fiber-reinforced polyester composites after water absorption decreased by 23.56%, 41.15%, and 31.07% for fiber orientations of $[0]_7$, $[0/\pm45/0]_S$, and $[(0/90)_3/0]$, respectively. In contrast, such properties based on oil-treated fiber decreased by 18.8%, 23.23%, and 21% for fiber orientations of $[0]_7$, $[0/\pm45/ 0]_S$, and $[(0/90)_3/0]$, respectively.

Bamboo fiber treated with oil gained flexibility as a result of the lignin components absorbing the oil and becoming stronger than untreated fibers [45–47]. As a result, oil-treated fibers had more strength than untreated fibers.

Based on the results obtained in the experiment, as the number of fiber orientations in the laminate increases, water absorption of the laminate is insignificant, but its impact strength increases. Oil-treated bamboo fiber composites were better at resisting water absorption than untreated fiber laminates. Since the oil absorbed by the fiber develops hydrophobic properties inside the bundle, the loss of impact energy in oil-treated bamboo fiber laminates of polyester composites is less. The detailed experimental results (impact and toughness) of oil-treated and untreated bamboo fiber-reinforced polyester composites before and after water absorption are shown in Table 5.6.

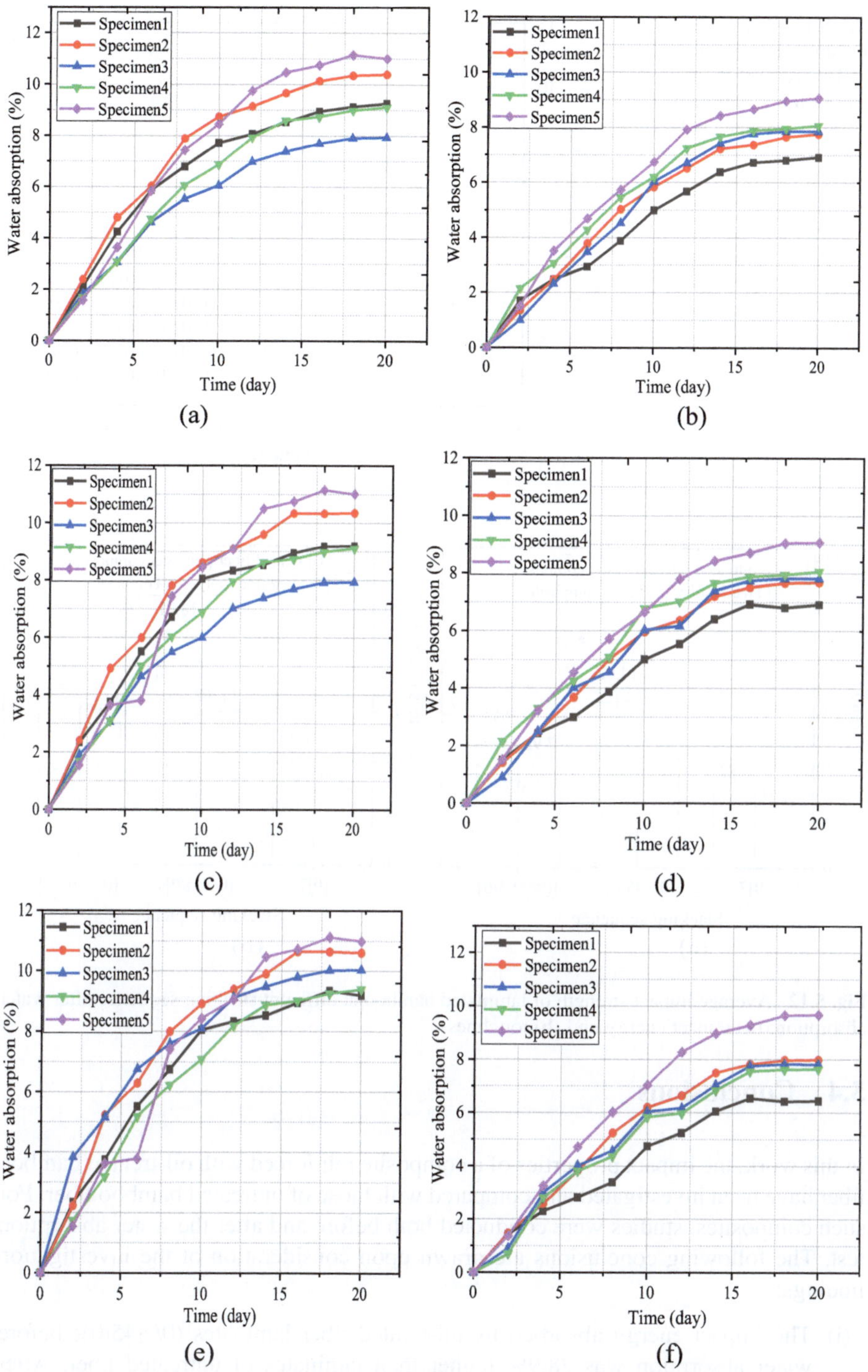

Fig. 5.10 Water absorption properties at different stacking sequence: **(a)** UBFP $[0_7]$; **(b)** OBFP $[0_7]$; **(c)** UBFP $[(0/90)_3/0]$; **(d)** OBFP $[(0/90)_3/0]$; **(e)** UBFP $[(0/\pm45/0)_S]$; and **(f)** OBFP $[(0/\pm45/0)_S]$ [41]

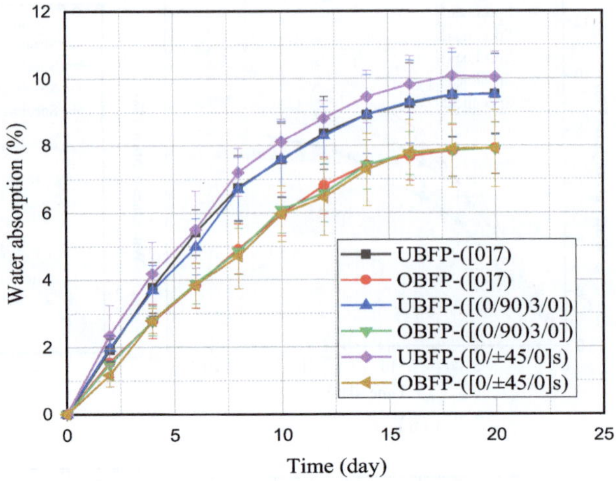

Fig. 5.11 Average moisture absorption of UBFP and OBFP composites at different stacking sequence [41]

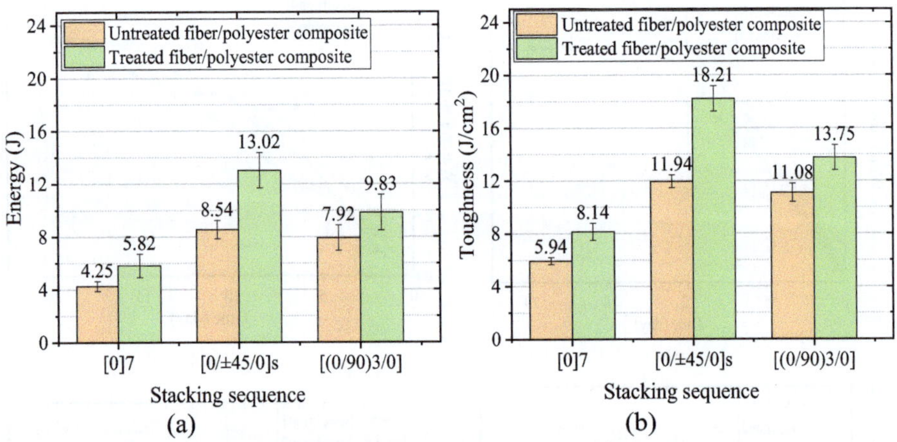

Fig. 5.12 Average impact strength of laminated bamboo at different stacking sequence after water absorption: (**a**) impact energy and (**b**) toughness

5.4 Conclusions

In this work, the impact properties of a composite reinforced with oil-treated bamboo fiber have been investigated and compared with those of untreated bamboo fiber. For such composites, studies were conducted both before and after the water absorption test. The following conclusions are drawn upon consideration of the investigation findings:

(i) The impact energy absorbed by oil-treated fiber laminates (0/±45/0)s before water absorption was 28.9% higher than laminates of untreated fiber. After water absorption, such laminates lose their impact strength by 23.23%.

Table 5.6 Impact properties of OBFP and UBFP before and after water absorption test

Materials		Specimen	Impact energy (J) $[0]_7$	Impact energy (J) $[0/\pm45/0]_S$	Impact energy (J) $[(0/90)_3/0]_S$	Toughness (J/cc) $[0]_7$	Toughness (J/cc) $[0/\pm45/0]_S$	Toughness (J/cc) $[(0/90)_3/0]_S$
OBFP	Before water absorption	1	7.04	16.13	12.54	10.8	19.29	24.82
		2	6.35	15.75	13.12	9.77	20.18	24.23
		3	8.11	14.00	12.25	12.4	18.85	21.50
		Average	7.17	16.96	12.64	11.0	23.53	19.44
		SD	0.89	1.14	0.44	1.37	1.75	0.67
	After water absorption	1	6.55	10.02	7.98	11.6	15.4	10.02
		2	5.74	14.23	10.5	5.35	20.88	15.75
		3	5.17	14.89	11.0	9.40	22.9	13.5
		Average	5.82	13.02	9.83	8.14	18.21	13.75
		SD	1.93	2.64	1.62	3.17	3.88	1.90
UBFP	Before water absorption	1	5.92	13.91	11.68	9.11	21.4	17.97
		2	5.8	14.6	10.83	8.92	22.48	16.66
		3	4.95	15.00	11.96	7.62	23.08	18.4
		Average	5.56	14.51	11.49	8.55	22.32	17.68
		SD	0.53	0.56	0.59	0.50	0.42	0.91
	After water absorption	1	3.30	8.28	7.75	5.04	9.37	11.90
		2	4.67	8.31	7.90	5.47	12.6	9.24
		3	4.78	9.05	8.10	7.31	13.85	12.1
		Average	4.25	8.54	7.92	5.94	11.94	11.08
		SD	0.82	0.44	0.18	1.21	2.31	1.60

(ii) Also, as the investigation shows, the number of fiber orientations and the impact energy absorbed by the laminate have a direct relationship.
(iii) Water absorption of the laminate does not depend on fibers orientation.
(iv) The failure mode of the laminated composite is due to the breakage of the fiber and matrix.
(v) Investigations on the shear, compressive, fatigue, vibration, fracture, and thermal properties are suggested for future studies.

References

1. Gon, D., Das, K., Paul, P., & Maity, S. (2013). Jute composites as wood substitute. *International Journal of Textile Science, 1*(6), 84–93.
2. Naveen, J., Jawaid, M., Amuthakkannan, P., & Chandrasekar, M. (2018). Mechanical and physical properties of sisal and hybrid sisal fiber-reinforced polymer composites. In *Mechanical and physical testing of biocomposites, fibre-reinforced composites and hybrid composites* (pp. 427–440). Woodhead Publishing.

3. Sivakandhan, C., Murali, G., Tamiloli, N., & Ravikumar, L. (2020). Studies on mechanical properties of sisal and jute fiber hybrid sandwich composite. *Materials Today: Proceeding, 21*, 404–407.
4. Ghamarian, N., Azmah Hanim, M. A., Penjumras, P., & Majid, D. L. (2016). *Effect of fiber orientation on the mechanical properties of laminated polymer composites*. Elsevier Ltd.
5. Huang, J. K., & Bin Young, W. (2018). The mechanical, hygral, and interfacial strength of continuous bamboo fiber reinforced epoxy composites. *Composites. Part B, Engineering, 166*(November), 272–283.
6. Indra Reddy, M., Anil Kumar, M., & Raju, C. R. B. (2018). Tensile and flexural properties of jute, pineapple leaf and glass fiber reinforced polymer matrix hybrid composites. *Materials Today: Proceedings, 5*(1), 458–462.
7. Dhakal, S., & Keerthi Gowda, B. S. (2017). An experimental study on mechanical properties of Banana polyester composite. *Materials Today: Proceedings, 4*(8), 7592–7598.
8. Sonar, T., Patil, S., Deshmukh, V., & Acharya, R. (2015). Natural fiber reinforced polymer composite material-a review. *IOSR Journal of Mechanical and Civil Engineerin*, 142–147.
9. Biswas, S., Shahinur, S., Hasan, M., & Ahsan, Q. (2015). Physical, mechanical and thermal properties of jute and bamboo fiber reinforced unidirectional epoxy composites. *Procedia Engineering, 105*, 933–939.
10. Pickering, K. L., Efendy, M. G. A., & Le, T. M. (2016). A review of recent developments in natural fibre composites and their mechanical performance. *Composites. Part A, Applied Science and Manufacturing, 83*, 98–112.
11. Mallick, P. K. (2007). *Fiber-reinforced composites: Materials, manufacturing, and design* (3rd ed.). Taylor and Francis Group, LLC.
12. Jawaid, M., & Abdul Khalil, H. P. S. (2011). Cellulosic/synthetic fibre reinforced polymer hybrid composites: A review. *Carbohydrate Polymers, 86*(1), 1–18.
13. Perremans, D., Trujillo, E., Ivens, J., & Van Vuure, A. W. (2018). Effect of discontinuities in bamboo fibre reinforced epoxy composites. *Composites Science and Technology, 155*, 50–57.
14. Osorio, L., Trujillo, E., Van Vuure, A. W., & Verpoest, I. (2011). Morphological aspects and mechanical properties of single bamboo fibers and flexural characterization of bamboo/epoxy composites. *Journal of Reinforced Plastics and Composites, 30*(5), 396–408.
15. Gratani, L., Crescente, M. F., Varone, L., Fabrini, G., & Digiulio, E. (2008). Growth pattern and photosynthetic activity of different bamboo species growing in the Botanical65 Garden of Rome. *Flora: Morphology, Distribution, Functional Ecology of Plants, 203*(1), 77–84.
16. Abdul Khalil, H. P. S., Bhat, I. U. H., Jawaid, M., Zaidon, A., Hermawan, D., & Hadi, Y. S. (2012). Bamboo fibre reinforced biocomposites: A review. *Materials and Design, 42*, 353–368.
17. Zakikhani, P., Zahari, R., Sultan, M. T. H., & Majid, D. L. (2014). Extraction and preparation of bamboo fibre-reinforced composites. *Materials and Design, 63*, 820–828.
18. Elanchezhian, C., Ramnath, B. V., Ramakrishnan, G., Rajendrakumar, M., Naveenkumar, V., & Saravanakumar, M. K. (2018). Review on mechanical properties of natural fiber composites. *Materials Today: Proceedings, 5*(1), 1785–1790.
19. Holbery, J., & Houston, D. (2006). Natural-fiber-reinforced polymer composites in automotive applications. *JOM, 58*(11), 80–86.
20. Lin, J., He, X., Hu, Y., Kuang, T., & Ceulemans, R. (2002). Lignification and lignin heterogeneity for various age classes of bamboo (Phyllostachys pubescens) stems. *Physiologia Plantarum, 114*(2), 296–302.
21. Li, T., Cheng, D. L., Wålinder, M. E., & Zhou, D. G. (2015). Wettability of oil heat-treated bamboo and bonding strength of laminated bamboo board. *Industrial Crops and Products, 69*, 15–20.
22. Mengstie, B. A., & Addisu, E. A. (2021). Effect of oil heat treatment on tensile and bending properties of Ethiopian Lowland Bamboo. In *Advances of science and technology: 8th EAI international conference, ICAST 2020*, Bahir Dar, Ethiopia, October 2–4, 2020, Proceedings, Part II 8 (pp. 204–215). Springer International Publishing.

23. Joshi, S. V., Drzal, L. T., Mohanty, A. K., & Arora, S. (2004). Are natural fiber composites environmentally superior to glass fiber reinforced composites. *Composites. Part A, Applied Science and Manufacturing, 35*(3), 371–376.
24. Wambua, P., Ivens, J., & Verpoest, I. (2003). Natural fibres: Can they replace glass in fibre reinforced plastics? *Composites Science and Technology, 63*(9), 1259–1264.
25. Varada Rajulu, A., Allah Baksh, S., Ramachandra Reddy, G., & Narasimha Chary, K. (1998). Chemical resistance and tensile properties of short bamboo fiber reinforced epoxy composites. *Journal of Reinforced Plastics and Composites, 17*(17), 1507–1511.
26. Chen, X., Guo, Q., & Mi, Y. (1998). Bamboo fiber-reinforced polypropylene composites: A study of the mechanical properties. *Journal of Applied Polymer Science, 69*(10), 1891–1899.
27. Schneider, C. A., Rasband, W. S., & Eliceiri, K. W. (2012). NIH image to ImageJ: 25 years of image analysis. *Nature Methods, 9*(7), 671–675.
28. David-West, O. (2019). Water absorption and effect of temperature on the characteristics of flax/hemp hybrid composites. *Materials Today: Proceedings, 31*, 4–8.
29. Yahaya, R., Sapuan, S. M., Jawaid, M., Leman, Z., & Zainudin, E. S. (2016). Effect of fibre orientations on the mechanical properties of kenaf–aramid hybrid composites for spall-liner application. *Defence Technology, 12*(1), 52–58.
30. Nahak, B., Gupta, M. K., & Kumar, A. (2019). Mechanical and water absorption properties of sisal composites: Effect of charcoal particles loading. *Materials Today: Proceedings, 18*, 3766–3774.
31. Arputhabalan, J., Palanikumar, K., Roche Adaikalaraj, S., & Sugan Priyan, M. (2019). Investigation of glass fiber influence on mechanical characteristics and resistance to water absorption of natural fiber reinforced polyester composites. *Materials Today: Proceedings, 16*, 843–852.
32. Devnani, G. L., Mittal, V., & Sinha, S. (2018). Mathematical modelling of water absorption behavior of bagasse fiber reinforced epoxy composite material. *Materials Today: Proceedings, 5*(9), 16912–16918.
33. Ashik, K. P., Sharma, R. S., & Guptha, V. L. J. (2018). Investigation of moisture absorption and mechanical properties of natural/glass fiber reinforced polymer hybrid composites. *Materials Today: Proceedings, 5*(1), 3000–3007.
34. Lila, M. K., Verma, A., & Bhurat, S. S. (2022). Impact behaviours of epoxy/synthetic fiber composites. In *Handbook of epoxy/fiber composites* (pp. 1–18). Springer.
35. Gupta, A., & Kumar, A. (2008). Potential of bamboo in sustainable development. *Asia Pacific Business Review, 4*(3), 100–107.
36. McMullen, P. (1984). Fibre/resin composites for aircraft primary structures: A short history, 1936–1984. *Composites, 15*(3), 222–230.
37. Embaye, K. (2003). *Ecological aspects and resource management of bamboo forests in Ethiopia* (Vol. 273). Swedish University of Agricultural Sciences.
38. Hyojin, K., Kazuya, O., Toru, F., & Kenichi, T. (2013). Influence of fiber extraction and surface modification on mechanical properties of green composites with bamboo fiber. *Journal of Adhesion Science and Technology, 27*(12), 1348–1358.
39. https://www.trl.com/astm_d256_izod_pendulum_impact_resistance_of_plastics/
40. Weast, R. C. (1972). Density of water (g/mL) vs. temperature (° C). *Handbook of Chemistry and Physics, 995371*, 6–7.
41. Mebratie, B. A., Ayele, B. S., & Meku, A. A. (2023). Investigating the tensile and flexural strength of sunflower oil treated Ethiopian Highland bamboo fibre reinforced polyester composites. *Advances in Bamboo Science, 3*, 100021. https://doi.org/10.1016/j.bamboo.2023.100021
42. Ab Ghani, M. H., & Ahmad, S. (2011). The comparison of water absorption analysis between counterrotating and corotating twin-screw extruders with different antioxidants content in wood plastic composites. *Advances in Materials Science and Engineering, 4*, 1.
43. Piao, X., Zhao, Z., Guo, H., Wang, Z., & Jin, C. (2022). Improved properties of bamboo by thermal treatment with wood wax oil. *Colloids and Surfaces A: Physicochemical and Engineering Aspects, 643*, 128807. https://doi.org/10.1016/j.colsurfa.2022.128807

44. Kushwaha, P. K., & Kumar, R. (2012). Influence of pre-impregnation treatment on bamboo reinforced Epoxy/UPE resin composites. *Open Journal of Composite Materials, 2*, 139–141.
45. Kaur, V., Chattopadhyay, D. P., & Kaur, S. (2013). Study on extraction of bamboo fibres from raw bamboo fibres bundles using different retting techniques. *Textiles and Light Industrial Science and Technology, 2*(4), 34–47.
46. Sood, M., & Dwivedi, G. (2018). Effect of fiber treatment on flexural properties of natural fiber reinforced composites: A review. *Egyptian Journal of Petroleum, 27*(4), 775–783.
47. Rassiah, K., Ahmad, M. M. H. M., & Ali, A. (2014). Mechanical properties of laminated bamboo strips from gigantochloa scortechinii/polyester composites. *Materials & Design, 57*, 551–559.

Chapter 6
Effects of Aggregate Size and Water-to-Cement Ratio on Fracture Energy of Normal Weight Concrete

Nikodimos A. Kasaw, Tesfaye A. Mohammed, and Habtamu A. Tadesse

6.1 Introduction

In the realm of global construction, concrete stands out as a versatile and frequently utilized building material. It is crafted by blending cement, water, additives, and both fine and coarse aggregate in precise proportions. Normal weight concrete finds extensive use in civil engineering structures due to its uncomplicated mix design, ready availability, and prolonged service life in normal and accidental conditions, workability, and commendable fire rating.

The study of the mechanical behavior of materials with cracks under applied loads is known as fracture mechanics. In the context of metals and concrete, fracture mechanics exhibits nonlinearity owing to the development of a substantial nonlinear zone at the fracture front. While metal fractures involve nonlinear zones associated with perfect yielding, hardening plasticity, and minimal fracture process zones (FPZ), concrete fractures predominantly fill the nonlinear zone with minimal to no plastic deformation. The region where the material undergoes softening damage is termed the fracture process zone (FPZ). Structural materials are categorized as brittle, quasi-brittle, or elastic-plastic based on the shape and dimensions of the

N. A. Kasaw (✉)
School of Civil and Hydraulics Engineering, Civil Engineering Department,
University of Gondar, Gondar, Ethiopia

T. A. Mohammed
Civil Engineering Department, Addis Ababa Science and Technology University,
Addis Ababa, Ethiopia

H. A. Tadesse
Faculty of Civil and Water Resources Engineering, Department of Structural Engineering,
Bahir Dar University, Bahir Dar, Ethiopia

K. Mequanint et al. (eds.), *Sustainable Development Research in Materials and Energy*, Green Energy and Technology,
https://doi.org/10.1007/978-3-031-69860-6_6

FPZ formed before fracture ahead of the crack tip. Concrete falls into the quasi-brittle materials category.

Within quasi-brittle materials, the energy dissipated for FPZ formation ahead of the crack tip is referred to as fracture energy (GF). This parameter, denoting the amount of energy required to create a unit area of crack surface, is pivotal in understanding concrete's fracture behavior. Concrete structures, during production and construction, often develop numerous microcracks and flaws, which, under various loading scenarios, can lead to unexpected failures. Predicting potential failure modes is crucial for preventing such mishaps.

The fracture energy of concrete serves as a valuable index in safe structural design and failure behavior modeling. By measuring the energy consumed in fracture propagation and the creation of new crack surfaces, the failure process can be examined. There are two types of fracture energies: total fracture energy (GF), representing the complete load displacement curve area, and initial fracture energy (Gf), representing the initial tangent area of the softening curve.

This research aims to investigate the fracture energy of locally manufactured normal weight concrete. Specifically, it seeks to demonstrate how altering the maximum aggregate size distribution and water-cement ratio in the mix design can affect the fracture matrix and lead to variations in fracture energy.

The study explores the influence of maximum aggregate size distribution and water-to-cement ratio on the total fracture energy of normal weight concrete. Laboratory tests involved four mix designs based on two nominal maximum coarse aggregate sizes (9.5 and 12.5 mm) and two water-cement ratios (0.5 and 0.6). Twelve notched beams were prepared, and fracture energy was examined using the work of fracture method, following RILEM TC50-FMC recommendations for three-point bending tests.

Fracture mechanics, when employed, enables the characterization of failure, identification of structural failure causes, selection of crack-resistant materials, and prediction of crack behavior. Recent studies underscore the indispensability of integrating fracture mechanics with strength criteria for designing safer and more effective concrete structures. Despite this, the fracture behavior of locally sourced concrete remains underexplored in the existing literature.

For quasi-brittle materials, fracture energy emerges as a potent index in failure behavior modeling and safe structural design. This research delves into the effects of coarse aggregate size and water-cement ratio on the fracture energy property of concrete produced from locally manufactured aggregates. The overarching objective is to scrutinize how these factors impact the fracture energy of locally manufactured normal weight concrete.

6.2 Literature Review

Numerous scholars have extensively delved into the fracture behavior of concrete from diverse perspectives, investigating the influence of matrix characteristics, inclusions, bond behavior, particle size distribution, volume fraction, and shape on composite materials like concrete. As a result, various studies have been conducted to uncover the fracture characteristics and cracking modes in different types of concrete.

In an experimental study conducted by Amparano et al. [1], the impact of aggregate content on concrete fracture behavior was explored using a size effect approach. Two key parameters defining concrete's fracture behavior, namely, fracture energy (GF) and effective fracture process zone (Cf), were examined. The overall quantity of aggregate and the ratio of coarse to fine aggregate emerged as crucial factors determining these characteristics. The study maintained a fixed volume ratio of coarse to fine aggregate at 2, focusing specifically on the influence of total aggregate content. Results from varying total aggregate content (45–75% of concrete volume) revealed significant insights.

With an increase in the volume fraction of aggregate, the compressive strength of concrete showed a decrease, contrary to conventional composite theories. Fracture energy GF displayed variations within a 25% range, suggesting that the impact of aggregate on GF is less pronounced compared to other composite types. GF decreased with rising aggregate content, reaching a minimum value at a 65% aggregate volume, and then exhibited an increasing trend. This implies that the critical volume fraction of aggregate, corresponding to the peak value of GF, may not exist for the concrete under study (with a water-to-cement ratio of 0.5 and fixed aggregate gradation).

Siregar et al. [32] further investigated the influence of various aggregate size distributions on the fracture behavior of high-strength concrete. Their study concluded that the fracture energy of high-strength concrete is significantly affected by aggregate size distribution, as indicated by the coarseness of the internal random grain structure. A 50% increase in measured values of GF was observed when the value changed from 0.32 to 0.22. Additionally, the study revealed that aggregate size distribution and the water-to-cement ratio impact the ductility of high-strength concrete, reflected in observed values of GF and KIC. The strength of the aggregate emerged as a limiting factor for the maximum value of GF in high-strength concrete.

Chen and Liu's research in 2004 highlighted the influence of both volume proportion of coarse aggregate and aggregate size distribution on the KIC and GF of normal and high-strength concrete. This influence was attributed to the physical properties of the crack path through the concrete, where microcracking, crack bridging, and aggregate interlocking contribute to increased energy absorption. The coarseness of the internal random grain structure, determined by aggregate grading, affected the direction of a crack during fracture, indicating that aggregate grade influences how concrete fractures. Their study suggested that larger aggregate could be employed to create high-strength concrete with reduced

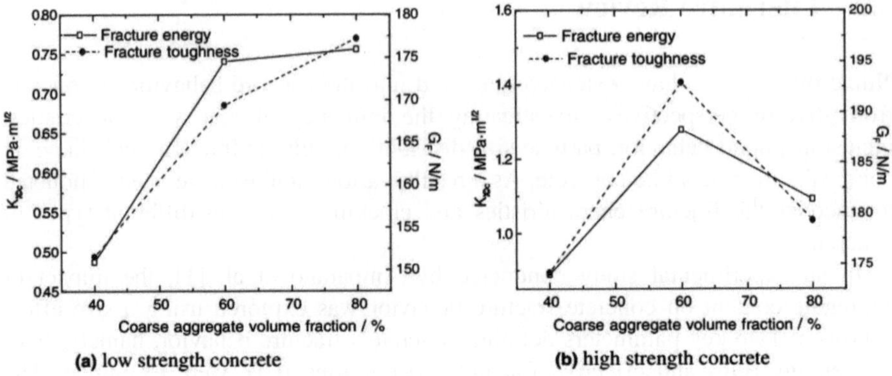

Fig. 6.1 Variation of G_F and K_{IC} of concrete with different coarse aggregate volume percent [10]

brittleness. Furthermore, they found that GF and KIC of low-strength concrete increased as aggregate volume increased from 40% to 80%. The investigation observed that with increasing aggregate size, the GF and KIC of notched concrete beams with specified dimensions rose. The hypothesis was proposed that using aggregate with lesser brittleness would result in high-strength concrete. Figure 6.1 below illustrates that the maximum GF and KIC values for high-strength concrete are achieved at 60% of aggregate volume.

The interfacial transition zone (ITZ) stands out as the most sensitive region in normal concrete, and enhancing this phase could lead to an increase in fracture energy. The characteristics of the ITZ play a crucial role in determining the cracking path. Research indicates that the size, volume, shape, surface, and texture of the aggregate significantly impact the properties of the ITZ.

In 2003, Elsharief, Cohen, and Olek conducted an experimental study examining the impact of aggregate size, age, and water-to-cement ratio (w/c) on the interfacial transition zone (ITZ) microstructure between normal weight aggregate and bulk cement paste. The research underscores the pivotal role of the w/c ratio in governing the thickness and microstructure of the ITZ. Notably, a reduction in the w/c ratio from 0.55 to 0.40 led to an ITZ exhibiting characteristics indistinguishable from those of the bulk paste, as revealed by backscattered electron images (BSE).

Furthermore, the study highlights the significant influence of aggregate size on ITZ properties. A decrease in aggregate size correlates with a reduction in ITZ porosity, emphasizing a noteworthy relationship between aggregate size and ITZ characteristics. This emphasizes the necessity of considering factors related to aggregates when aiming to optimize the ITZ microstructure for enhanced concrete performance.

6.3 Methodology

The primary objective of this study's experimental program was to examine the influence of water-to-cement ratio and aggregate size distribution on the fracture energy of normal weight concrete. The concrete mixes were formulated following the ACI 211.1 method for normal strength concrete. Subsequently, both compressive strength and three-point bending tests were conducted. Additionally, a comparative analysis was carried out by utilizing the mechanical properties necessary for an established empirical model to predict fracture energy. This analysis involved a thorough comparison between the experimentally determined fracture energy and the predictions provided by existing empirical models. The specific details of the experimental program are outlined in the subsequent sections.

6.3.1 Materials

6.3.1.1 Cement

The cement used was a grade of 42.5 N ordinary Portland cement (OPC) made by Derba cement factory (Ethiopia) throughout the experimental investigation.

6.3.1.2 Aggregates

In all the concrete mixes used in this study, the fine aggregate was natural river sand with a fine modulus of 2.75. The sand underwent a preliminary screening process using a sieve, followed by washing and drying before undergoing tests to examine its properties.

For the coarse aggregate in specimen preparation, crushed basaltic stone sourced from Bahir Dar, Ethiopia, was employed. To minimize the impact of water absorption during mixing and accurately assess the influence of aggregates on concrete properties, the aggregates were maintained in a saturated surface dry condition. The coarse aggregate underwent a process of washing and drying, followed by screening with sieves of sizes 19.5 mm, 12.5 mm, 9.5 mm, and 4.75 mm. Subsequently, the coarse aggregates were blended in accordance with the grading requirements specified by ASTM.C33-07 [2].

To investigate the effects of water-to-cement ratio and maximum size of coarse aggregate on the fracture energy of normal weight concrete, four mixes (NWC1, NWC2, NWC3, and NWC4) were designed. These mixes had water-to-cement ratios of 0.5 and 0.6 and maximum coarse aggregate sizes (dnmax) of 9.5 mm and 12.5 mm. The design of the concrete mixes adhered to the ACI 211.1 method for normal strength concrete. Detailed mix proportions can be found in Table 6.1.

Table 6.1 Mixture composition of concrete

	NWC1	NWC2	NWC3	NWC4
d_{nmax}(mm)	9.5	9.5	12.5	12.5
w/c ratio	0.5	0.6	0.5	0.6
Cement(kg/m^3)	400	333.33	400	333.33
Water (kg/m^3)	200	200	200	200
Coarse aggregate (kg/m^3)				
(12.5 mm–16 mm)	–	–	95	95
(9.5 mm–12.5 mm)	142.5	142.5	380	380
(4.75 mm–9.5 mm)	807.5	807.5	475	475
Fine aggregate (kg/m^3)	780	780	780	780

6.3.2 Experimental Test Specimen Preparation and Test Procedure

An experimental initiative was conducted to investigate the effects of maximum coarse aggregate size and water-to-cement ratio on fracture parameters in normal concrete. This study utilized 12 notched beam specimens following the dimensions recommended by RILEM TC50-FMC. Four concrete mixes, identified as NWC1, NWC2, NWC3, and NWC4, were formulated, incorporating variations in water-cement ratios and nominal maximum coarse aggregate sizes. The proportions other than these variables were kept constant for a comparative analysis, as outlined in Table 6.1.

To meet the stringent horizontal requirements of the casted three-point beam specimens, meticulously crafted wood molds were employed for concrete casting. Prior to casting, both molds and steel sheets were coated with a thin layer of oil. A consistent 3-mm-wide vertical notch was initiated on the beam's tensile face during casting, maintaining a notch depth-to-beam depth ratio of 0.4 (αo = 0.4d). After 4 h of casting, the steel sheets were removed, and all specimens were demolded approximately 24 h later. Subsequently, the specimens underwent a 28-day water soaking period before testing.

Following the recommendations of RILEM TC50-FMC [28], the specimen dimensions were proportional to dmax. Three standard specimens were cast for each mix to determine both compressive strength and fracture energy. The size of the beam specimen depended on the maximum aggregate size, dmax. For dmax less than or equal to 16 mm, the fracture behavior of the hardened concrete was analyzed using 100 mm × 100 mm × 840 mm beam specimens subjected to a three-point bend test according to RILEM TC 50-FMC [28]. The dimensions of the beam—depth (d), width (b), and total length (L)—were 100 mm, 100 mm, and 840 mm, respectively, with a support span (S) of 600 mm. Additionally, the cube compression strength (fc) was tested at 28 days using specimens of dimensions 150 mm × 150 mm × 150 mm, with notches strategically placed in the middle of the length of the beam during casting (Fig. 6.2).

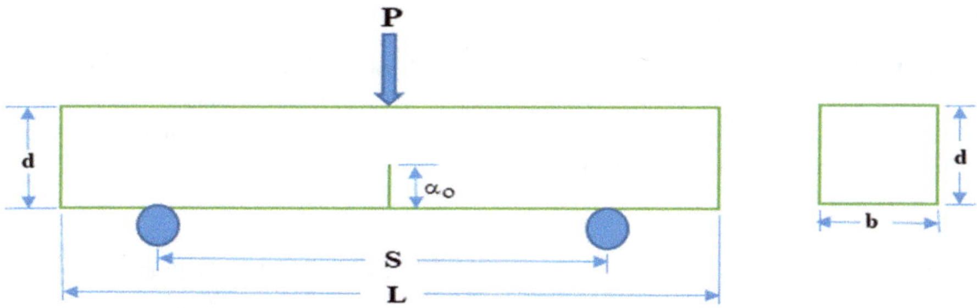

Fig. 6.2 Schematic three-point bending test configuration

Table 6.2 The test parameter description

Test parameters	Levels	Description
Water-to-cement ratio	1	0.5
	2	0.6
Aggregate size	1	$d_{n\max}$=9.5 mm
	2	$d_{n\max}$=12.5 mm

To evaluate fracture energy using the work of fracture method (WFM), a series of tests were executed on notched beams exposed to three-point bending. Each mixture underwent testing with three notched beams, maintaining consistent dimensions of 100 × 100 × 840 mm (width, depth, length), featuring a 40-mm notch depth. The preparation of these notched beams was meticulous. All beams were subjected to testing using a computer-controlled electrohydraulic servo universal testing machine (WAW-1000E) with a maximum load capacity of 1000kN, operating under displacement control. A constant displacement rate of 0.2 mm/s was applied to the beam specimens to ensure thorough tracking of post-peak ductility. Additionally, for each experimental mix, three cubes measuring 150 × 150 × 150 mm were cast to determine compressive strength (fc).

6.3.3 Test Parameter Description

The test parameters description and result format for the three-point bending test are summarized in Tables 6.2 and 6.3 respectively.

Table 6.3 Test result format for 3-PBT

Type of test	W/C ratio	trials	Aggregate size (d_{nmax})	
			9.5 mm	12.5 mm
Three-point bending test (fracture energy in N/m)	0.5	1	✓	✓
	0.5	2	✓	✓
	0.5	3	✓	✓
Type of test	**W/C ratio**	**Trials**	**Aggregate size (d_{nmax})**	
			9.5 mm	**12.5 mm**
Three-point bending test (fracture energy in N/m)	0.6	1	✓	✓
	0.6	2	✓	✓
	0.6	3	✓	✓

6.4 Results and Discussion

6.4.1 Analysis of Fracture Parameters from WFM

According to RILEM 50-FMC [28], the total fracture energy (G_F) by using WFM is determined by calculating the area under load displacement curve. The test results of 3-PBT are shown below in Table 6.4.

6.4.2 Load Versus Displacement Graph of the Notch Beam Test Specimen

The three-point bending test were performed on WAW-1000D hydraulic universal testing machine. The loading rate was displacement controlled with a rate of 0.2 mm/s. It was carefully examined in displacement controlled loading rate of 0.2 mm/s to fully trace the post-peak load displacement curve. The load displacement curve for NWC1, NWC2, NWC3, and NWC4 is shown in Figs. 6.3, 6.4, 6.5, and 6.6, respectively.

6.4.3 Effects of Water-to-Cement Ratio on Fracture Energy

Several investigations have delved into the effects of the water-to-cement ratio on fracture parameters, with a particular focus on fracture energy (GF). Carpinteri and Brighenti's [7] study, examining the influence of the water-to-cement ratio ranging from w/c = 0.45 to w/c = 0.75 on both plain and fiber-reinforced concrete, identified that the maximum GF value occurs at w/c = 0.55.

Table 6.4 Test results for 3-PBT

Type of test	W/C ratio	Trials	Aggregate size (d_{nmax})	
			9.5 mm	**12.5 mm**
Three-point bending test (fracture energy in N/m)	0.5	1	239.5	329.167
	0.5	2	197.167	331.667
	0.5	3	230	341.333
Type of test	**W/C ratio**	**Trials**	**Aggregate size (d_{nmax})**	
			9.5 mm	**12.5 mm**
Three-point bending test (fracture energy in N/m)	0.6	1	250	279.33
	0.6	2	259.667	289
	0.6	3	248	–

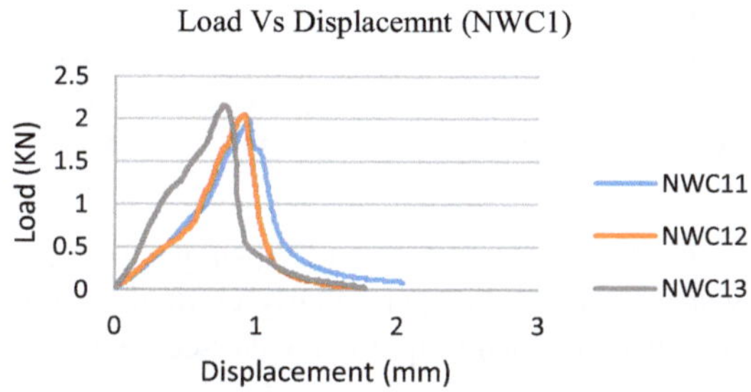

Fig. 6.3 Load versus displacement plot of NWC1

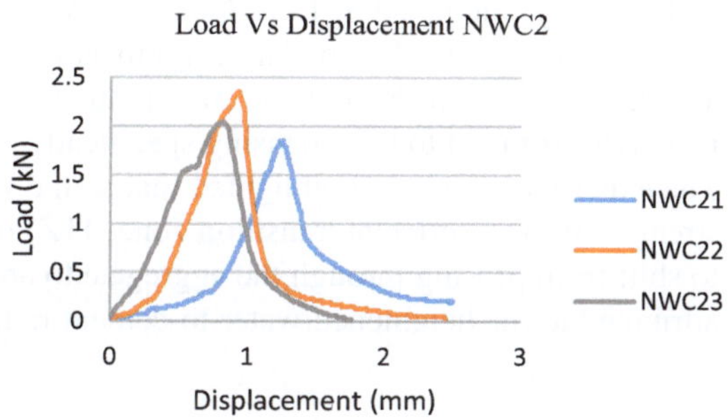

Fig. 6.4 Load versus displacement plot of NWC2

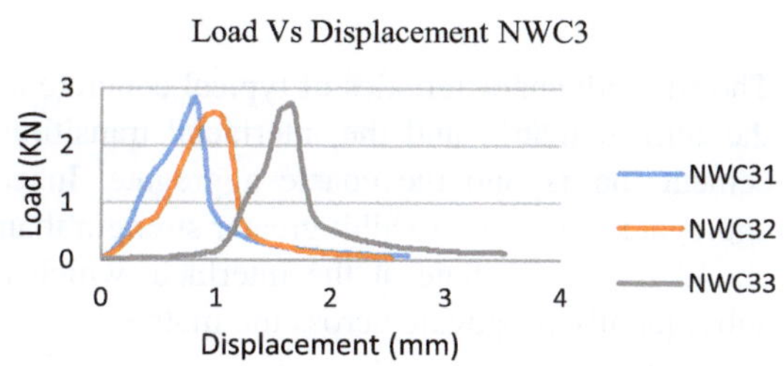

Fig. 6.5 Load versus displacement plot of NWC3

Fig. 6.6 Load versus displacement plot of NWC4

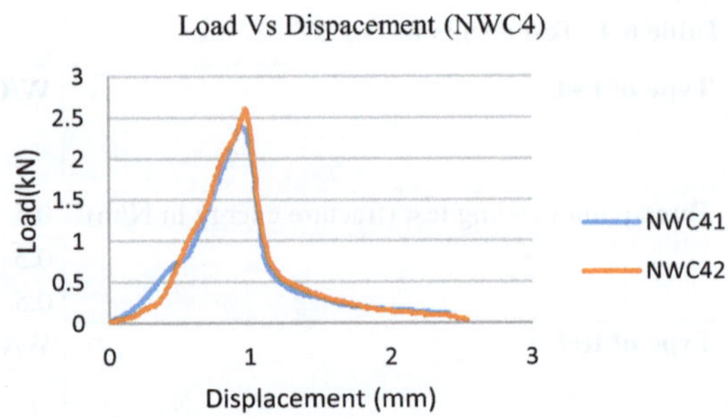

Table 6.5 Summary of compressive and 3-PBT test results

Mix ID.	f_c (Mpa)	G_F(N/m)			Average G_F (N/m)
		Beam 1	Beam 2	Beam 3	
NWC1	49.13	239.5	197.167	230	222.22
NWC2	40.22	250	259.667	248	252.556
NWC3	47.7	329.167	331.667	341.333	334.05
NWC4	44.35	279.33	289	–	284.167

Findings from Petersson [25] suggested a decrease in GF with an elevation in the water-cement ratio, particularly evident beyond a ratio of 0.4. While GF appears relatively constant up to a water-cement ratio of 0.5, a pronounced decline is observed beyond this threshold. Conversely, Khatieb [20] proposed that increasing the water-to-cement ratio could enhance fracture energy (Table 6.5).

In a study conducted by Lotfi-Omran, Nikbin, and Sandrmomtazi [30], the water-to-cement ratio was identified as a significant influencer of total fracture energy GF. The research revealed a notable 27% reduction in GF as the water-to-cement ratio increased from 0.4 to 0.7, focusing specifically on heavyweight magnetite concrete. The experimental study highlighted that as the water-to-cement ratio increased, the strength of the interfacial transition zone (ITZ) decreased, causing the fracture path to shift from passing through the aggregate to around the aggregates. This shift was attributed to the heightened water-to-cement ratio.

6.4.4 Effects of Aggregate Size on Fracture Energy

The strength characteristics of typical concrete are contingent upon the attributes of the cement matrix and the interfacial transition zone (ITZ) existing between the cement matrix and the coarse aggregate. In concrete with normal strength, the aggregates generally exhibit greater strength than the cement matrix. Consequently, cracks tend to initiate at the interface, which represents the weakest phase, and subsequently propagate across the matrix.

As the crack encounters the more robust aggregate phase, it deflects, bypassing the aggregate and leading to its debonding. The crack follows an elongated path as it navigates around the coarse aggregate, particularly noticeable with larger aggregate sizes. Moreover, the presence of coarse aggregate fosters interlocking among aggregates, demanding more energy for crack propagation. Larger aggregates not only impact crack propagation but also wield influence over the post-peak behavior of bending specimens.

Research outcomes suggest several observations as the maximum size of the aggregate increases. Ductility tends to rise, indicating an augmented ability for deformation before failure. Concurrently, the slope of the post-peak segment of the load displacement curve diminishes, signifying a change in the material's stiffness after reaching the peak load. Additionally, the tail of the load displacement curve ascends, indicating a delayed and gradual decrease in load after reaching the peak point. These effects collectively underscore the intricate interplay between aggregate size and the mechanical behavior of concrete specimens, encompassing both crack propagation and post-peak performance.

6.4.5 Comparison of the Results of the Present Research with the Proposed Empirical Relations

The analysis involves a comparison between the fracture energy (GF) values derived from the current experimental study on normal weight concrete and those predicted by several established models, including [3, 8, 9, 17, 22]. Table 6.6 succinctly

Table 6.6 Proposed empirical equations by researchers to determine GF

Researchers	Proposed equations	Remarks
Bažant and Giraudon [3]	$G_f = \alpha_o \left(\frac{f'_c}{0.051}\right)^{0.46} \left(1 + \frac{d_{max}}{11.27}\right)^{0.22} \left(\frac{w}{C}\right)^{-0.3}$ $G_F = 2.5\alpha_o \left(\frac{f'_c}{0.051}\right)^{0.46} \left(1 + \frac{d_{max}}{11.27}\right)^{0.22} \left(\frac{w}{C}\right)^{-0.3}$	$\alpha_o = 1$ for round aggregates $\alpha_o = 1.44$ for crushed or angular aggregate
Nikbin et al. [22]	$G_f = \alpha_o (G_f)_{chart} \times C_{fc} \times C_{Age} \times C_{W/C}$ $G_F = 2.46\alpha_o (G_f)_{chart} \times C_{fc} \times C_{Age} \times C_{W/C}$	$C_{fc} = 0.261\left(\frac{f_c}{50}\right) + 0.7516$ $C_{Age} = 0.1951\left(\frac{Age}{28}\right) + 0.8132$ $C_{W/C} = -0.4735\left(\frac{W/C}{0.4}\right) + 1.4903$ $\alpha_o = 0.9$ for round aggregates $\alpha_o = 1.44$ for crushed or angular aggregate
CEB-FIP [8]	$G_F = \left(0.0469 d_{max}^2 - 0.5 d_{max} + 26\right)\left(\frac{f_c}{10}\right)^{0.7}$	
CEB-FIP [9]	$G_F = 73(f_{cm})^{0.18}$	$f_{cm} = f_{ck} + 8$ (Mpa)
JSCE [17]	$G_F = 10(d_{max})^{0.33}(f'_c)^{0.33}$	

Notes: G_F (N/m) = fracture energy; G_f (N/m) = size-independent fracture energy; f'_c (MPa) = compressive strength of cylindrical concrete specimen; f_c (MPa) = compressive strength of cubic concrete specimen; f_{ck} is the specific characteristic compressive strength (MPa); w/c = water-to-cement ratio; d_{max} = maximum aggregate size

Table 6.7 The experimental results and existing empirical relations by researchers to determine the fracture energy

Mix id.	f'_c (Mpa)	G_F (N/m) Exp. result	Bazant and Becq-Giraudon [3]	Nikbin et al. [22]	CIB-FIP (1990)	CIB-FIP [9]	JSCE [17]
NWC1	39.304	222.22	110.796	133.817	70.254	146.005	77.122
NWC2	32.176	252.556	95.971	111.838	61.36	141.918	72.35
NWC3	38.16	334.05	113.005	141.491	76.62	145.509	83.041
NWC4	35.48	284.167	103.465	121.051	72.812	143.95	81.069

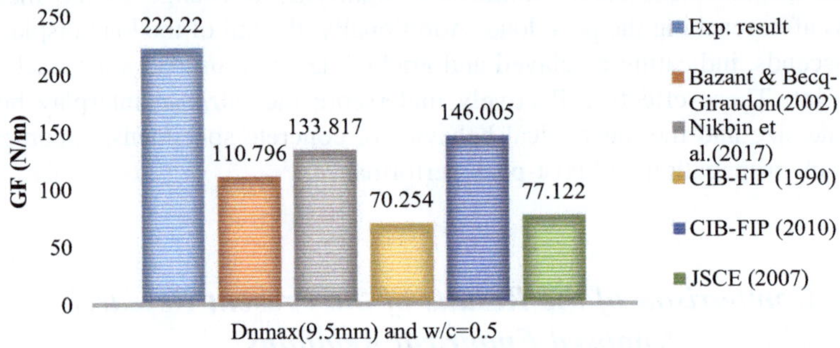

Fig. 6.7 Experimental and empirical estimated fracture energy of NWC1

presents some of the equations proposed by these researchers for determining the fracture energy GF, specifically the size-independent fracture energy Gf derived from the effective parameters of mix design and compressive strength (Table 6.7).

The following figures show a comparative analysis of experimentally estimated fracture energy values by work of fracture method with five existing empirical models for normal weight concrete (Figs. 6.7, 6.8, 6.9, and 6.10).

6.5 Conclusions

In this investigation, the influence of aggregate size and water-to-cement ratio on the fracture energy of normal weight concrete was explored through experimental analysis using the work of fracture method. Additionally, a comparative assessment was conducted between experimentally determined fracture energy values and those predicted by five existing empirical models. The key findings are summarized as follows:

Effect of Water-to-Cement Ratio The experimentally estimated fracture energy values demonstrated that with an increase in the water-to-cement ratio from 0.5 to 0.6, the energy required for crack propagation rose from 222.22 N/m to

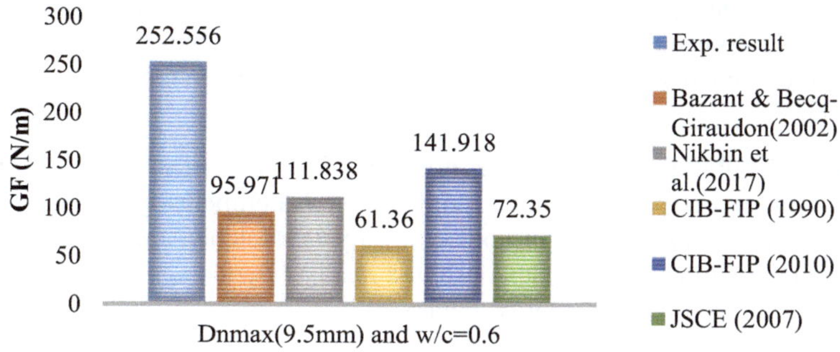

Fig. 6.8 Experimental and empirical estimated fracture energy of NWC2

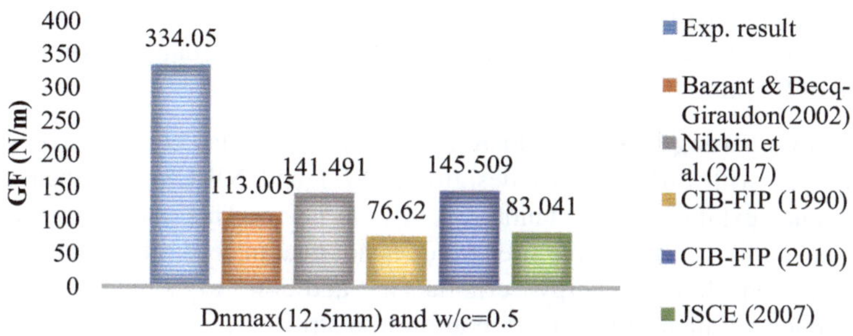

Fig. 6.9 Experimental and empirical estimated fracture energy of NWC3

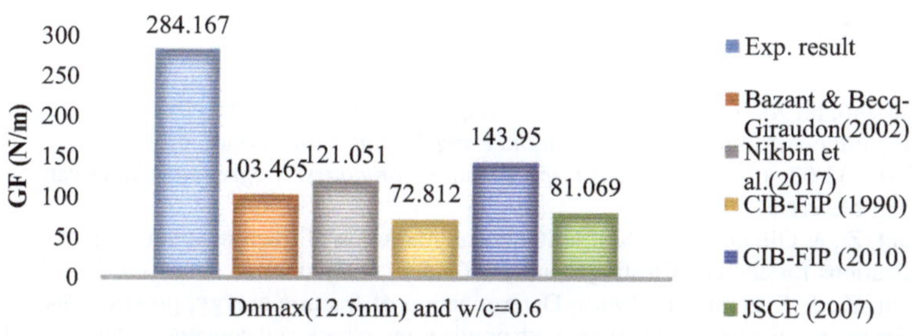

Fig. 6.10 Experimental and empirical estimated fracture energy of NWC4

252.556 N/m for a nominal maximum aggregate size distribution of 9.5 mm. Conversely, for a 12.5-mm nominal maximum aggregate size distribution, the fracture energy decreased from 334.05 N/m to 284.167 N/m.

Impact of Aggregate Size The experimental results revealed that increasing the nominal maximum aggregate size from 9.5 mm to 12.5 mm led to a 50.34% increase in fracture energy when the water-to-cement ratio was fixed at 0.5 and a 12.52% increase when the water-cement ratio was 0.6. This suggests that larger aggregates result in greater energy absorption due to the intricacies of crack propagation and fracture dimensions.

Optimal Mix Design Among the tested normal weight concrete samples, NWC3 (with a water-to-cement ratio of 0.5 and a nominal maximum aggregate size of 12.5 mm) exhibited superior performance in both fracture energy and compressive strength.

Empirical Model Predictions The predicted fracture energy values from five empirical models consistently underestimated the experimentally obtained fracture energy values of normal weight concrete samples. Notably, the empirical model proposed by Nikbin et al. [22] demonstrated relatively good agreement with experimentally estimated fracture energy values.

Water-to-Cement Ratio Relationship Empirical relations proposed by Bažant and Giraudon [3] and Nikbin et al. [22] suggested an inverse relationship between water-to-cement ratio and fracture energy for all maximum aggregate size distributions. However, the experimental results in this study indicated a direct relationship between water-to-cement ratio and fracture energy for a nominal maximum aggregate size of 9.5 mm.

Compressive Strength Versus Fracture Energy Similar compressive strength was achieved with different mix designs (NWC1 and NWC3), yet their fracture energy values exhibited a substantial 50.34% variation. This underscores the conclusion that similar compressive strength does not guarantee similar fracture energy. Therefore, considering energy criteria in addition to strength criteria is recommended to enhance the safety margin of a structure.

References

1. Amparano, F., Xi, Y., & Roh, Y. (2000). Experimental study on the effect of aggregate content on fracture behaviour of concrete. *Engineering Fracture Mechanics, 65-84.*
2. ASTM.C33-07. (2004). *Standard specification for coarse aggregate.* American Society of Testing Materials.
3. Bažant, Z., & Giraudon, E. (2002). Statistical prediction of fracture parameters of concrete and implications for choice of testing standard. *Cement and Concrete Research, 529–556.*
4. Bazant, Z., & Kazemi, M. (1990). Determination of fracture energy, process zone length and brittleness number from size effect, with application to rock and concrete. *International Journal of Fracture, 44*, 111–131.
5. Bazant, Z., & Pfeiffer, P. (1987). Determination of fracture energy from size effect and brittleness number. *ACI Material Journal, 84*(6), 463–480.
6. Bazant, Z., & Planas, J. (1997). *Fracture and size effect in concrete and other Quasibrittle materials.* CRC Press.
7. Carpinteri, A., & Brighenti, R. (2010). Fracture behaviour of plain and fiber-reinforced concrete with different water content under mixed mode loading. *Materials and Design, 2032–2042.*
8. CEB-FIP. (1990). Fib model code for concrete structures 1990.
9. CEB-FIP. (2010). Fib model code for concrete structures.
10. Chen, B., & Liu, J. (2004). Effect of aggregate on the fracture behaviour of high strength concrete. *Construction and Building Materials, 18*, 585–590.
11. Elices, M., & Rocco, C. (2008). Effect of aggregate size on the fracture and mechanical properties of a simple concrete. *Engineering Fracture Mechanics, 75*, 3839–3851.

12. Elices, M., Guinea, G., & Planas, J. (1997). On the measurement of concrete fracture energy using three-point bend tests. *Material and Structures, 30*, 375–376.
13. Elsharief, A., Cohen, M. D., & Olek, J. (2003). Influence of aggregate size, water cement ratio and age on the microstructure of the interfacial transition zone. *Cement and Concrete Research, 33*, 1837–1849.
14. Habtamu, A., Temesgen, W., & Alemayehu, G. (2021). Application of potential based cohesive model for analysis of concrete fracture. In *Advances of science and technology: 9th EAI international conference, ICAST* (pp. 15–26), Hybrid Event, Bahir Dar, Ethiopia, August 27–29, 2021, Proceedings, Part II.
15. Hillerborg, A. (1985). The theoretical basis of a method to determine the fracture energy GF of concrete. *Material and Structures, 18*(4), 291–296.
16. Hillerborg, A., Modeer, M., & Petersson, P. (1976). Analysis of crack formation and crack growth in concrete by means of fracture mechanics and finite elements. *Cement and Concrete Research, 6*(6), 773–781.
17. JSCE. (2007). Standard Specifications for Concrete Structures (15), 1–503.
18. Karamloo, M., Mazloom, M., & Payganeh, G. (2016). Effects of maximum aggregate size on fracture behaviour of self-compacting light weight concrete. *Construction and Building Materials, 123*, 508–515.
19. Khalilpour, S., Banisad, E., & Dehestai, M. (2019). A review on concrete fracture energy and effective. *Cement and Concrete Research, 120*, 294–321.
20. Khatieb, M. (2016). Experimental evaluation of concrete fracture energy and its dependency on relevant parameters. *International Journal of Applied Engineering Research, 11*, 10348–10352.
21. Nakamura, H., Nanrib, T., Miura, T., & Roya, S. (2018). Experimental investigation of compressive strength and compressive fracture energy of longitudinally cracked concrete. *Cement and Concrete Composites, 93*, 1–18.
22. Nikbin, I., Rahimi, S., & Allahyari, H. (2017). A new empirical formula for prediction of fracture energy of concrete based on the artificial neural network. *Engineering Fracture Mechanics, 186*, 466–482.
23. Park, K. (2005). *Concrete fracture mechanics and size effect using a specialized choisive zone model*. University of Illinois at Urbana-Champaign.
24. Park, K. (2009). *Potential-based fracture mechanics using cohesive zone and virtual internal bond modeling*. University of Illinois at Urbana-Champaign.
25. Petersson, P. (1980). Fracture energy of concrete: Practical performance and experimental. *Cement Concrete Research, 10*, 91–101.
26. Petersson, P. (1981). *Crack growth and development of fracture zones in plain concrete and similar materials*. Division of Building Materials, LTH, Lund University.
27. Rao, A., & Rao, G. (2014). Fracture mechanics of fiber reinforced concrete: An overview. *International Journal of Engineering Innovations and Research, 517.*
28. Rilem, D. (1985). Determination of the fracture energy of mortar and concrete by means of the three-point bend tests on notched beams. *Material and Structures, 18*, 285–290.
29. RILMFMT-89. (1990). Size-effect method for determining fracture energy and process zone size of concrete. *Material and Structure, 23*(6), 461–465.
30. Sandrmomtazi, A., Lotfi-Omran, O., & Nikbin, I. (2019). On the fracture parameters of heavy-weight magnetite concrete with different water-cement ratios through three methods. *Engineering Fracture Mechanics, 219*, 106615.
31. Shi, Z. (2009). *Crack analysis in structural concrete: Theory and applications*. Butterworth-Heinemann.
32. Siregar, A. P. N., Rafiq, M. I., & Mulheron, M. (2017). Experimental investigation of the effects of aggregate size distribution on the fracture behaviour of high strength concrete. *Construction and Building Materials, 150*, 252–259.
33. Standard, A. C. I. (1996). Standard practice for selecting proportions for normal, heavyweight, and mass concrete. ACI Man. Concr. Pract, pp. 1–38.

Chapter 7
Investigation of Synthetic Versus Natural Fiber on Interlocking Compressed Stabilized Earth Blocks: A Comparison of Plastic Bottle Fiber and "Teff" Straw

Selamawit Fetene, Asregidew K. Woldesenbet, Rahel A. Negatu, and Belaynesh N. Alamirew

7.1 Introduction

7.1.1 Background

The African population is expected to grow at a rate of 2.52% on average by 2020 [2]. This growth is about six times that of the rate projected for developed countries. This rate is even higher for Eastern African nations, with an approximate estimate of 2.72% per year [1]. This rapid pace of population in the continent has greatly increased the demand for shelter, infrastructure, and public services. As a result, the issue of shelter and low-cost house construction appears to be at the top of the long list of urbanization problems and is considered as the first and most important economic activity in the process of rapid urbanization [2].

Ethiopia is one of the East African nations, where the housing situation of its towns and cities is a manifestation of the challenges of urbanization. Overcrowding of dwelling units results from a high rate of population growth and migration to urban areas which creates a high demand for housing [3]. According to a report by Habitat for Humanity [4], only 30% of the current housing stock in Ethiopia is in fair condition, while the remaining 70% needs total replacement. Like other cities of Ethiopia, Bahir Dar encounters a plethora of problems including acute and ever-

S. Fetene · R. A. Negatu (✉) · B. N. Alamirew
Faculty of Civil and Water Resource Engineering, Bahir Dar Institute of Technology, Bahir Dar University, Bahir Dar, Ethiopia
e-mail: Rahel.Ayalew@bdu.edu.et; Belaynesh.Nibret@bdu.edu.et

A. K. Woldesenbet
Ethiopian Institute of Architecture, Building Construction and City Development, Addis Ababa University, Addis Ababa, Ethiopia

K. Mequanint et al. (eds.), *Sustainable Development Research in Materials and Energy*, Green Energy and Technology,
https://doi.org/10.1007/978-3-031-69860-6_7

worsening housing shortage, insufficient solid and liquid waste management, shortage of potable water, growing unemployment, and poverty [5]. Thus, there is a need to find alternative solutions to replace poorly built housings that are affordable for low- and middle-income residents and meet the ever-increasing housing demand. Today, the high cost of materials, labor, and transportation have adversely affected the construction industry in building low-cost housing in developing economies. This has forced the industry to revisit ancient building methods of utilizing natural, locally abundant, and sustainable materials. Natural building materials offer several environmental benefits including simple production and quick processes without the need for highly skilled labor and high energy using raw materials from plant wastes and construction materials [6]. Some of these natural, low-cost building materials include earth, lime, chalk, wheat, and rice straws that are used across the globe to build various components of a house such as adobe walls, straw bale insulation, rammed earth floors, and compressed soil construction. There are so many researches done so far related to cement-stabilized earth blocks, but this study assesses the use of plastic and "teff" fibers for interlocking compressed stabilized earth blocks (ICSEB).

7.2 Materials and Methods

7.2.1 Materials

In this study three soil samples, Portland cement, water, plastic bottle fiber, and teff straw were used for the preparation of interlocking cement-stabilized earth blocks. Based on the soil characteristics and the test results that followed, the soil cement ratio was determined as 1:18 which is 5.5% of cement by weight for all mix. Plastic fibers is produced by collecting empty water plastic bottles and strip into rope-like structure with a homemade plastic bottle striper with 35 mm length and 3 mm width like the teff straw. After drying, the soil samples was crushed and passed through a sieve of 75 μm, in order to ensure that the stabilizer or reinforcing additives could be uniformly distributed throughout the mix. Physical properties of three soil samples from three different locations were tested based on their test procedures, and the result is presented in Table 7.1. Based on the field and laboratory soil test results, Gudo Bahir soil had been selected, and all blocks were produced using soil collected from Gudo Bahir area in Bahir Dar City.

Empty water packaging plastic bottles were collected from local cafeterias and hotels. Then, cut the bottom part of the bottle by using scissors and strip into a rope-like structure with a homemade plastic bottle striper which has three holes to cut the bottle into 3-mm width as shown in Figs. 7.1 and 7.2.

As shown in Fig. 7.3, teff straw was collected from the local market and washed with water in order to remove any debris and cattle dung. The straw was also cut into pieces by scissors with an approximate aspect ratio of the PET fiber (35 mm length and 3 mm width).

Table 7.1 Physical properties of three soil samples collected from different locations

No	Physical properties	Values of Gudo Bahir (GBR)	Values of Abay Mado (AMD)	Values of Geter Menged (GMD)
1	Optimum moisture content	23.08	31.11	29.49
2	Maximum dry density	1.47	1.43	1.43
3	Silt content	7.50	6.36	13.50
4	Clay content	19.50	20.91	25.60
5	Sand content	70.00	72.73	60.90
6	Liquid limit	53.74	55.63	56.53
7	Plastic limit	33.96	31.82	32.92
8	Plasticity index	19.78	23.81	23.61
9	Shrinkage test	4.25	1.12	1.75
10	Unconfined compression strength of the soil = qu	154.57	95.97	81.37
11	Shear strength of the soil = qu/2	77.28	47.98	40.68

Fig. 7.1 Rough-edged plastic fiber

7.2.2 Ingredients Proportion

In this study there are five types of mixture including (1) ICSEB without fiber, (2) ICSEB with plastic fiber, (3) ICSEB with teff straw, (4) ICSEB with teff straw top-up (5) ICSEB with rough-edged plastic fiber. The size of ICSEB in this study was 300 × 150 × 120 mm.

Fig. 7.2 Plastic fiber

Fig. 7.3 Teff straw preparation

7.2.3 Sample Preparation and Curing

Prepare all ingredients and dry mixing was thoroughly done before wet mixing with water to sufficient workability. Mixing of straw, plastic fiber, cement, water and soil was done by hand on the bucket, and manual pressing machine was used for block

Fig. 7.4 Manual pressing machine with the mold

preparation as shown in Fig. 7.4. Plastic fiber and straw spread over the soil-cement mixture and diligently mixed. For each mixture, six blocks were prepared and a total of 150 blocks were casted. To achieve maximum strength, mud blocks need a period of damp curing under plastic shelter, where they were kept moist as shown in Fig. 7.5. Two curing periods were practiced in this study, which are 7 and 28 days. Finally compressive strength, water jet or pressure test, comparison of weight reduction, absorption test, or submersion were conducted at 7 day and 28 days.

7.3 Test Result and Discussion

To simplify the analysis, the researcher symbolized the mix in the following way as indicated in Table 7.2.

7.3.1 Compressive Strength

From the results shown in Fig. 7.6, soil cement "Teff" straw top-up (SCSTU) achieved the maximum mean result for both the 7th and 28th curing periods; i.e., 2.95 and 4.08 MPa; respectively. Whereas rough-edged plastic reinforced block

Fig. 7.5 Curing of blocks

Table 7.2 Mix symbol and their descriptions of ICSEB

No	Mix symbol	Descriptions
1	(SC)	Soil cement only
2	(SCT)	Soil cement and "teff" straw (straw type 1)
3	(SCP)	Soil cement plastic fiber (plastic type 1)
4	(SCSTU)	Soil cement "teff" straw and top-up (straw type 2)
5	(SCREP)	Soil cement rough-edged plastic fiber (plastic type 2)

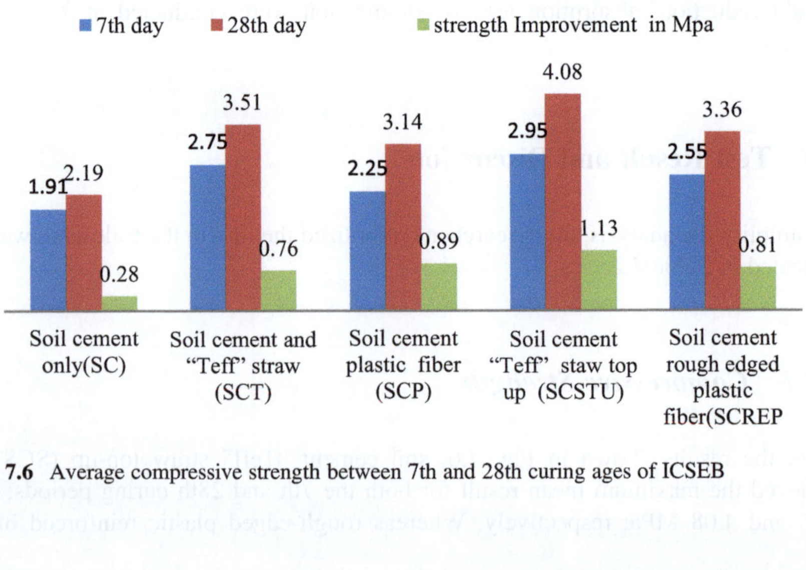

Fig. 7.6 Average compressive strength between 7th and 28th curing ages of ICSEB

(SCREP) type recorded the second maximum result of 2.55 and 3.36 MPa for the 7th and 28th days, respectively, this shows the tough behavior of blocks due to the presence of dispersed straw. The strength improvements were recorded as 14.66% to 39.56% for the 7th and 28th curing ages among all the five types of blocks. Previous studies show that the addition of fiber in ICSEB increases the compressive strength by 68% up to 72% as compared to unstabilized soil blocks [7].

7.3.2 Water Absorption

The plain block, i.e., ICSEB without fiber water absorption, was 10.90% on the seventh day of curing age. Nevertheless, the block severely damaged few minutes after removed from the bath, whereas the finding of all types of ICSEB counts less than 2% of water absorption as compared with the ICSEB without fiber. Blocks on the 28th day of curing time absorb less water as compared with their 7th day of curing time. Therefore, it is possible to conclude that water absorption was influenced by the curing ages of the ICSEBs.

From the results shown in Fig. 7.7, "teff" straw-reinforced ICSEB influenced the water absorption amount of ICSEB as compared with the plastic-reinforced ICSEB category (type 1) of 7th and 28th curing days since natural fibers absorb more water than plastic fibers. However, type 2 category block of "teff" straw shows an incremental absorption as compared with the same category of plastic fiber, i.e., from 6.26 to 9.6 of its 7th to 28th of its age, whereas the plastic fiber block decreased its absorption from 9.86 to 8.59 from fiber (type 1) and in the 7th day of type 1 into 5.35 of 7th days of type 2. The reason for type 2 "teff" straw, i.e., cement-soil "teff" straw top-up ICSEB to increase water absorption, was due to the topping up straw; it is exposed directly to the external environment during the experimental period, and the straw could up take more water than those straw buried inside the mud. Previous

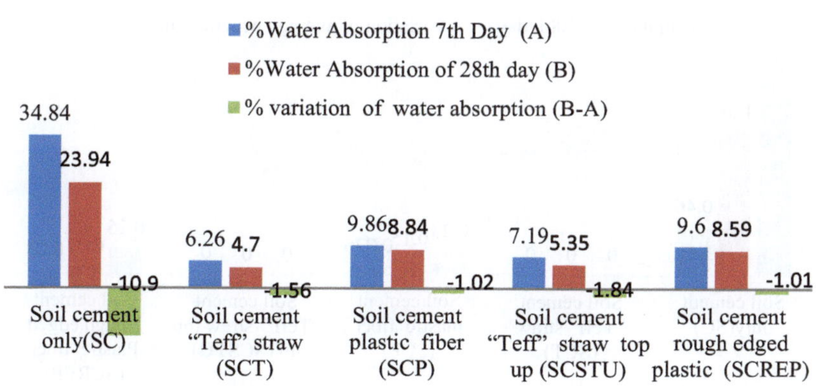

Fig. 7.7 Comparison of water absorption on 7th and 28th curing periods

studies shows that the water absorption test indicated that samples reinforced with 0.5% and 1.0% (by weight) fiber showed least water absorption [7].

7.3.3 *Water Jet Test*

Figure 7.8 shows the 28th day curing age stands well the water jet test than the 7th day ICSEBs. In contrast, the soil cement, i.e., ICSEB without fiber, was the one eroded severely. Its erosion test result was recorded as 4.65 cm (1.83 inch) which was far from the standard and can be considered as severe. The "teff" straw-reinforced ICSEB showed no sign of penetration at all. The type 1 plastic-reinforced ICSEB showed 0.23 cm (0.09 inch) on the 7th day and 0.15 cm (0.06 inch) on the 28th day. Type 2 plastic-reinforced ICSEB showed less penetration as compared with type 1. The comparison between 7th day and 28th day of curing time shows the penetration had decreased from 1.83 cm (0.73 inch) to 1.39 cm (0.55 inch) for soil cement blocks. Therefore, we can conclude that fiber significantly influences water penetration and curing ages also play a vital role in enhancing the capacity of the erosion resistance of ICSEB. Since the straw stitching effect in ICSEB can resist water jet. From the results obtained, it can be understood that all sets of plastic fiber stabilized and cement soil only blocks could not meet the 0.10 cm/minute (0.04 inch/min) standard except the "teff" straw-stabilized blocks. Based on previous studies, the results of water jet indicate that the sample reinforced with 2% fiber showed no erosion [7].

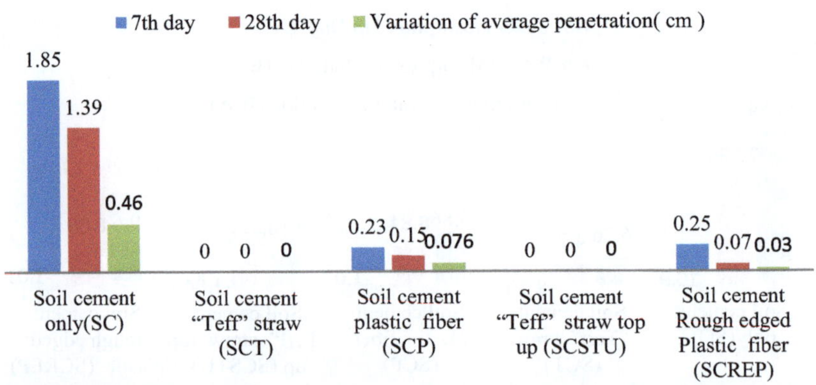

Fig. 7.8 Comparison of 7th and 28th curing ages of water jet test result of ICSE

7.3.4 Weight Reduction

The weight reduction between the two curing ages was not significant. The minimum expected weight loss from a 7.5-kg block was 0.375 kg or 375 gm. Therefore, all types of blocks recorded a weight loss below and around the minimum recommended setting as presented in Fig. 7.9. Especially, type 1 plastic- and "teff" straw-reinforced ICSEB lost 340 gm and 320 gm of weight, respectively, whereas the other three types of block, viz., plain ICSEB and rough plastic and "teff" straw-reinforced ICSEB, show much more below the recommendations. Therefore the test result shows that curing age did not have any significant effect on weight reduction of ICSEB.

The plastic-reinforced ICSEB weighted a little bit more from both type 1 and type 2 blocks of "teff" straw. The percentage range of initial weight difference was 26–28% which was 260–280 gm by weight. As Fig. 7.9 presented, the "teff" straw had less weight than the plastic fiber-reinforced blocks; top-up "teff" straw had a slight weight gain as compared with the non-top-up "teff" straw blocks. In the meantime anchored plastic-reinforced blocks also recorded a slight increase in weight gain. Though the weight differences were not significant as per the total weight of the block, and taking the variance in weight into consideration type 1 is preferable. Therefore, "teff" straw fiber showed less weight loss than other ICSEBs.

7.3.5 Test Result Analysis

A two-way ANOVA was performed to determine if there was a statistically significant difference in compressive strength, water absorption, water jet test, and weight reduction of the five types of cement-stabilized earth blocks and the two curing ages, i.e., 7th and 28th.

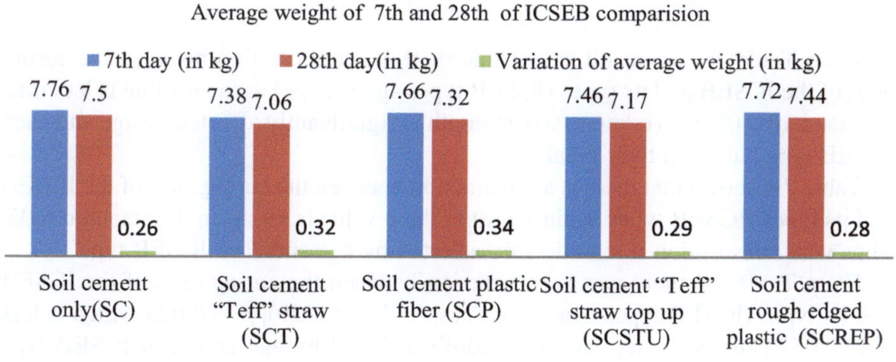

Fig. 7.9 Average weight of 7th and 28th of ICSEB comparison

Table 7.3 Tests of blocks against their age on compressive strength

Dependent variable: compressive strength						
Source	Type 3 sum of squares	Df	Mean square	F	Sig.	Partial Eta squared
Corrected model	24.70[a]	9	2.719	124.293	0.000	0.957
Intercept	493.755	1	439.755	22571.997	0.000	0.998
CSEB types	14.301	4	3.575	163.444	0.000	0.929
Age	9.017	1	9.017	412.218	0.000	0.892
ICSEB types*age	1.152	4	0.288	13.160	0.000	0.513
Error	1.094	50	0.022			
Total	519.318	60				
Corrected Total	25.563	59				

[a]R squared = 0.957

Table 7.4 Tests on the interaction of ICSEB types against age on water absorption. Tests of blocks against their age on water absorption

Source	Type 3 sum of squares	Df	Mean square	F	Sig.	Partial Eta squared
Corrected model	2549.019[a]	9	283.224	118.999	0.000	0.982
Intercept	4259.970	1	4259.970	1783.837	0.000	0.989
CSEB types	2359.203	4	589.801	246.976	0.000	0.980
Age	79.935	1	79.935	33.472	0.000	0.626
ICSEB types*age	109.881	4	27.470	11.503	0.000	0.697
Error	47.762	20	2.388			
Total	6856.751	30				
Corrected Total	2596.781	29				

[a]R squared = 0.982 (adjusted R squared = 0.973)

From Table 7.3 the result shows that there is an interaction between the curing ages of the ICSEB and its types (ICSEB types against age) as the p value is less than the standard 0.05; also compressive strength is significantly affected by age and each ICSEB type based on their p value.

Table 7.4 shows that there is an interaction between the curing ages of the ICSEB and its types (ICSEB types against age) as the p value is less than the standard 0.05. Thus water absorption is significantly affected by age and each ICSEB type.

From Table 7.5 there was an interaction between the curing ages of the ICSEB and its types (ICSEB types against age) as the $F = (4, 20)$, $p = 0.028$ which is less than 0.05. Thus water jet was significantly affected by age and each ICSEB type (Tables 7.6 and 7.7).

Table 7.5 Tests on the interaction of ICSEB types against age on water jet

Dependent variable: water jet test						
Source	Type 3 sum of squares	of Df	Mean square	F	Sig.	Partial Eta squared
Corrected model	1.879[a]	9	0.209	73.091	0.000	0.97
Intercept	0.672	1	0.672	235.240	0.000	0.922
ICSEB	1.826	4	0.457	159.803	0.000	0.97
Age	0.014	1	0.014	4.930	0.000	0.198
ICSEB types*age	0.039	4	0.010	3.419	0.000	0.406
Error	0.057	20	0.003			
Total	2.608	30				
Corrected total	1.936	29				

[a]R squared = 0.970 (adjusted R squared = 0.957)

Table 7.6 Tests on the interaction of ICSEB types against age on weight reduction

Dependent variable: absorption test						
Source	Type 3 sum of squares	Df	Mean square	F	Sig.	Partial Eta squared
Corrected model	1.418[a]	9	0.158	13.015	0.000	0.854
Intercept	1664.628	1	1664.628	137458.962	0.000	1.000
ICSEB	0.741	4	0.185	15.305	0.000	0.754
Age	0.672	1	0.672	55.492	0.000	0.735
ICSEB types*age	0.005	4	0.001	0.105	0.979	0.021
Error	0.242	20	0.012			
Total	1666.289	30				
Corrected total	1.661	29				

[a]R squared = 0.854 (adjusted R squared = 0.789)

The test for interaction was not significant, $F\ (4, 20) = 0.0105$, $p = 0.979$. Since there was no significant interaction, we tested the main effects. The main effect for ICSEB type was significant, $F\ (4, 20) = 15.305$, $p = 0.000$. The main effect for age was also significant, $F\ (1, 20) = 55.492$, $p = 0.000$.

7.4 Cost Analysis of Interlocking Cement-Stabilized Earth Block

To calculate the unit cost analysis of cement-stabilized earth block, the following assumptions were taken into consideration.

Table 7.7 Unit cost analysis of ICSEBs

Material cost					Labor cost				Equipment cost			
Item	Unit	Qty.	Rate (birr)	Cost per unit (birr)	Labor by trade	No.	Daily cost (birr)	Cost per piece (birr)	Types of equipment	No.	Daily rental (birr)	Cost per piece (birr)
Cement	Qt.	0.004	2150	8.6	Machine operator	1	350	0.7	Pressing machine	1	600	1.2
Soil	M^3	0.075	360	27	Assistance	2	300	0.6	Hand tools	2	40	0.08
Straw	kg	0.079	4	0.32								
Water	M^3	0.001	8	0.01								
Total				35.9(A)				1.3(B)				1.28(C)

A unit production cost of one "teff" straw ICSEBs was = A + B + C = **38.5** birr/ Block

Assumptions
Date: July 2023
Material

PPC cement price	2150 birr/Qt.
Soil	360 birr/m^3
Straw	0.4 birr/kg

(One plastic bag "teff" straw (ጭድ weighs 25 kg and costs 100 birr in Bahir Dar)

Water	8.00 birr/m^3

Labor

Machine operator	10,500 birr/month
Assistant	300 birr/day

Machine rental cost

Manual pressing machine daily rent	0.600 birr/day
Hand tools	40 birr/day

Capacity

Machine daily production capacity	0.500 piece/day
Labor production capacity	0.500 piece/day

7.5 Conclusions and Recommendations

7.5.1 Conclusions

The following conclusions are drawn from the experiments conducted and the analyses made.

- Among the three soil samples, Gudo Bahir (GBR) soil had been selected for ICSEB making based on the field and laboratory tests.
- The proportion of materials used in producing 300 mm × 150 mm × 120 mm size of ICSEB was 7.5 kg of soil from Gudo Bahir, 412.5 gm PPC cement, 0.5% of "teff" straw, and 17.3% of water by weight of the block.
- Curing ages had an influence on the compressive strength of all the five types of ICSEBs because the attained compressive strength values ranged between 1.91 and 2.95 MPa at the 7th day and 2.19 to 4.08 MPa at the 28th day.

- Blocks on the 28th day curing time take less water as compared with their 7th curing time. Therefore, water absorption was affected by the curing ages of the ICSEBs.
- The weight reduction between the two curing ages was not significant. When the minimum expected weight loss from a 7.5 kg block was 0.375 kg or 375 gm, all types of blocks recorded a weight loss below and around the minimum recommended setting.
- Blocks with plastic fiber reinforcement attained low strength of 3.6 MPa by type 2 (SCERP) compared with "teff" straw blocks (SCSTU) within the same category and at their 7th and 28th curing ages which happened to be 4.08 MPa.
- The "teff" straw had an appreciable influence on compressive strength than the plastic fiber because "teff" straw-reinforced blocks have shown a compressive strength from 0.37 to 0.72 MPa improvements than the plastic fiber-reinforced blocks. This shows that using "teff" straw as reinforcement can increase ICSEBs strengths.
- "Teff" straw had stronger resistance to water jet/erosion/than the plain and plastic fiber blocks.
- Soil cement and "teff" straw (SCT) and soil cement rough-edged plastic fiber (SCREP) had absorb less water than the other types of ICSEB,

7.5.2 *Recommendations*

1. After conducting the experiments presented in this investigation and reviewing the related literatures, the following design recommendations are made: "teff" straw top-up blocks outsmart the other types of blocks in this study. The compressive strength of "teff" straw top-up block result was 4.08 MPa; the absorption found was 9.6% below the recommended setting of 15%, and water jet result of this ICSEB was also nil or no sign of penetration at its 28th day of curing age. The performance of these blocks indicates that the products are good options for low-cost house construction.
2. Soil cement plastic fiber-reinforced blocks were also recommended as an alternative ICSEBs wall material for those who have interest to substitute the "teff" straw with waste plastics, as it is cheap resource and in effect reduces environmental pollution. From the experimental point of view, it had 3.6 MPa which was more than the British standard of 2.8 MPa, the absorption rate was also 8.6% below the recommended stetting of 15%, and water jet test result of this ICSEB was 0.076 cm/min on its 28th day of curing age which was less than 0.01 cm/min of the standard.

References

1. United Nations. (2017). Department of Economic and Social Affairs, Population Division Population Facts Report -"The impact of population momentum on future population growth." No. 2017/4.
2. Akmel, A., & Mandado, M. (2014). Assessment of Problems of Access to Residential Housing by Government Employees: The Case of Sodo Town, Wolaita Zone, Snnpr, Ethiopia. Doctoral dissertation. Haramaya University.
3. Tebarek, L. (2013). "Inner city housing and urban development-induced displacement": Impact on poor female-headed households in Arada sub city of Addis Ababa, Ethiopia. *Journal of Sustainable Development in Africa, 15, 131*(2).
4. UN-Habitat Affordable land and housing in Africa. United Nation Human Settlements Program. Addis Ababa, Ethiopia: UN Habitat: Vol.3. (2011).
5. Solomon, M., & Ruth, M. L (2004). Feasibility Study for the Application of Community led Infrastructure Finance Facility (CLIFF), Operation in Ethiopia, Homeless International.
6. Walker, P. J. (2004). Strength and erosion characteristics of earth blocks and earth block masonry. *Journal of Materials in Civil Engineering, 16*, 497–506.
7. Shrma, V., Marwaha, M., & Vinayak, K. (2016). Enhancing durability of adobe by natural reinforcement for propagating sustainable mud housing. *International Journal of Sustainable Built Environment, 5*, 141–155.

Chapter 8
The Use of Sisal Fiber for the Reduction of Plastic Shrinkage Cracking in Concrete

Rahel A. Negatu, Belaynesh N. Alamrew, and Asregdew K. Woldesenbet

8.1 Introduction

8.1.1 Background

One of the most widely used construction materials in the construction industry of Ethiopia is concrete. For the past several decades, the design of concrete has typically been based on strength since it can be easily measured for quality control and quality assurance purpose. Although the requirements for strength are frequently met, the durability performance is not always satisfied. One of the major problems that is frequently observed with respect to durability is shrinkage cracking.

Concrete is typically subjected to different types of volume changes in their service life. Once mortar and concrete are exposed to dry conditions, the materials begin to lose water and that leads to shrinkage.

Shrinkage is a volume reduction that occurs due to moisture loss, temperature change, and chemical reactions. Typically, plastic shrinkage, drying shrinkage, thermal shrinkage, and autogenous shrinkage are the main types of shrinkage [4].

Plastic shrinkage cracking is especially problematic since it occurs when the concrete is still in a young state and has not gained sufficient strength to resist the tensile stress that develops. Early age shrinkage cracking in concrete can occur due

R. A. Negatu (✉) · B. N. Alamrew (✉)
Faculty of Civil and Water Resources Engineering, Bahir Dar Institute of Technology, Bahir Dar University, Bahirdar, Ethiopia
e-mail: Rahel.Ayalew@bdu.edu.et; Belaynesh.Nibret@bdu.edu.et

A. K. Woldesenbet
Ethiopian Institute of Architecture, Building Construction and City Development, Addis Ababa University, Addis Ababa, Ethiopia

K. Mequanint et al. (eds.), *Sustainable Development Research in Materials and Energy*, Green Energy and Technology,
https://doi.org/10.1007/978-3-031-69860-6_8

to different reasons that include improper construction practice, poor mix design, and insufficient curing.

According to ACI [1], plastic shrinkage cracking is frequently associated with hot weather which is serious when concreting is done in arid climates. It occurs in concretes exposed to hot weather, primarily in flat slabs, beams, and footings. The main driving force behind the phenomenon is believed to be rapid and excessive loss of water, which mainly takes place in the form of surface water evaporation.

Nonmetallic fibers added to concrete have a considerably reducing effect on early-age cracking which have better results compared to ordinary mesh reinforcement. It is also found that longer fibers have a better effect than shorter fibers, probably due to better bond strength between concrete and fiber. The optimum fiber length is found to be somewhat larger than the maximum aggregate size [3].

Sisal fiber is the most widely used natural fiber that is obtained from the leaves of the sisal plant which can be very easily cultivated. Each leaf contains a number of long, straight fibers which can be removed in a process known as decortication. Because of diverse agroecologies, Ethiopia is the 11th sisal-producing country in the world with an annual production of 700 tons (FAOSTAT, 2017). Most of the sisal produced in Ethiopia comes from Hawassa State Farm. However, wild varieties of sisal that grow abundantly are also growing in Metekel area.

Thus, this study was initiated to reduce plastic shrinkage cracking by using fiber and evaluate the effect on the compressive and flexural strength of concrete.

8.2 Materials and Methods

8.2.1 Materials

8.2.1.1 Sisal Fiber

Sisal fiber was collected from the local market at Bahir Dar where it was sourced from west Gojjam and extracted from a local tree called biska. The fiber used in the concrete mix was 0%, 0.1%, 0.2%, 0.3%, and 0.4% [2] in volumetric proportion with 25 mm and 40 mm lengths for all mixes. Figure 8.1 shows sisal tree and extracted sisal fiber.

8.2.1.2 Fine and Coarse Aggregate

Well-graded crushed aggregate and sand collected from quarry sites located around Bahir Dar were used and sieve analysis, unit weight, specific gravity, and silt content were done based on standard requirements [5].

Fig. 8.1 Sisal tree and extracted fiber

8.2.2 *Test Methods*

8.2.2.1 Preparation of Trial Mix

The ACI method was used to determine the mix proportions and the mix design for C-25 concrete grade. Twelve trial mixes were prepared, such as the control mix without fiber, a volumetric proportion of fiber 0.1%, 0.2%, 0.3% and 0.4% each with 25-mm and 40-mm fiber length. Experiments were conducted to select fiber mixtures that may reduce plastic cracking while still maintaining workability and strength.

8.2.2.2 Plastic Shrinkage Cracking Test

The restrained plastic shrinkage test was performed in accordance with the ASTM 1579 "standard test method for evaluating plastic shrinkage cracking of restrained fiber reinforced concrete using steel inserts" with slight modifications. Figure 8.2 shows that the restrained slab mold is made from steel based on ASTM 1579 standards.

8.2.3 *Compressive Strength Test*

The specimens for compressive strength test with 0%, 0.1%, 0.2%, 0.3%, and 0.4% fiber dosage each with 25 mm and 40 mm length were casted and cured in a water tank until they were tested according to (BS.1881)-111 "Method for normal curing

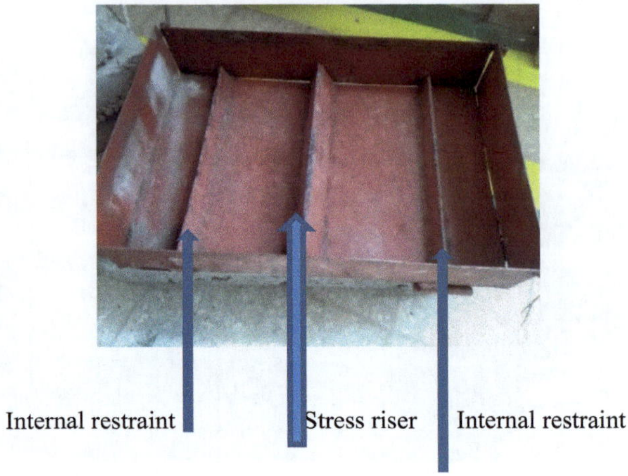

Fig. 8.2 Restrained slab mold

of test specimens." Compressive strength tests were conducted on the 28th day for all concrete mixes.

8.2.4 Flexural Strength Test

Specimens for flexural strength test with 0%, 0.1%, 0.2%, 0.3%, and 0.4% fibers dosage each with a length of 25 mm and 40 mm of C-25-grade concrete were casted and cured in a water bath until they were tested using ASTM C 192 "Practice for Making and Curing Concrete Test Specimens in the Laboratory."

8.3 Result and Discussion

8.3.1 Environmental Chamber

An environmental chamber was used for this research based on the procedure described by ASTM 1579 standard as shown in Fig. 8.3. The environmental conditions used in this investigation were temperature of 36±3°C, relative humidity of 30±10, and wind speed of 24 km/h. Temperature and wind speed were controlled during the test. The electronic balance and thermometer were used to measure weight losses and temperatures of the concrete and the atmosphere, respectively.

Fig. 8.3 Environmental chamber

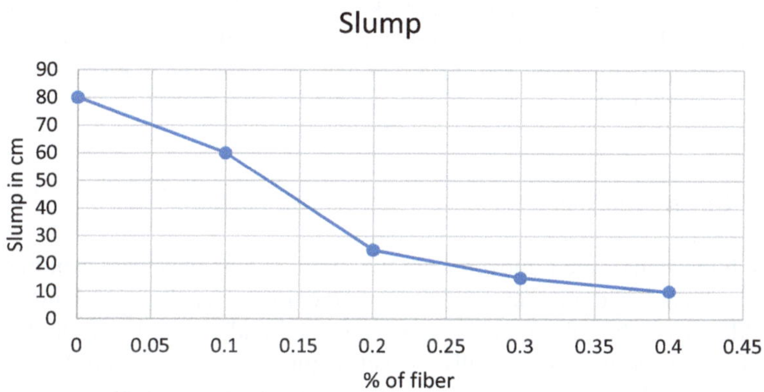

Fig. 8.4 Effect of fiber on workability of concrete mix

8.3.2 Workability Test

The workability test was conducted through an inverted slump cone with the aid of internal vibration. Figure 8.4 indicates that the addition of fiber at concentrations ranging from 0% to 0.4% decreased the slump value from 80 mm to 10 mm, respectively. This result shows that a low volume of fibers has little impact on the workability, while as the fiber volume increases, slump decreases which might be due to the cohesive properties of the concrete with fiber. Therefore the use of proper admixture which does not affect other properties except workability might be a better solution for higher fiber-cement ratios.

8.3.3 *Unit Weight of Concrete*

The density (unit weight) of concrete was measured using (ASTM) C 138 "Standard Test Method for Density (Unit Weight), Yield, and Air Content (Gravimetric) of concrete," and the result is presented in Table 8.1.

8.3.4 *Plastic Shrinkage Cracking Test*

After casting the samples on the restrained slab mold, the specimens were placed at a controlled temperature. Wind speed was set up and observed periodically, and weight loss and plastic shrinkage cracking were monitored continuously for the first 6 h. at 30-min intervals. Then after specimens were stored in the chamber for an additional 18 h. After a total of 24 h from casting, the slab was removed from the chamber, and the cracks were measured.

On the plain concrete, fine hairline cracks could be observed above the stress riser approximately 30 min after the concrete was placed in the environmental chamber. On fiber-reinforced mixtures, the surface of concrete appeared to be free of cracking for a longer period of time.

8.3.4.1 Image Analysis

After 24 h the slabs were removed from the setup, and images were captured by the camera and analyzed the image by using the image analysis technique in the MATLAB. After that, the images were processed and edited to get a clear crack profile. It was used to automatically extract features of interest from digital images and provide rapid and accurate interpretation of the image contents.

For this purpose programing language has been prepared in MATLAB, and crack lengths were measured by uploading and processing the image as shown in Fig. 8.5.

Based on the image analysis results, four cracks were observed on the concrete surface, and the maximum crack was recorded in the control mix (without fiber) with values of 49.8 mm and 1.6 mm, second 44.4 mm and 1.43 mm, third 29.2 mm and 0.6 mm, the fourth one 15.6 mm and 0.6 mm, length and width, respectively (Fig. 8.6).

Table 8.1 Unit weight of concrete

Sample of fiber (%)	Unit weight (kg/m3)
Control	2553
0.1	2545
0.2	2550
0.3	2550
0.4	2552

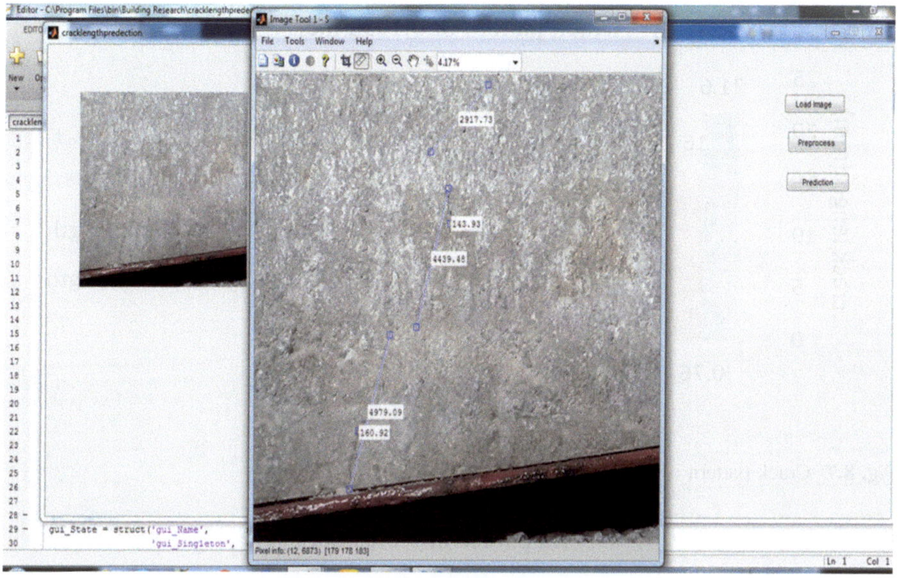

Fig. 8.5 Results from image analysis in MATLAB

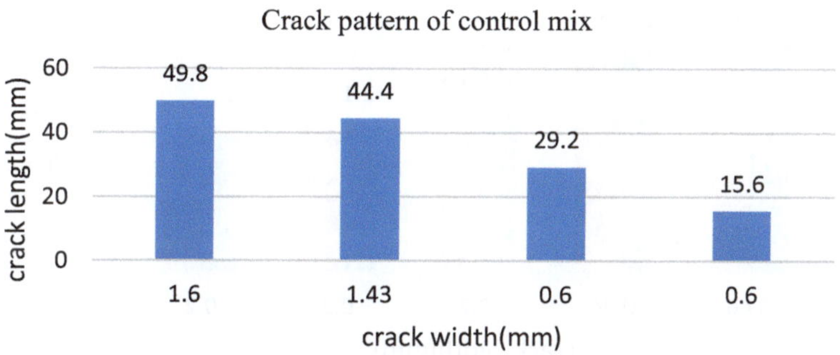

Fig. 8.6 Crack pattern of control mix (without fiber)

In concrete with 0.1% fiber mix and 25-mm fiber length, the maximum crack length was 21.6 mm, while the crack length of 0.1% fiber mix and 40-mm fiber length was 16.2 mm as indicated in Fig. 8.7. These results clearly showed that crack length and width were reduced with the addition of 0.1% fiber (in volume) compared to without fiber. Moreover, the results indicated that the concrete mix of 0.1% fiber (in volume) with 40-mm fiber length reduced the crack lengths better than at 25-mm fiber length. The inclusion of sisal fibers is effective in delaying first crack appearance which may be attributed to the high elastic modulus of fibers compared to the cementitious matrix, in addition to its ability to provide bridging forces across the cracks. Based on a previous study, the addition of 0.2% and 0.1% of sisal fibers

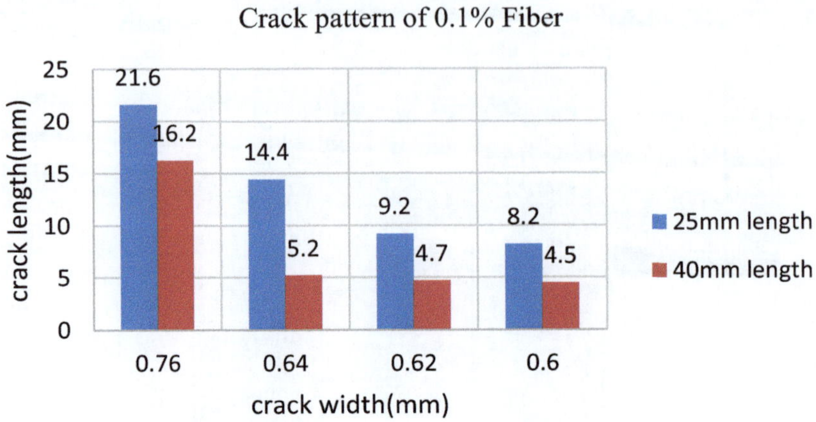

Fig. 8.7 Crack pattern of 0.1% fiber

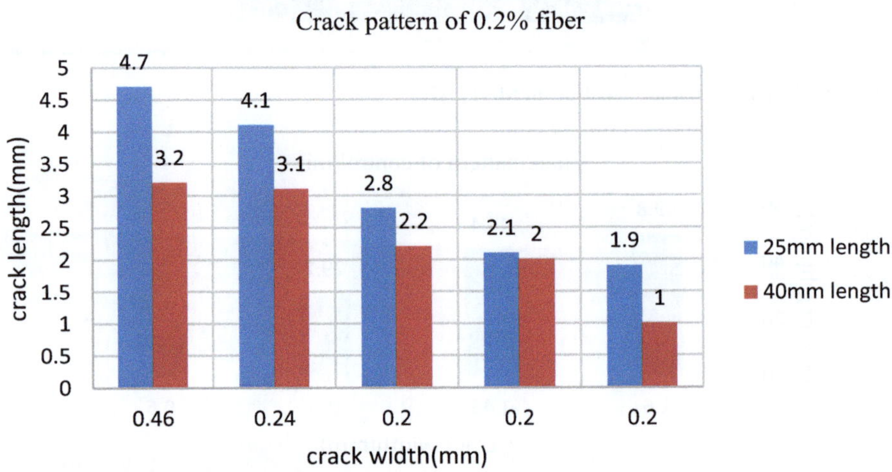

Fig. 8.8 Crack pattern of 0.2% fiber

(25 mm long) resulted in the reduction plastic shrinkage of the mortar mixes by 29.6% and 24.1%, respectively [6].

As indicated in Fig. 8.8, on concrete with 0.2% fiber mix and 25-mm fiber length, the maximum crack length was 4.7 mm, while the crack length of 0.2% fiber mix and 40-mm fiber length was 3.2 mm. These results clearly showed that crack length and width were better reduced with the addition of 0.2% fiber (in volume) compared to with the addition of 0.1% fiber. Moreover, the results indicated that the crack length was reduced in 40 mm length of fiber than in 25 mm length. Filho R. et al. observed that adding a 0.2% volume fraction of 25-mm sisal fiber to the mortar matrix greatly reduced free plastic shrinkage. Sisal fibers give restraint to the sliding of the matrix by means of frictional resistance. The addition of 0.2% sisal fibers reduced the

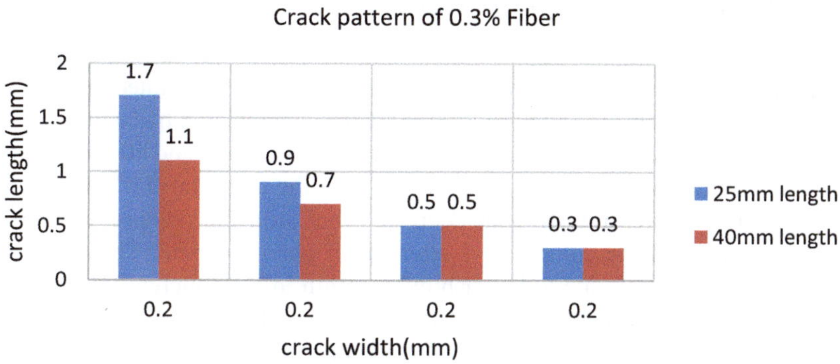

Fig. 8.9 Crack pattern of 0.3% fiber

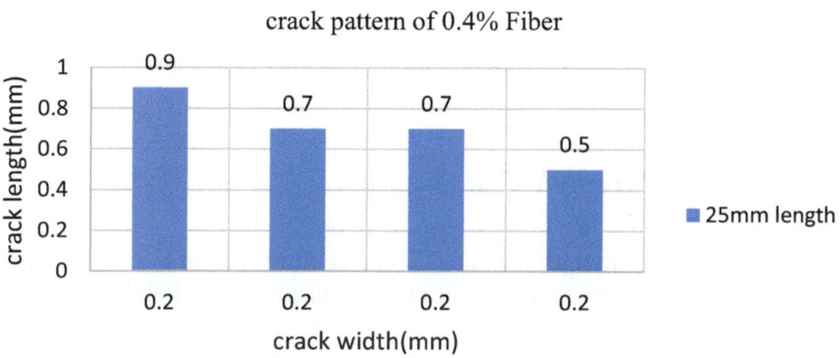

Fig. 8.10 Crack pattern for 0.4% fiber

restrained plastic shrinkage by delaying the initial cracking and effectively controlling the crack development in the composite in its plastic state [6].

As indicated in Fig. 8.9, on concrete with 0.3% fiber mix and 25-mm fiber length, the maximum crack length was 1.7 mm, while the crack length of 0.3% fiber mix and 40-mm fiber length was 1.1 mm. These results clearly showed that crack length and width were better reduced with the addition of 0.3% fiber (in volume) compared to with the addition of 0.2% fiber. Moreover, the results indicated that the crack length was more reduced in 40 mm length of fiber than in 25 mm length.

As indicated in Fig. 8.10, on concrete with 0.4% fiber mix and 25-mm fiber length, the maximum crack length was 0.9 mm, while the crack length of 0.4% fiber mix and 40-mm fiber length was almost crack-free element.

Based on the findings of the study, it can be concluded that with the addition of fiber from 0.1% to 0.4% by volumetric proportion, the plastic shrinkage cracks were reduced compared to plain concrete. As observed from the results recorded in 0.4% fiber addition, there is crack-free concrete surface for 40 mm length of fiber, and for 25 mm length of fiber there is a crack-free concrete surface for 0.5% fiber.

8.3.5 Compressive Strength Test

The results of compressive strength tests at 28 days are illustrated in Fig. 8.11. According to the results, the addition of fiber in the concrete mix had a little effect on its compressive strength. At 0.1% fiber mix addition of 25 mm and 40 mm lengths of fiber, the compressive strength increases, but with the addition of 0.2% and above fiber mix, the compressive strength decreases, but still, the compressive strength remains greater than the compressive strength of plain concrete. There is a slight modification of strength due to better bond strength in the concrete matrix, but the main reason for the addition of fibers to concrete is not for the strength gains. The reason for the development of fiber-reinforced concrete is for enhancing the flexural toughness of the material. It effectively prevents cracks due to shrinkage and creep. Slightly modified strength and better flexural strength of concrete is the result of fiber infusion [8].

But by adding 0.4% fiber mix with 40 mm length, the compressive strength decreases; this is due to the dispersion problem of fiber at this specific dosage.

8.3.6 Flexural Strength Test

The results of flexural strength tests at 28 days are illustrated in Fig. 8.12; there was an increase in flexural strength with an increase in fiber. The trend shows that the addition of fiber between 0.1% and 0.3% percentages in both lengths (25 mm and 40 mm) into concrete increases the flexural strength with maximum flexural strength achieved at 0.3%. This may be due to the improved interfacial bond between paste and filler. The addition of fiber beyond 0.4% shows a decline in flexural strength [7].

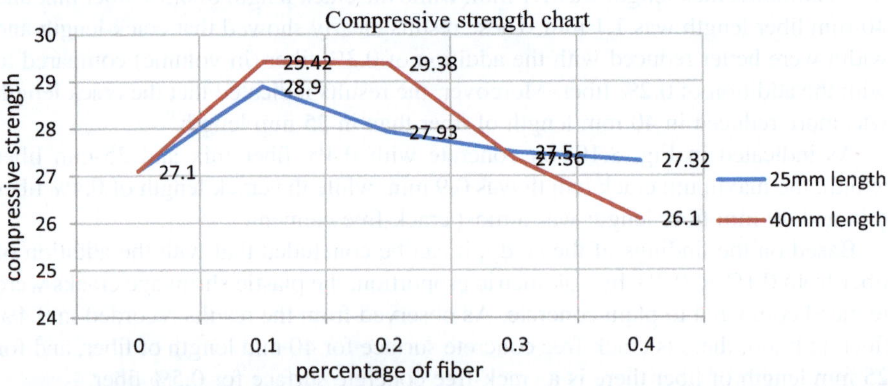

Fig. 8.11 Compressive strength of each mix at 28 days

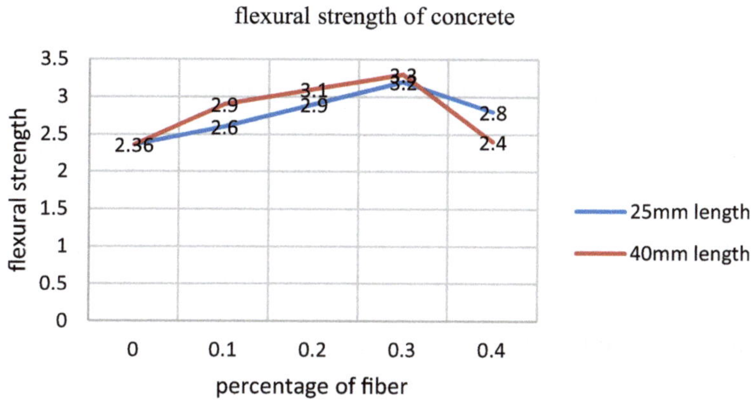

Fig. 8.12 Flexural strength of each mix at 28 days

Table 8.2 Test between subject effects in crack length

Dependent variable: crack length					
Source	Type 3 sum of squares	Df	Mean square	F	Sig.
Corrected model	10597.076[a]	9	1177.453	2956.32	0.000
Intercept	6456.560	1	6456.560	16210.973	0.000
Volume	10545.976	4	2636.494	6619.644	0.000
Length	2550	1	21.862	54.892	0.000
Volume* length	2552	4	7.31	18.353	0.000
Error	7.966	20	0.398		
Total	17061.602	30			
Corrected total	10605.042	29			

[a]R squared = 0.999

8.3.7 Data Analysis

Statistical analysis was applied to the conducted laboratory test results. SPSS V 20 is used to analyze two-way ANOVA based on dependent variables (crack length, compressive strength, and flexural strength) and independent variables (volume of fiber and length of fiber).

About 99.9% of the variability in the crack length is explained by the volume of fiber, length of fiber, and volume* length interaction. As shown in Table 8.2, each p-value above is less than the significance value (0.05) for both main factors (volume of fiber and length of fiber) and interaction effect. This shows that the volume of fiber, length of fiber, and interaction effect significantly affect the crack length.

About 69.2% of the variability in compressive strength is explained by the volume of fiber, length of fiber, and volume* length interaction (Table 8.3). Each p-value is greater than the significance value for volume, which shows that the compressive strength is significantly affected by the volume of fiber. But the other

Table 8.3 Tests of between-subjects effects in compressive strength

Dependent variable: compressive strength					
Source	Type 3 sum of squares	Df	Mean square	F	Sig.
Corrected model	10597.076[a]	9	1177.453	2956.32	0.000
Intercept	6456.560	1	6456.560	16210.973	0.000
Volume	10545.976	4	2636.494	6619.644	0.000
Length	21.862	1	21.862	54.892	0.000
Volume* length	29.238	4	7.31	18.353	0.000
Error	7.966	20	0.398		
Total	17061.602	30			
Corrected Total	10605.042	29			

[a]R squared = 0.999

Table 8.4 Tests of between-subjects effects in flexural strength

Dependent variable: flexural strength					
Source	Type 3 sum of squares	Df	Mean square	F	Sig.
Corrected model	3.120	9	.347	3.152	.016
Intercept	235.200	1	235.200	2138.182	0.000
Volume	2.67	4	.668	6.068	0.002
Length	.012	1	.012	.109	0.745
Volume * length	.438	4	.110	.995	0.433
Error	2.200	20	.110		
Total	240.520	30			
Corrected total	5.32	29			

factor (length of fiber) and the interaction effect do not significantly affect the compressive strength since the p-value (0.05) is less than the significance value.

About 58.6% of the variability in flexural strength is explained by the volume of fiber, length of fiber, and volume* length interaction (Table 8.4). Each p-value is greater than the significance value for volume showing that the flexural strength is significantly affected by the volume of fiber. But the other factor (length of fiber) and the interaction effect are not significantly affecting the flexural strength because the p-value (0.05) is less than the significance value for both length of fiber and interaction effect.

8.4 Conclusions and Recommendations

8.4.1 Conclusions

1. Plastic shrinkage cracks were reduced by 56–96% compared to plain concrete by the addition of sisal fibers at concentrations ranging from 0.1% to 0.4% and

25-mm fiber length. On the other hand, plastic shrinkage cracking was reduced by 67.4–97.7% compared to plain concrete when concrete was mixed with a 40 mm length addition of fiber at a concentration ranging from 0.1% to 0.3% in volumetric proportion. Fiber with 40 mm length is better than 25 mm length for crack reduction since 40 mm length had more surface area than 25 mm at a given volume fraction.
2. Addition of sisal fiber in the concrete mix in the range of 0.1–0.4% volume increases the compressive strength of concrete from 6.2% to 0.8%, respectively, for 25 mm length. With the addition of fiber for 40 mm length with 0.1–0.3% of fiber, the compressive strength increased by 7.8–0.95%. However, with the addition of 0.4% fiber, the compressive strength decreased by 3.7% which may be due to the dispersion problem.
3. The flexural strength of the concrete increased by about 3.45.2% with the addition of 0.1–0.4% in 25 mm length compared to the control concrete. With the addition of 40 mm length in a volume of 0.1–0.3%, the flexural strength increased by 17–27%.
4. The optimum fiber volume for the compressive and flexural strength is 0.3%, but at a fiber content of 0.4% both the compressive strength and flexural strength decrease may be due to fiber dispersion problems above 0.4% fiber volume.
5. The slump was decreasing with the addition of sisal fibers. The more the fiber-cement ratio, the more the decrease in slump due to the cohesion property of concrete making materials and fiber.

8.4.2 Recommendations

1. It can be recommended that mixing sisal fiber to large exposed concrete elements have the potential to reduce excessive plastic shrinkage cracking considering the limitation of the study.
2. As the slump of fiber-reinforced concrete decreased with increasing the volume and length of fiber, the use of proper admixture which does not affect other properties except workability is recommended for higher fiber-cement ratios.

References

1. ACI, c. Hot weather concreting (1999).
2. Aire, C., Mendoza, C., & Davila, P. (2011). Polypropylene fibers reinforced concrete: optimization on plastic shrinkage cracking'. Proceedings of the Second International Conference on Future Concrete. ACTS Advanced Construction Technology Services. Dubai.
3. Balaguru, P. (1994). Contribution of fibers to crack reduction of cement composites during the initial and final setting period. *Materials Journal, 91*(3), 280–288.
4. Banthia, N., Azzabi, M., & Pigeon, M. (1994). Restrained shrinkage cracking in fibre-reinforced cementitious composites. *Materials and Structures, 26*, 405–413.
5. BS. British standard (1983).

6. Filho, R. D. T., & Sanjuan, M. A. (1999). Effect of low modulus sisal and polypropylene fiber on the free and restrained shrinkage of mortars at early age. *Cement and Concrete Research, 29*, 1597–1604.
7. Mora, J., Gettu, R., Olazábal, C., Martín, M. A., & Aguado, A. (2000). Effect of the incorporation of fibers on the plastic shrinkage of concrete. In P. Rossi & G. Chanvillard (Eds.), *Fifth international RILEM symposium on fibre-reinforced concrete (FRC)* (pp. 705–714). RILEM Publications SARL.
8. Boshoff, W. P., & Combrinck, R. (2013). Modeling the severity of plastic shrinkage cracking in concrete. *Cement and Concrete Research, 48*, 29–34. https://doi.org/10.1016/j.cemconres.2013.02.003

Part II
Renewable and Green Energy

Chapter 9
Synthesis of Novel Polyaniline-Nanosphere Composite for the Detection of Ammonia in Poised Meat

Tadele Ageru Alemu, Delele Worku Ayele, Ababay Ketema Worku, Minbale Admas Teshager, Praveen C. Ramamurthy, and Dhakshnamoorthy Mani

9.1 Introduction

Food spoilage like spoilage of meat is the cause of food-borne diseases/food-borne illness due to unexpected accumulation of gases such NH_3 that could lead to very danger diseases/sharp poisonous such as brain and nerve damage, chronic arthritis, and hemolytic uremic syndrome, HUS, resulting in kidney failure or long-term diseases like cancer [33], WFP. Spoilage of meat has a close correlation with the increased production of nitrogen-containing compounds such as ammonia (NH_3) as one hazardous gas that causes contamination and brings health problems when humans are exposed to it for the first time at a concentration of about 55 ppm [4, 6, 18, 25]. Food-borne diseases are not only health problems but also responsible for a significant amount of economic loss due to wastage in the food and catering industries. This problem seeks a best gas sensor technology method having fast meat freshness detection ability. There have been different types of conservative methods which have given different kinds of sensors for gas detection. These conservative

T. Ageru Alemu
Faculty of Materials Science and Engineering, Jimma Institute of Technology, Jimma University, Jimma, Ethiopia

D. W. Ayele (✉) · M. A. Teshager · D. Mani
Department of Chemistry, College of Science, Bahir Dar University, Bahir Dar, Ethiopia

A. K. Worku
Bahir Dar Energy Center, Bahir Dar Institute of Technology, Bahir Dar University, Bahir Dar, Ethiopia

P. C. Ramamurthy
Department of Materials Engineering, Indian Institute of Science, Bangalore, Karnataka, India

K. Mequanint et al. (eds.), *Sustainable Development Research in Materials and Energy*, Green Energy and Technology,
https://doi.org/10.1007/978-3-031-69860-6_9

methods have depended on metal oxide semiconductor (MOS) sensors, but they have their own limit of detection which begin in the range from 1 to 1000 ppm, and the system temperature falls in the ranges from 200 to 500°C as reported [36]. Thus, this high-operating temperature which is the cause of high-power consumption makes MOS sensor limited in applications [41]. Specifically, MOS sensors cannot be used in handheld devices where the power is supplied by a battery [41].

Conducting polymers such as polyaniline (PANI) having π-conjugated system were discovered and attracted by researchers since it has excellent and tunable properties as well as its less-priced, frivolous, ease of processing, resistance to corrosion, and excellent electrical, mechanical, and optical properties [5, 9, 14, 21]. Besides conducting polymers (CP) polyaniline (PANI) becomes attractive for bioelectronic and electronic interfaces with it intrinsic biocompatibility, fast good self-conductivity activities, intrinsic redox reaction, and reversible doping-de-doping properties [14, 20, 22, 31, 32, 39]. These make polyaniline (PANI) applicable in the production of sensors having short response time, room temperature operation, high sensitivity, and high selectivity, and they can be tuned in both chemical and physical properties under different possibilities [14, 19, 23, 26, 35, 42]. When we see researches' report, there is not enough reports regarding PANI-nanosphere composite especially in the area of sensing ammonia (NH_3) gas, and particularly not enough researches' report for PANI-nanosphere composite electrochemical sensor-based meat spoilage/poisoning sensing. Thus, polyaniline (PANI)-nanosphere composite electrochemical sensor-based materials has been found to be very interesting in the area of sensing activities that can participate in several advantages for the detection of ammonia (NH_3) gas in spoilage meat due to its good stability, redox properties, and chemical and biochemical performances as well as sensing abilities [7, 10–14, 19, 23, 26, 30, 35, 42, 43].

In this work, polyaniline-nanosphere composite electrochemical sensor was successfully synthesized by taking the appropriate dopant and polymerization initiator to monomer through the chemical oxidative polymerization method [3] (via a self-assembly process) at lower temperatures (0–5°C) and has been characterized using FT-IR, XRD, UV-Vis, TGA, Zetasizer(DLS), SEM, TEM, and CV devices.

9.2 Experimental Methods

9.2.1 Materials

Aniline (99.9%, Sigma-Aldrich), azobisisobutyronitrile (AIBN, 99.5%), p-toluene sulfonic acid, (PTSA, 98%), N-methyl-2-pyrrolidone (NMP, 98.5%), acetone (98%), diethyl ether (98%), azobisisobutyronitrile (AIBN, 99.5%), and ethanol (99%)

9.2.2 Production of Polyaniline-Nanosphere

At first, polyaniline was produced via chemical oxidative polymerization process [1, 3, 16]. The 5-g aniline and 1-g p-toluene sulfonic acid (as an acidic dopant and as a stabilizer to control stereo-regularity) were mixed in a 500-ml flask. Then, 15 mL of N-methyl-2pyrrolidone (NMP) solvent was added to the mixture which was kept on the ice bath (below 5°C) under vigorous stirring conditions for 10 min. Next to this, the polymerization free-radical initiator (5 g of azobisisobutyronitrile, AIBN) was added to the reaction mixture and stirred for 24 h at 5°C to complete the oxidative polymerization reaction. As a result, the dark green solution was obtained, and then 60 mL of acetone, 60 mL of diethyl ether, and 20 mL of ethanol were added to the solution. The solution was filtered, and the precipitate (PANI) was washed with de-ionized water until the filtrate becomes colorless. The oligomeric impurities have been removed from the precipitate by washing with methanol and acetone [1, 3, 16]. The PANI-nanosphere composite precipitate was dried in a vacuum oven at 60° C for 12 h and ground into very fine powder.

9.2.3 Characterization of PANI-Nanosphere Composite

Functional groups' fingerprint results were obtained from a FT-IR spectrometer (spectrum one, Perkin Elemer, Waltham, MA, USA) characterization with an ATR transmission mode in the wavelength range of 400–4000 cm^{-1}. The semicrystalline structure of PANI-nanosphere composite was characterized by the X-ray diffraction (XRD) with Cu Kα radiation ($\lambda = 0.1541$ nm), an accelerating potential of 40 kV, and current of 30 mA at a scanning rate of 0.5°/min on an X-ray diffractometer (Rigaku MultiFlex ZD3609N). The surface morphology and nanosphere particle arrangement of PANI-nanosphere composite was investigated by field emission scanning electron microscope (FESEM, Carl Zeiss Ultra 55 and JSM-IT300) at an accelerating voltage of 10 kV with energy and angle selective backscattered electron (EsB) detector and Transmission electron microscopy (TEM) (Model; JSM Tecnai F20). The absorbance of each sample solution was studied by UV-Vis spectrophotometer (model V-750 UV-Visible spectrophotometer) in a wavelength range of 200–800 nm. The thermal decomposition of samples was determined by a Thermogravimetric analyzer, TGA (model: STA409PC Netzsch, Germany). The size distributions and zeta potentials of the samples were determined by Zetasizer, dynamic light scattering (Zetasizer Ver.7.13. MAL 1031019). A computer-controlled, Instrument model: CHI660D electrochemical workstation potentiostat (CH Instruments, TX, USA) was used for voltametric measurements. It was a conventional three-electrode electrochemical cell with Ag/AgCl(3MKCl) as reference electrode, Pt wire as counter electrode, and GCE as working electrode that was applied for electrochemical characterization (anodic and cathodic reaction properties, cyclic stability, reproducibility, selectivity, and performance) of

PANI-nanosphere composite electrochemical ammonia (NH_3) gas sensor at IISC, Bangalore, India.

9.3 Results and Discussion

Figure 9.1a displays the FTIR spectrum of PANI-nanosphere composite. Thus, the absorption peaks at 2935 cm^{-1} and 2302 cm^{-1} are attributed to N-H stretching of the quinoid and C=C stretching of the benzenoid rings present in PANI-nanosphere composite in the order already mentioned [34]. The absorption peaks at 3642.88 cm^{-1} and 1815 cm^{-1} correspond to N-H stretching vibration of the amine and the C-H stretching band of cyclic alkene. The characteristic peaks at 1393 cm^{-1} and 667 cm^{-1} are assigned to asymmetric C-H stretching and plane C-H bending vibrations of benzenoid rings in the order already mentioned [2, 37]. A sharp peak at 663.94 cm^{-1} corresponds to the C=C bending vibration of PANI-nanosphere composite [37]. The XRD pattern of PANI-nanosphere composite is shown in Fig. 9.1b. The prominent semi-diffraction peaks at 2θ (15.18,20.91, 25.43, and 32.4) indicate the semicrystalline structure of PANI-nanosphere composite [27, 38]. It confirms that the formed PANI-nanosphere composite is appropriate for energy storage materials (supercapacitors) and sensor applications.

Figure 9.2a displays the UV-visible absorption spectrum of PANI-nanosphere composite. The absorption band at 270.43 nm, 378.74 nm, and 568.74 nm correspond to π-π* transition of the benzenoid ring and π-polaron transition (due to the charge transfer between the benzenoid and quinoid units) in the order already mentioned [27, 34, 38, 40]. Figure 9.2b displays the thermogravimetric analysis curve of PANI-nanosphere nanocomposites. Figure 9.2b indicates its thermal stability before 35°C specially at room temperature but after 35°C, step-wise decomposition (instability) regions due to weight losses of the contents of

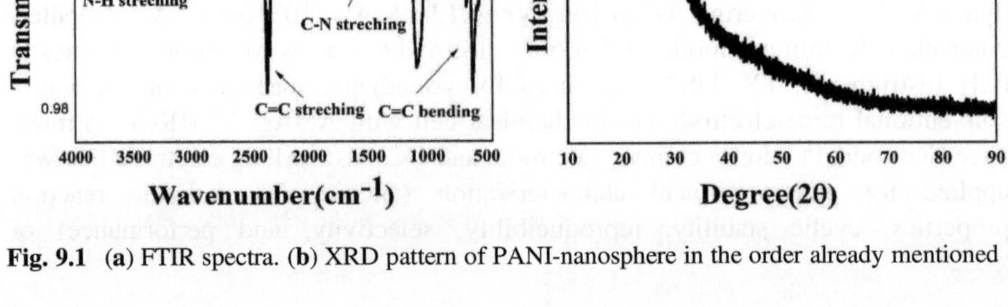

Fig. 9.1 (**a**) FTIR spectra. (**b**) XRD pattern of PANI-nanosphere in the order already mentioned

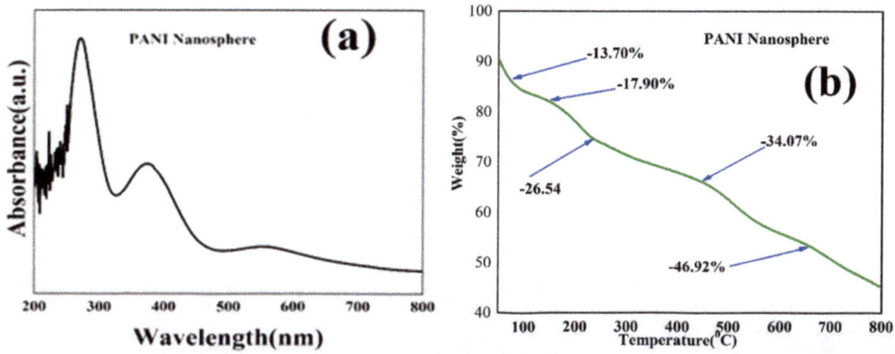

Fig. 9.2 (**a**) UV-visible absorption spectra. (**b**) Thermogravimetric analysis of PANI-nanosphere composite in the order already mentioned

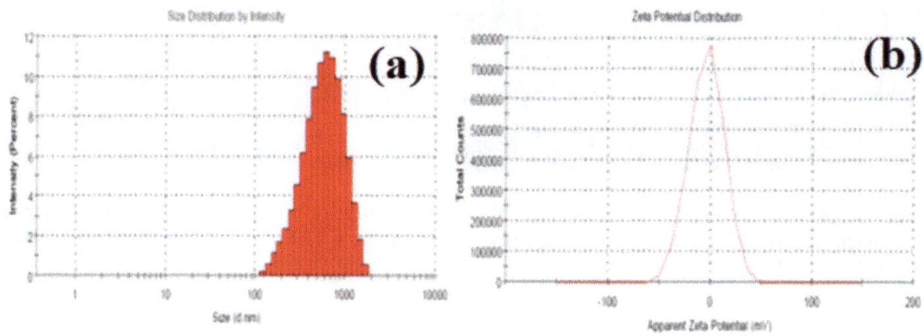

Fig. 9.3 (**a**) Size distribution report. (**b**) Zeta potential of PANI-nanosphere composite in the order already mentioned

polyaniline-nanosphere composite: (i) the weight loss observed below 78.70°C (−13.70%) correspond to the loss of water molecules/moisture, which is due to the hygroscopic nature of the polymer nanosphere; (ii) the second weight loss (−17.90% occurred at 152.77°C) that could be related to removal of dopant ions/tosylic acid, TsOH, ions and decompositions; (iii) the third weight loss (−26.54% occurred at 234.41 °C) that could be related to atmosphere switched from N_2 to O_2); (iv) the fourth weight loss (−34.07% occurred at 449.57 °C), which might be related to combustion of carbon parts; and (v) the five and above stages weight losses (−46.92% occurred at 657.69 °C), which could be related to inert inorganic residue of ash. Furthermore, polyaniline-nanosphere composite has good thermal stability specially at room temperature that makes its goodness for the detection of ammonia (NH_3) gas from poised meat.

The size distribution and zeta potential reports for PANI-nanosphere composite are shown in Fig. 9.3a and b. The size distribution and zeta potential values are given in Table 9.1. In principle, the size of a single unit of the nanosphere/nanomaterial is between 1 and 100 nm (at least one dimension). As per the standard BS ISO 22412:

Table 9.1 The size distribution report and zeta potential values of PANI-nanosphere composite

Sample type	Size distribution reports		Zeta potential reports	
PANI-nanosphere	Specification	Values	Specification	Values
	Size (d.nm)	499.2	Zeta potential	−4.23
	PdI	0.212	Conductivity (μS/cm)	23.4
	Intercept	0.872	Area (%)	100
	Intensity (%)	100		

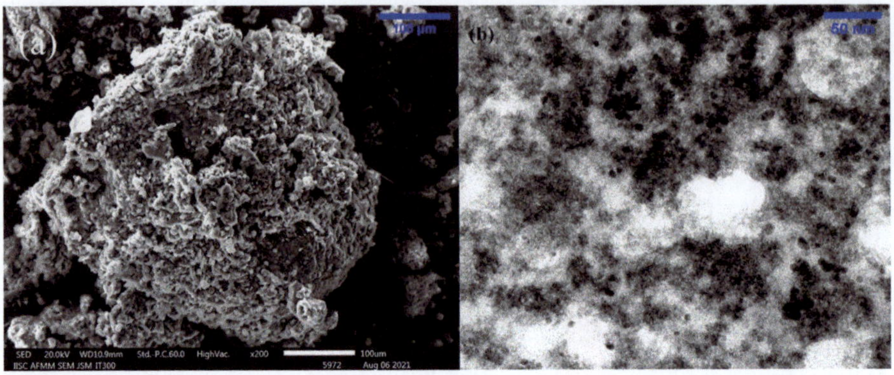

Fig. 9.4 (**a**) SEM image. (**b**) TEM Image of PANI-nanosphere (electron beam acceleration energy and magnification are 20 kV and ×1300, respectively) using a TEM device (Tecnai F20) at IISC AFMM, Bangalore, India

2017, ISO, 22412, Second Edition 2017-02 (from 1 nm to 100 μm range), the size distribution of PANI-nanosphere in Fig. 9.3a displays a narrow and sharp peak, and the obtained polyaniline-nanocomposite particle size result (z-average = 499.2 nm and PdI =0.184) is under considerable. This indicates that PANI-nanosphere composite has excellent consistency in particle size, shape, and mass of the particles. Figure 9.3b shows the plots of apparent zeta potential vs. total counts of PANI-nanosphere composite [28]. The zeta potential value (−4.23) as well as conductivity (23.4 μS/cm) indicated in Table 9.1 and Fig. 9.3b assured that there is stability and conductivity. The negative potential indicates the net charge of the scattering materials to the negative slipping plane. All these characteristics make it for good ability in the detection of ammonia from poised meat.

The surface morphology and the internal nanosphere arrangement of the PANI-nanosphere composite was analyzed by SEM and TEM [15, 24]. Figure 9.4a (SEM surface morphology result) displays the presence of PANI in the nanosphere composite powder sample in 100 μm scale, while the TEM image of PANI-nanosphere composite is displayed in Fig. 9.4b in 50 nm scale. Bothe the SEM and TEM images confirm good porous structure of PANI-nanosphere composite, and it can be suitable for electrochemical sensor applications in the detection of ammonia (NH_3) gas in poised meat sample.

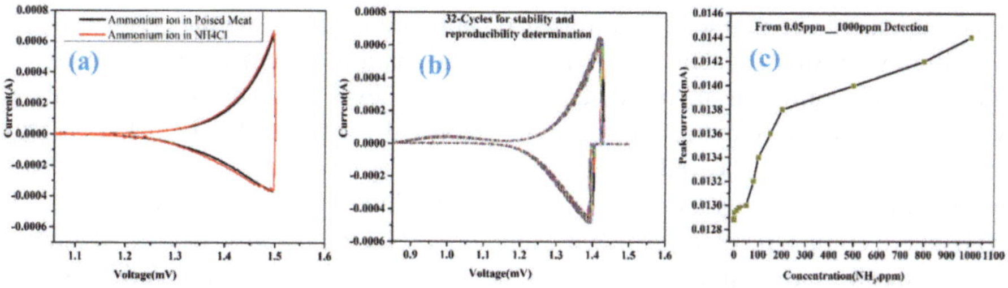

Fig. 9.5 (**a**) Red voltammogram curve for ammonium ion in NH_4Cl solution and black voltammogram curve for ammonium ion in decomposed meat solution. (**b**) Voltammograms of 32 cycles for ammonium ion in decomposed meat solution for stability and reproducibility determination. (**c**) Peak currents vs concentrations (0.05_1000ppm) determination

The electrochemical activities of the redox processes of PANI-nanosphere composite electrochemical ammonia (NH_3) gas sensor with ammonium ion in the detection of ammonia in poised meat were carried out using three electrodes electrochemical cell cyclic voltammetry (GCE as working electrode) in 0.1MKCl as supporting electrolyte (as an indicator or probe solution) in the presence of 0.1 M HCl(P^H < 3, acidic medium) under the setups of 50 mV/S as scan rate, 1e^{-4} sensitivity (A/V), 0.001 sample interval (V), and 2-second quiet time. As clearly shown in Fig. 9.5a–c, there have been changes and produced currents after concept potential (almost 1.24 mV for all activities) due to the PANI-nanosphere composite electrochemical ammonia (NH_3) gas sensor on GCE that senses ammonia (NH_3) in the electrochemical solution. As a result, anodic and cathodic current peak values occurred at effective surface of GCE due to the generated PANI-nanosphere composite film on GCE after being incorporated into it and immersed in the solution and have been converted into signals (voltammograms). Polyaniline (PANI) has played a great role in electronic transport proving a conducting pathway for electron transfer. As illustrated in Fig. 5a, the anodic peak current value for ammonium ion detection in NH_4Cl (red color curve) is 6.631E^{-4}A at1.496 mV, while for ammonium ion detection in poised meat (black color curve) is 6.585E^{-4} A at1.495 mV and the cathodic peak current value for ammonium ion detection in NH_4Cl (red color curve) is −3.691E^{-4} A at1.497 mV, whereas for ammonium ion detection in poised meat (black color curve) is −3.835E^{-4} A at1.495Mv. So it is possible to conclude that the synthesized PANI-nanosphere composite electrochemical ammonia (NH_3) gas sensor is ammonia (NH_3) selective in the decomposed poised meat. The cyclic stability and its reproducibility of the synthesized PANI-nanosphere composite electrochemical ammonia (NH_3) gas sensor have been checked by running 32 cycles after 7 days of decomposition of the meat sample. Thus, the 32-cycle voltammograms indicated in Fig. 9.5b have almost same anodic peak current values (almost 6.3202E^{-4} at 1.423 mV for all voltammograms) and same cathodic peak current values (almost −3.839E^{-4} at 1.394 mV for all voltammograms). This can assure the cyclic stability and repeatability/reproducibility of the synthesized sensor. Figure 5c illustrates the response of PANI-nanosphere composite electrochemical ammonia (NH_3) gas sensor operated at room temperature, 25°C, good dependence on the concentration. As a highly sensitive sensor, it can measure and exhibit extremely small amount to ammonia concentration having detecting very low level (0.05 ppm) in the analyte, which is considerably better than currently reported literatures [8, 17, 24, 29]. This ammonia gas sensor showed excellent detection performance and high selectivity against the components of decomposed meat. This result indicates that PANI-nanosphere composite electrochemical ammonia (NH_3) gas sensor is a good ammonia gas-sensing novel material for detecting ammonia (NH_3) accumulation, which can be applied for checking of the level of ammonia in the given meat sample. The detection response changes of the sensor to different ammonia (NH_3) gas concentrations (0.05_1000 ppm) is shown in Fig. 9.5c. The peak current increases in a direct relation to the ammonia (NH_3) gas concentrations.

9.4 Conclusion

New and noble PANI-nanosphere composite material, which can be taken as recent advances toward the use of electrochemical ammonia (NH_3) gas sensor, has been synthesized successfully through chemical oxidative polymerization by taking the appropriate dopant and polymerization initiator to monomers. This possesses good electrical conductivity, thermal stability, semicrystalline, and electrochemical properties. The size and zeta potential of the PANI-nanosphere composite particles (499.2 nm and −4.23) have been determined by the Zetasizer. The quinoid units along the polymer chain being affected, such that strong interactions between nanosphere particles and quinoidal sites of PANI, were observed from FT-IR data. Semicrystalline nature of PANI-nanosphere composite has been confirmed from XRD analysis. The morphological study (SEM result:100 μm and TEM result: 50 nm) confirmed the formation and distribution of nanosphere particles in PANI-nanosphere polymer surface. The characteristic absorption peaks of the synthesized material analysis (UV-visible spectrum) were very appropriate and depictable. Due to better chemical kinetics, chemical activity, shorter ionic diffusion path lengths, high specific surface area, and surplus of active sites for electrochemical reactions, the polyaniline nanosphere electrochemical sensor has exhibited excellent performance in the detection of ammonia (NH_3) gas in poised meat sample. So, polyaniline nanosphere composite has been found to have interesting features that make it quite a suitable and promising electrochemical sensor for the detection of ammonia (NH_3) gas applications.

Acknowledgments Authors have acknowledged Jimma University (Ethiopia), Debre Markos University (Ethiopia), and Indian Institute of Science (Bangalore, India) for providing financial support and laboratory facilities to carry out the research work.

Funding Statement This work was supported by the Ethiopian Ministry of Science and Higher Education and Indian Institute of Science, IISC, Bangalore, India.

Data Availability Statement The data that we have used to support the findings of the study have been included within the article.

Conflict of Interest The authors declare no competing financial interest.

References

1. Abu-Thabit, N. Y. (2016). Chemical oxidative polymerization of polyaniline: A practical approach for preparation of smart conductive textiles. *Journal of Chemical Education, 93*(9), 1606–1611.
2. Ahad, I. Z. M., Harun, S. W., Gan, S. N., & Phang, S. W. (2018). Polyaniline (PAni) optical sensor in chloroform detection. *Sensors and Actuators B: Chemical, 261*, 97–105.
3. Alemu, T. A., Ayele, D. W., Worku, A. K., Teshager, M. A., & Ramamurthy, P. (2023). Highly sensitive detection of bacteria (E. Coli) Endotoxin using novel PANI-benzimidazole-Ag nanocomposite by DMMB dye displacement assay. *Materials Research Express*.
4. Bendahan, M., Lauque, P., Lambert-Mauriat, C., Carchano, H., & Seguin, J.-L. (2002). Sputtered thin films of CuBr for ammonia microsensors: Morphology, composition and ageing. *Sensors and Actuators B: Chemical, 84*(1), 6–11.
5. Carrasco, P. M., Grande, H. J., Cortazar, M., Alberdi, J. M., Areizaga, J., & Pomposo, J. A. (2006). Structure–conductivity relationships in chemical polypyrroles of low, medium and high conductivity. *Synthetic Metals, 156*(5–6), 420–425.
6. Chougule, M., Pawar, S., Patil, S., Raut, B., Godse, P., Sen, S., & Patil, V. B. (2011). Polypyrrole thin film: Room temperature ammonia gas sensor. *IEEE Sensors Journal, 11*(9), 2137–2141.
7. Ding, H., Wan, M., & Wei, Y. (2007). Controlling the diameter of polyaniline nanofibers by adjusting the oxidant redox potential. *Advanced Materials, 19*(3), 465–469.
8. Fu, H., Jiang, Y., Ding, J., Zhang, J., Zhang, M., Zhu, Y., & Li, H. (2018). Zinc oxide nanoparticle incorporated graphene oxide as sensing coating for interferometric optical microfiber for ammonia gas detection. *Sensors and Actuators B: Chemical, 254*, 239–247.
9. Gospodinova, N., & Terlemezyan, L. (1998). Conducting polymers prepared by oxidative polymerization: Polyaniline. *Progress in Polymer Science, 23*(8), 1443–1484.
10. Gu, H., Guo, J., Wei, H., Zhang, X., Zhu, J., Shao, L., et al. (2014a). Magnetoresistive conductive polymer-tungsten trioxide nanocomposites with ultrahigh sensitivity at low magnetic field. *Polymer, 55*(3), 944–950.
11. Gu, H., Guo, J., Yan, X., Wei, H., Zhang, X., Liu, J., et al. (2014b). Electrical transport and magnetoresistance in advanced polyaniline nanostructures and nanocomposites. *Polymer, 55*(17), 4405–4419.
12. Gu, H., Huang, Y., Zhang, X., Wang, Q., Zhu, J., Shao, L., et al. (2012). Magnetoresistive polyaniline-magnetite nanocomposites with negative dielectrical properties. *Polymer, 53*(3), 801–809.
13. Gu, H., Wei, H., Guo, J., Haldolaarachige, N., Young, D. P., Wei, S., & Guo, Z. (2013). Hexavalent chromium synthesized polyaniline nanostructures: Magnetoresistance and electrochemical energy storage behaviors. *Polymer, 54*(21), 5974–5985.
14. Hu, Q., Wang, Z., Chang, J., Wan, P., Huang, J., & Feng, L. (2021). Design and preparation of hollow NiO sphere-polyaniline composite for NH3 gas sensing at room temperature. *Sensors and Actuators B: Chemical, 344*, 130179.

15. Imani, A., & Farzi, G. (2015). Facile route for multi-walled carbon nanotube coating with polyaniline: Tubular morphology nanocomposites for supercapacitor applications. *Journal of Materials Science: Materials in Electronics, 26*, 7438–7444.
16. Kandulna, R., & Choudhary, R. (2017). Robust electron transport properties of PANI/PPY/ZnO polymeric nanocomposites for OLED applications. *Optik, 144*, 40–48.
17. Khan, H., Malook, K., & Shah, M. (2017). Highly selective and sensitive ammonia sensor using polypyrrole/V 2 O 5 composites. *Journal of Materials Science: Materials in Electronics, 28*, 13873–13879.
18. Kim, K.-H., Pal, R., Ahn, J.-W., & Kim, Y.-H. (2009). Food decay and offensive odorants: A comparative analysis among three types of food. *Waste Management, 29*(4), 1265–1273.
19. Li, S., Lin, P., Zhao, L., Wang, C., Liu, D., Liu, F., et al. (2018). The room temperature gas sensor based on polyaniline@ flower-like WO3 nanocomposites and flexible PET substrate for NH3 detection. *Sensors and Actuators B: Chemical, 259*, 505–513.
20. Li, T., Wang, J., Zhang, L., Yang, J., Yang, M., Zhu, D., et al. (2017). "Freezing", morphing, and folding of stretchy tough hydrogels. *Journal of Materials Chemistry B, 5*(29), 5726–5732.
21. Lim, V., Kang, E., Neoh, K., Ma, Z., & Tan, K. (2001). Determination of pyrrole–aniline copolymer compositions by X-ray photoelectron spectroscopy. *Applied Surface Science, 181*(3–4), 317–326.
22. Lin, S., Yuk, H., Zhang, T., Parada, G. A., Koo, H., Yu, C., & Zhao, X. (2016). Stretchable hydrogel electronics and devices. *Advanced Materials, 28*(22), 4497–4505.
23. Luo, G., Xie, L., He, M., Jaisutti, R., & Zhu, Z. (2021). Flexible fabric gas sensors based on reduced graphene-polyaniline nanocomposite for highly sensitive NH 3 detection at room temperature. *Nanotechnology, 32*(30), 305501.
24. Matindoust, S., Farzi, A., Baghaei Nejad, M., Shahrokh Abadi, M. H., Zou, Z., & Zheng, L.-R. (2017). Ammonia gas sensor based on flexible polyaniline films for rapid detection of spoilage in protein-rich foods. *Journal of Materials Science: Materials in Electronics, 28*, 7760–7768.
25. Matindoust, S., Farzi, G., Nejad, M. B., & Shahrokhabadi, M. H. (2021). Polymer-based gas sensors to detect meat spoilage: A review. *Reactive and Functional Polymers, 165*, 104962.
26. Matsuguchi, M., Nakamae, T., Fujisada, R., & Shiba, S. (2021). A highly sensitive ammonia gas sensor using micrometer-sized core–shell-type spherical polyaniline particles. *Sensors, 21*(22), 7522.
27. Neelgund, G. M., & Oki, A. (2011). A facile method for the synthesis of polyaniline nanospheres and the effect of doping on their electrical conductivity. *Polymer International, 60*(9), 1291–1295.
28. Ramaye, Y., Dabrio, M., Roebben, G., & Kestens, V. (2021). Development and validation of optical methods for zeta potential determination of silica and polystyrene particles in aqueous suspensions. *Materials, 14*(2), 290.
29. Ramesan, M., & Santhi, V. (2018). Synthesis, characterization, conductivity and sensor application study of polypyrrole/silver doped nickel oxide nanocomposites. *Composite Interfaces, 25*(8), 725–741.
30. Saravanan, C., Shekhar, R. C., & Palaniappan, S. (2006). Synthesis of polypyrrole using benzoyl peroxide as a novel oxidizing agent. *Macromolecular Chemistry and Physics, 207*(3), 342–348.
31. Shi, Y., & Yu, G. (2016). Designing hierarchically nanostructured conductive polymer gels for electrochemical energy storage and conversion. *Chemistry of Materials, 28*(8), 2466–2477.
32. Shi, Y., Zhou, X., Zhang, J., Bruck, A. M., Bond, A. C., Marschilok, A. C., et al. (2017). Nanostructured conductive polymer gels as a general framework material to improve electrochemical performance of cathode materials in Li-ion batteries. *Nano Letters, 17*(3), 1906–1914.
33. Snook, G. A., Kao, P., & Best, A. S. (2011). Conducting-polymer-based supercapacitor devices and electrodes. *Journal of Power Sources, 196*(1), 1–12.
34. Tamboli, M. S., Kulkarni, M. V., Patil, R. H., Gade, W. N., Navale, S. C., & Kale, B. B. (2012). Nanowires of silver–polyaniline nanocomposite synthesized via in situ polymerization and its novel functionality as an antibacterial agent. *Colloids and Surfaces B: Biointerfaces, 92*, 35–41.

35. Verma, A., Gupta, R., Verma, A. S., & Kumar, T. (2023). A review of composite conducting polymer-based sensors for detection of industrial waste gases. *Sensors and Actuators Reports, 5*, 100143.
36. Wang, C., Yin, L., Zhang, L., Xiang, D., & Gao, R. (2010). Metal oxide gas sensors: Sensitivity and influencing factors. *Sensors, 10*(3), 2088–2106.
37. Wasu, M., & Raut, A. (2014). Synthesis and characterization of polyaniline based conducting polymers. *Journal of Chemistry, 4*(2), 90–97.
38. Wu, H., Wang, Q., Fei, G. T., Xu, S. H., Guo, X., & De Zhang, L. (2018). Preparation of hollow polyaniline micro/nanospheres and their removal capacity of Cr (VI) from wastewater. *Nanoscale Research Letters, 13*, 1–9.
39. Yuk, H., Lu, B., & Zhao, X. (2019). Hydrogel bioelectronics. *Chemical Society Reviews, 48*(6), 1642–1667.
40. Zengin, H., Aksin, G., Zengin, G., Kahraman, M., & Kilic, I. H. (2019). Preparation and characterization of conductive polyaniline/silver nanocomposite films and their antimicrobial studies. *Polymer Engineering & Science, 59*(S1), E182–E194.
41. Zhang, J., Liu, X., Neri, G., & Pinna, N. (2016). Nanostructured materials for room-temperature gas sensors. *Advanced Materials, 28*(5), 795–831.
42. Zhang, Y., Zhang, J., Jiang, Y., Duan, Z., Liu, B., Zhao, Q., et al. (2020). Ultrasensitive flexible NH3 gas sensor based on polyaniline/SrGe4O9 nanocomposite with ppt-level detection ability at room temperature. *Sensors and Actuators B: Chemical, 319*, 128293.
43. Zhao, Y., Wei, H., Arowo, M., Yan, X., Wu, W., Chen, J., et al. (2015). Electrochemical energy storage by polyaniline nanofibers: High gravity assisted oxidative polymerization vs. rapid mixing chemical oxidative polymerization. *Physical Chemistry Chemical Physics, 17*(2), 1498–1502.

Chapter 10
Investigating the Thermal Performance of Different Solar-Assisted Dryers for Agricultural Produce: A Review

Hailemariam Mulugeta Wassie, Muluken Zegeye Getie, Bimrew Tamrat Admasu, Mulat Shitye Alem, Molla Asmare Alemu, Lamesgin Addisu Getnet, Atrsaw Jejaw Bezie, Aschale Getnet Alemu, Assefa Beyene Ayalew, and Tewodros Simeneh Ayalewu

10.1 Introduction

Food demand is anticipated to increase (40–50%) by 2030 [1], and maintaining a balance between food supply and consumption is extremely difficult. Drying is a key activity for food preservation that involves decreasing the moisture in food to avoid bacterial growth [2]. There are various mechanisms that cause moisture to evaporate near the surface, such as diffusion of liquid and vapor, capillary and gravity forces, shrinkage, and pressure gradients [3].

According to reports, improper preservation techniques result in the loss of a large number of cereals (10–30%) and fruits (50–70%) [4]. Small-scale farmers use open-sun drying methods to dry agricultural products [4]. For optimal drying of food items especially for fruits and vegetables, the drying medium temperature should be in a range of 45–75°C [5]. Agricultural food products should be dried quickly to a safe moisture content while preserving product quality by removing moisture from

H. M. Wassie
Faculty of Mechanical and Industrial Engineering,
Bahir Dar Institute of Technology, Bahir Dar University, Bahir Dar, Ethiopia

Bahir Dar Energy Centre, Bahir Dar institute of Technology,
Bahir Dar University, Bahir Dar, Ethiopia

M. Z. Getie (✉) · B. T. Admasu · M. S. Alem · L. A. Getnet · A. J. Bezie · A. G. Alemu
A. B. Ayalew · T. S. Ayalewu
Faculty of Mechanical and Industrial Engineering,
Bahir Dar Institute of Technology, Bahir Dar University, Bahir Dar, Ethiopia

M. A. Alemu
Bahir Dar Energy Centre, Bahir Dar institute of Technology,
Bahir Dar University, Bahir Dar, Ethiopia

K. Mequanint et al. (eds.), *Sustainable Development Research in Materials and Energy*, Green Energy and Technology,
https://doi.org/10.1007/978-3-031-69860-6_10

them under controlled temperature and humidity settings [6]. Another option for drying is industrial drying, which is more expensive. It consumes a considerable amount of coal, which causes global warming [7].

To address the aforementioned issues, many researchers have designed, tested, and investigated solar dryers, which provide higher dried product quality, decrease losses and bacterial contamination, use less space for drying, reduce labor requirements, and reduce drying time.

Controlled solar dryer is the best and more sustainable means of food drying than open-sun drying (OSD) [8, 9]. Direct, indirect, and hybrid solar dryers have several drawbacks in their drying performance, such as uneven drying rates, longer drying times, and lower drying efficiency. Moreover, the drying system is affected by sudden changes as temperature, insulation, cloud cover, and wind speed [10]. Both design and manufacturing limitation and inappropriate solar collector material also limit the thermal performance of the whole drying systems. As a result, recent efforts to improve the rate of drying and dryer efficiency have focused on and tested a variety of solutions, including concentrating solar radiation, incorporating phase-change materials [9], improving solar collectors design, integrating a heat pumping system, integrating solar photovoltaic systems, combining automatic sun-tracking mechanism, and using a geothermal flat plate. Furthermore, supplemental recovery dryers, hybrid solar-biomass dryers using cogeneration (Co-Gen) technology, solar-infrared drying, solar photovoltaic and evacuated tube collectors, and solar-assisted fluidized-bed drying play vital roles in the advancement of the technology [1].

10.1.1 Solar Dryer Working System

As shown in Fig. 10.1, heat is transferred from its source to the dried product, while mass (moisture) is transported from the food to be dried to the surrounding air. A solar dryer's primary functions are to increase the moisture-content of the air by lowering its relative humidity and mounting the vapor pressure of moisture in the

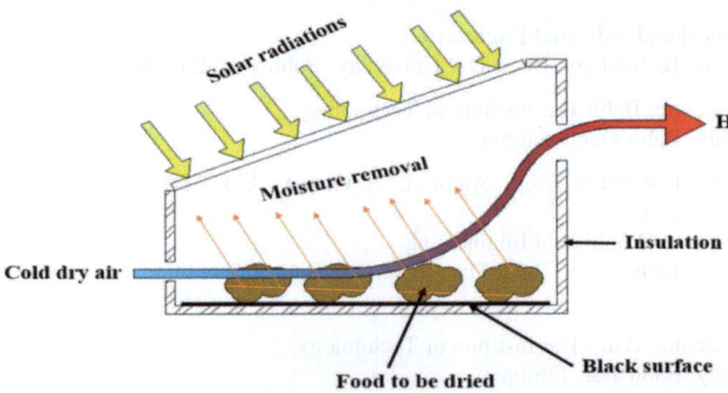

Fig. 10.1 Principle of solar dryer [11]

product. In solar-assisted drying method, hot air draws the water content from the substance to be dried. The temperature of the dried air determines how much moisture is removed; warm air absorbs moisture more easily than cold air [11].

10.1.2 Solar Thermal Collectors

Nowadays, many varieties of solar thermal energy collectors are used in drying of different agricultural and industrial products. Solar thermal energy collection can be done using either concentrating or non-concentrating technologies [12]. The non-concentrated solar collectors can be either flat plate collectors, which is applicable for a temperature range of up to 90°C and evacuated tube solar collectors applicable for a temperature range of up to 150°C [13, 14].

For the case of medium- and high-temperature (400–500°C) operation, concentrating solar collectors are the best option to achieve the desired result in solar heating, industrial processes, solar cooling, and electric power generation [15]. Figure 10.2 depicts the various types of solar thermal collectors used in the drying process.

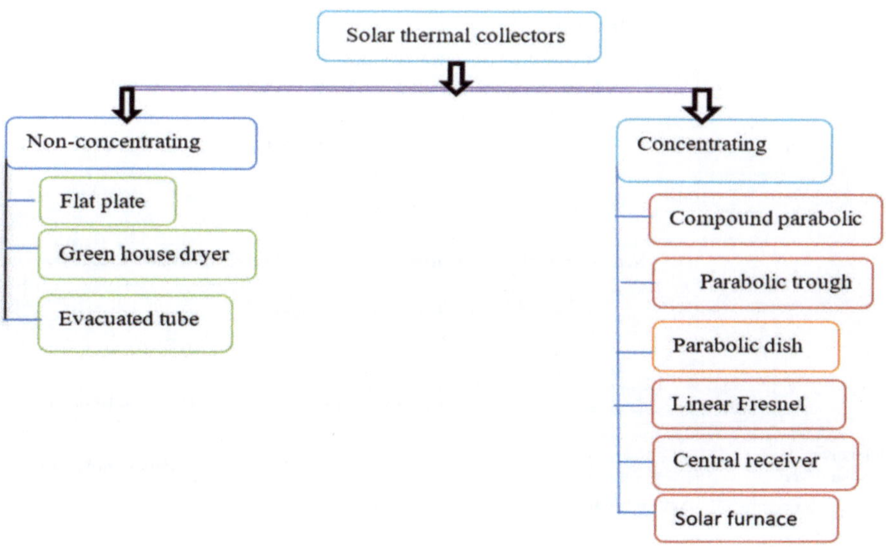

Fig. 10.2 Types of solar thermal collector types

10.2 Kinds of Solar Dryers

Based on their size, system design, and mode of operation (passive, active, or hybrid), agricultural solar dryers can be categorized [8]. Figure 10.3 depicts solar dryers that concentrate on heat transfer and air movement (natural or forced convection) (direct or indirect).

10.2.1 Forced Solar Dryer

A forced convection drying system is the one which utilizes external power such as fan or blower to promote heat transfer into the drying chamber to enhance the circulation of air within the drying cabinet [16]. The fan also ensures that the dryer maintains the required air flow rate, resulting in more consistent moisture evaporation from the goods. As per the size of the dryer, a large-sized fan or multiple fans are required to achieve the desired heating effect as shown in Fig. 10.4.

10.2.2 Passive Solar Energy Dryer

Instead of using mechanical or electrical power to force air through the product, these dryers use buoyancy forces to circulate the air through the crop being dried.

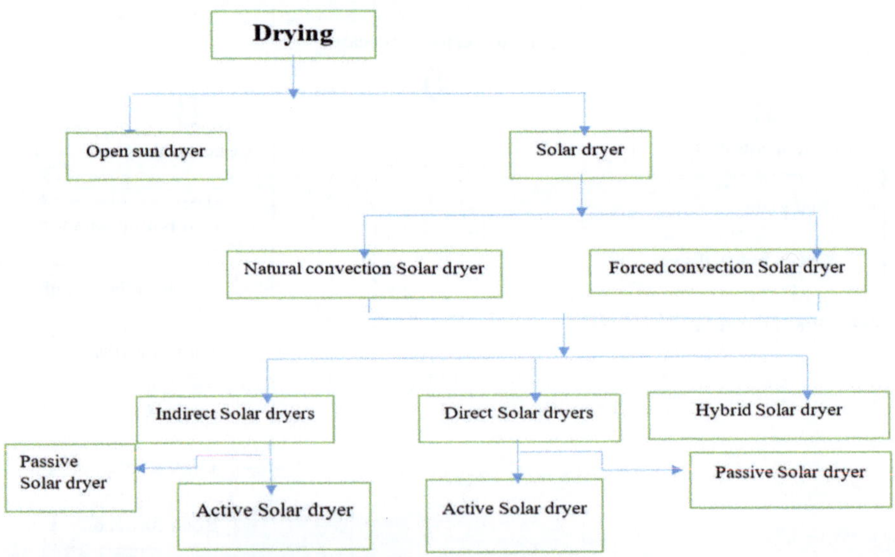

Fig. 10.3 Different categories of solar drying system

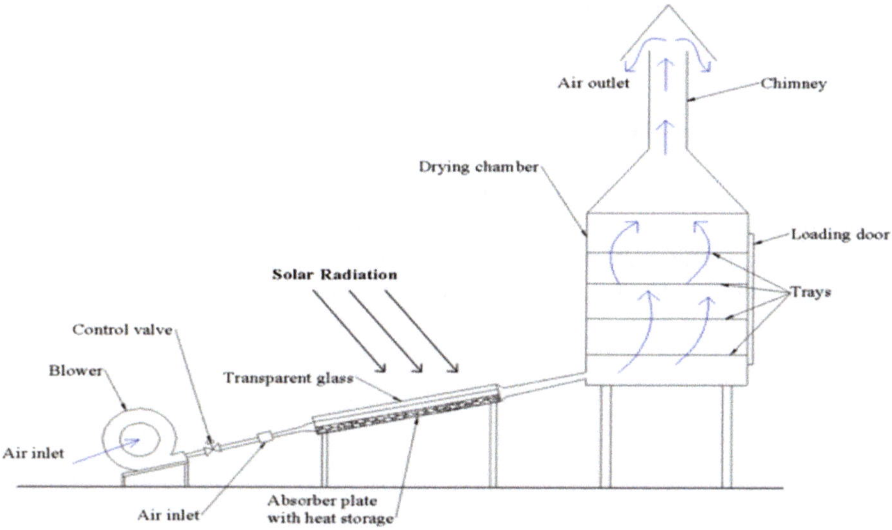

Fig. 10.4 Active solar dryer [16]

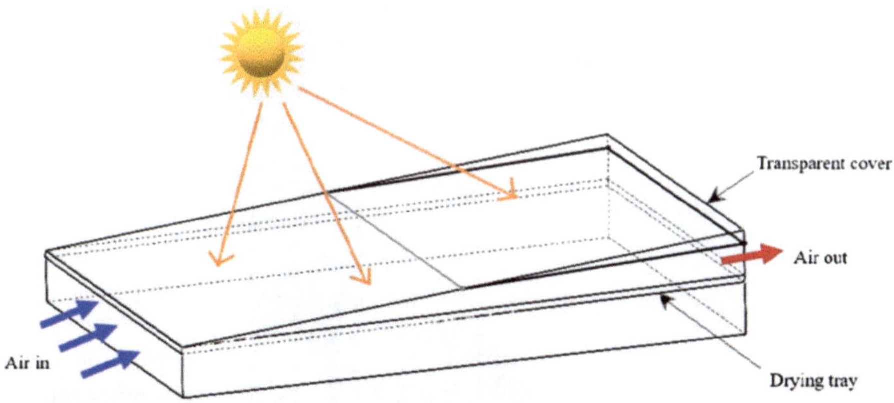

Fig. 10.5 Direct solar dryer [17]

The reason for this is because wind pressure brought on by density differences causes solar-heated air to be forced through the sample as it is drying. Figure 10.5 also illustrates such events.

10.2.2.1 Direct-Type Solar Dryer

A direct solar dryer uses a chamber to dry products without exposing them to sunlight, as seen in Fig. 10.2. Due to insufficient drying or crop burning from exposure to direct sunlight, the dried output from direct solar dryers may be lower quality [17].

10.2.2.2 Indirect Solar Dryer

In Musembi et al. [18], an indirect solar drier that is intended for usage in intermediate latitudes was designed, developed, and tested. So as to dry slices of fresh apple, 2.5 mm thick, a solar collector with a polycarbonate cover, absorber plate made of iron-sheet, a drying chamber, and a chimney compartment were used (Figs. 10.6 and 10.7).

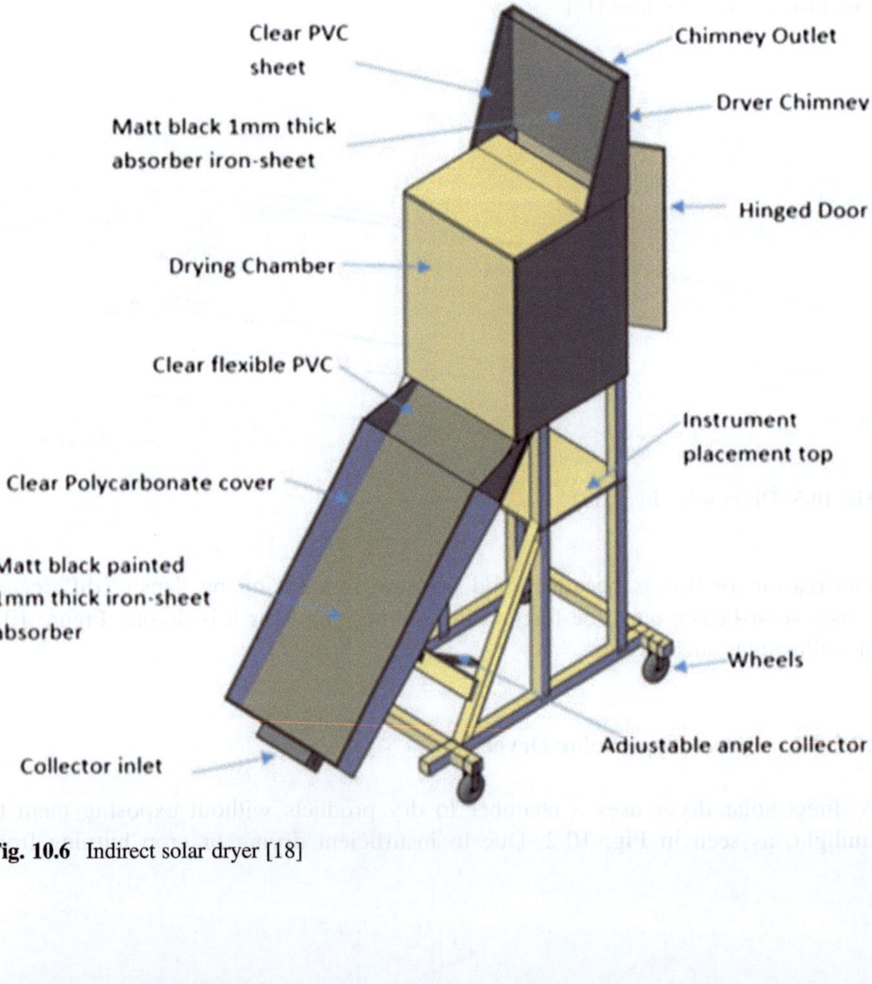

Fig. 10.6 Indirect solar dryer [18]

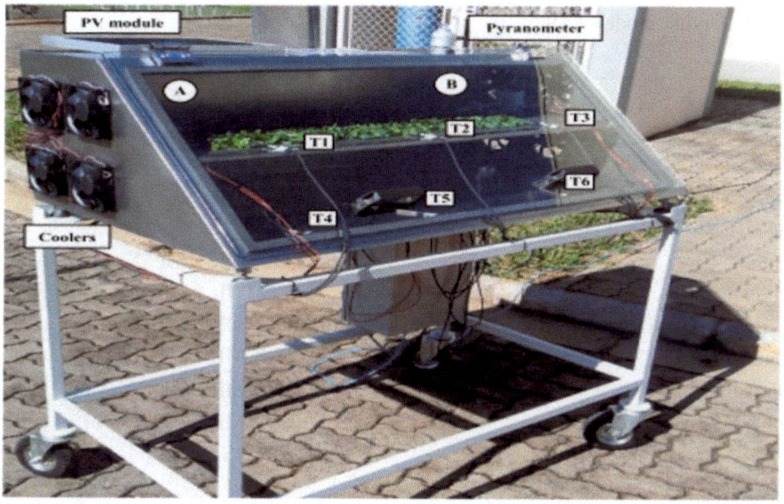

Fig. 10.7 Direct solar dryer [19]

10.3 Basic Drying Theory and Performance Evaluation Parameter

10.3.1 Thin-Layer Drying Model

In general, by Fick's second law, it is possible to generate semitheoretical models. These models work well in a broad range of temperature, moisture content, relative humidity, and air flow velocity (Tables 10.1 and 10.2).

10.3.2 Validation the Model Statistically

To suit the drying data, the thin-layer drying model could be used to analyze the drying kinetics of agricultural products. The best semitheoretical drying models are chosen using statistical parameters, and goodness of fit is used as a comparison criterion (Table 10.3).

10.4 Solar Dryer for Different Agricultural Product

Ferraresi et al. [19] determine that the mean efficiency and specific energy consumption for the natural convection drying process were 34.2% and 18.3 kWh/kg and 38.3% and 16.4 kWh/kg for the forced convection drying process, respectively.

Table 10.1 Drying models [20]

Name	Equation
Newton	$\mathrm{MR} = e^{(-\mathrm{kt})}$
Pages	$\mathrm{MR} = e^{(-\mathrm{kt}^n)}$
Modified page	$\mathrm{MR} = e^{(-(\mathrm{kt})^n)}$
Henderson and Pabis	$\mathrm{MR} = e^{(-(\mathrm{kt})^n)}$
Modified Henderson and Pabis	$\mathrm{MR} = \mathrm{ae}^{(-\mathrm{kt})} + \mathrm{be}^{(-\mathrm{gt})} + \mathrm{be}^{(-\mathrm{gt})}$
Logarithmic	$\mathrm{MR} = \mathrm{ae}^{(-\mathrm{kt})} + \mathrm{c}$
Two-term	$\mathrm{MR} = \mathrm{ae}^{(-\mathrm{kot})} + \mathrm{be}^{(-\mathrm{k1t})}$
Two-term exponential	$\mathrm{MR} = \mathrm{ae}^{(-\mathrm{kt})} + (1 - \mathrm{a})\ \mathrm{e}^{(-\mathrm{kat})}$
Diffusion approach	$\mathrm{MR} = \mathrm{ae}^{(-\mathrm{kt})} + (1 - \mathrm{a})\ \mathrm{e}^{(-\mathrm{kbt})}$
Midilli and Kucuk	$\mathrm{MR} = \mathrm{ae}^{(-\mathrm{kt}^n)} + \mathrm{bt}$
Wangh and Singh model	$\mathrm{MR} = \mathrm{a} + \mathrm{at} + \mathrm{bt}^2$

Table 10.2 Performance evaluation parameter [21]

Parameter	Equation
Moisture content wet base	$\mathrm{MCwb} = \frac{W_{i-w_f}}{W_i} \times 100\%$
Moisture content on a dry basis	$\mathrm{Mcdb} = \frac{Mc_{\mathrm{wb}}}{100 - Mc_{\mathrm{wb}}}$
Mass of water removed	$\mathrm{mw} = \mathrm{Mp}\ \frac{(M_i - M_f)}{(100 - M_f)}$
Heat energy required to evaporate moisture	$\mathrm{Q_m} = \mathrm{m_w}L$
The moisture ratio	$\mathrm{MR} = \frac{M_t - M_e}{M_{i-}M_e}$
Drying rate	$\mathrm{DR} = \frac{M_{t+\mathrm{dt}} - M_t}{\mathrm{dt}}$
Moisture loss	$\mathrm{M_L} = (W_i - W_f)$
Drying efficiency	$\eta d = \frac{W \times L}{I_b \times t \times A_a}$
Equilibrium moisture content	$\mathrm{Me} = \frac{M_i M_f - (M_m)^2}{M_i + M_f - 2M_m}$
Effective moisture diffusion coefficient	$\mathrm{MR} = \frac{8}{\pi^2}\left[\exp\left(\frac{-\pi^2 D_{\mathrm{eff}}\ t}{4L^2}\right)\right]$

According to the drying kinetics, there was a constant rate phase and a declining rate period for both operating settings. Periods of constant rate were applied to straight lines with an R^2 of 0.999. The falling rate phases were most accurately predicted by the Page and Overhauls models, with R^2 ranging from 0.932 to 0.999. Effective diffusivity values were 1.15×10^{-8} m^2/s for forced convection runs and 5.15×10^{-9} m^2/s for natural convection runs. When drying by natural convection, the specific energy consumption and solar dryer's average efficiency were 18.3 kWh/kg and 34.2%, respectively; when drying by forced convection, they were 38.3% and 16.4 kWh/kg.

A thin-layered drying kinetics of Granny Smith apples using thermogravimetric at a constant temperature range of 20–50°C and interval of 5°C analyzed by [24]. The Wang-Singh model accurately describes the drying process that takes

Table 10.3 Statistical validation for the selected model

Statistical parameters	Model equation	References
Coefficient of determination (R^2)	$R2 = 1 - \frac{\sum_{i=1}^{N}(MR_{pre.i} - MR_{exp.i})^2}{\sqrt{\left[\sum_{i=1}^{N}(\overline{MR}_{pre.i} - MR_{exp.i})^2\right]}}$	[22]
Reduced chi-square (X^2)	$X2 = \frac{\sum_{i=1}^{N}(MR_{exp,i} - MR_{pre,i})2}{N-n}$	[23]
Mean bias error (MBE)	$MBE = \frac{1}{N}\sum_{i=1}^{N}(MR_{pre,i} - MR_{exp,i})$	[22]
Root mean square error (RMSE)	$RMBE = \left[\frac{1}{N}\sum_{i=1}^{N}(MR_{pre,i} - MR_{exp,i})^2\right]^{1/2}$	[20]
Sum square error (SSE)	$SSE = \frac{1}{N}\sum_{i=1}^{N}(MR_{exp,i} - MR_{prid,i}$	[20]

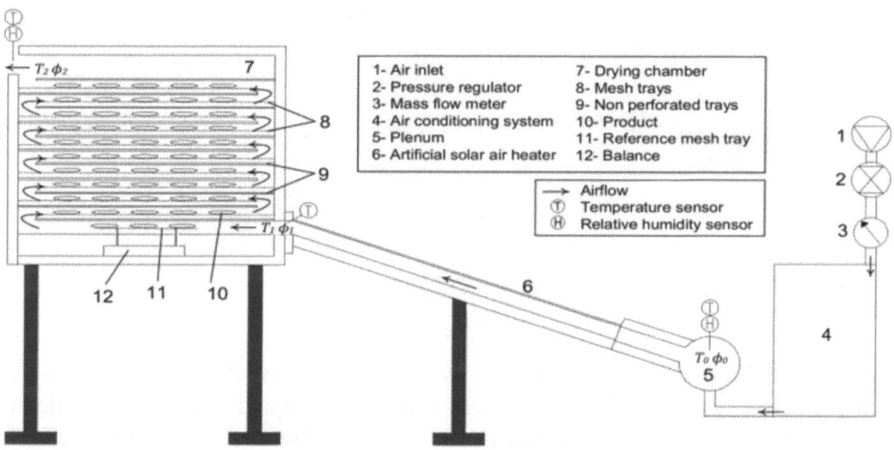

Fig. 10.8 Indirect solar dryer diagram [24]

place at a constant drying temperature. The model for solar dryer has been validated by experimental TGA measurements with maximum deviations of 1.5% in drying time. The suggested process was tested and then applied to the indirect sun dryer drying of Granny Smith apples. As shown in Fig. 10.8, the model prediction and experimental observations of apples drying in a lab-scale solar dryer in different drying settings showed good agreement with only minor deviations (less than 10%) caused by sample thermal inertia.

Lakshmi et al. [25] used a forced convection dryer operating in mixed modes and a paraffin wax thermal energy storage device (Curcuma cassia) to dry the black turmeric slices. The samples were dried for 18.5 h in a solar dryer with a moisture content ranging from 73.4% to 8.5%, as indicated in Fig. 10.9. After that, they were

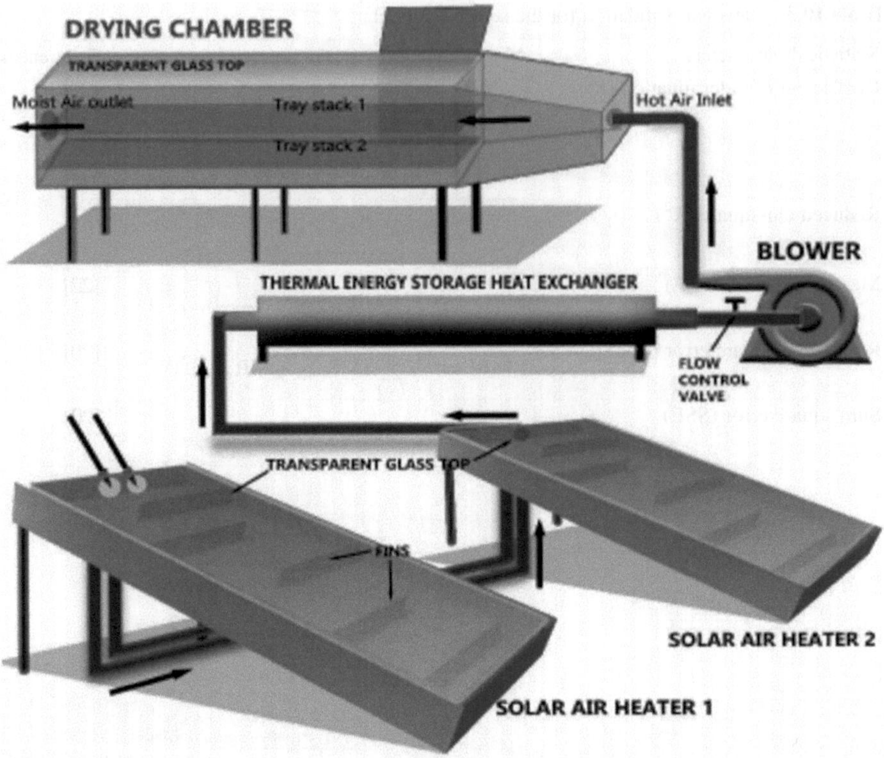

Fig. 10.9 Mixed-type active solar dryer with TESS [25]

dried for 46.5 h outside in the sun. The two-term model and the Page model are the most effective drying models when used in the open-sun and solar dryer. While the overall efficiency of the solar dryer was 12.0%, that of the solar collector was 25.6%.

Ssemwanga et al. [26] examined the HIP and SPE dryers' fruit drying capabilities with pineapples, and the outcomes were contrasted with those of the open-sun drying (OSD) standard. The food was dried in 10, 18, and 30 h using the SPE, HIP, and OSD procedures, respectively. The enhanced HIP dryer exhibited a drying efficiency that was equivalent to that of the SPE dryer and 18% better than that of the OSD method. As a consequence, upgrading the greenhouse cabinet and replacing the solar collector plate with many metallic solar concentrators significantly improves the HIP dryer's drying performance. So, the HIP dryer is recommended as an alternative to the OSD method for broad use as illustrated in Fig. 10.10.

Erick César et al. [27] examined a natural convection, hybrid-mode dryer that was used for the analysis illustrated as shown in Fig. 10.11. The temperature inside the drying cabinet during the testing varies from 65 to 70°C in MSD and 55 to 60°C in ISD at solar noon. According to the findings, the tomato was dried for 26 h in the ISD and 17 h in the MSD. The solar collector's efficiency ranged from 52.3% to

Fig. 10.10 Modified hybrid indirect passive (HIP) drying system [26]

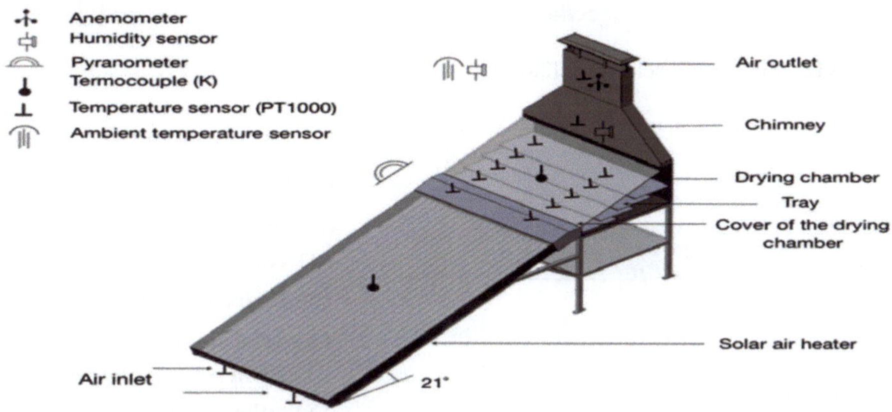

Fig. 10.11 Passive, mixed-type solar dryer [27]

55.45%, whereas the average efficiency was 10.66% in MSD mode and 8.8% in ISD mode. The drying efficiency was 5.47% and 4.48% for the MSD mode and ISD model, respectively.

Abubakar et al. [28] construct a mixed-mode solar dryer and assess its performance. Hot drying air from a separate solar collector is delivered via a grain bed in the dryer, as shown in Fig. 10.12. The drying room temperature reached 74% for almost 3 h after 12 o'clock. The drying rate was 0.62 kg/h, with a 57.5% system efficiency. The dryer's drying rate indicates its ability to dry food products to an optimal moisture level quickly.

Vásquez et al. [29] designed a dynamic model of a solar dryer system for agricultural products having a TESS using PCM as a medium. As illustrated in Fig. 10.13, the dryer's mathematical model was divided into three partitions: the PV, the energy storage, and the drying cabinet. These theoretical models were verified by using experimental results from previous studies on kiwifruit and mushroom drying.

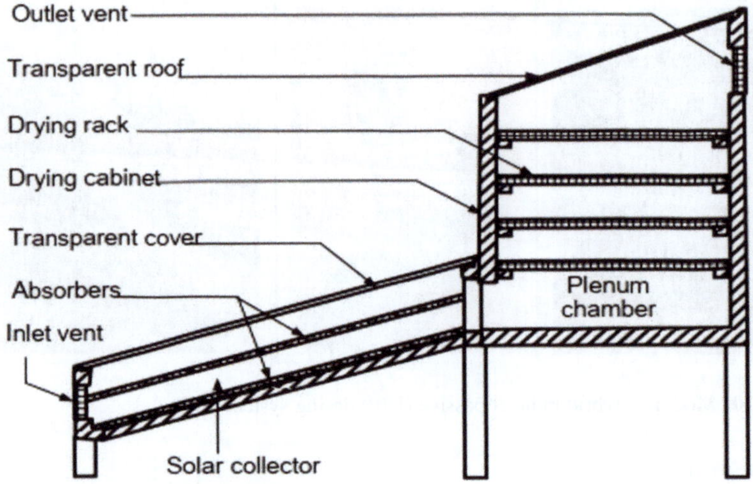

Fig. 10.12 Mixed-mode solar dryer [28]

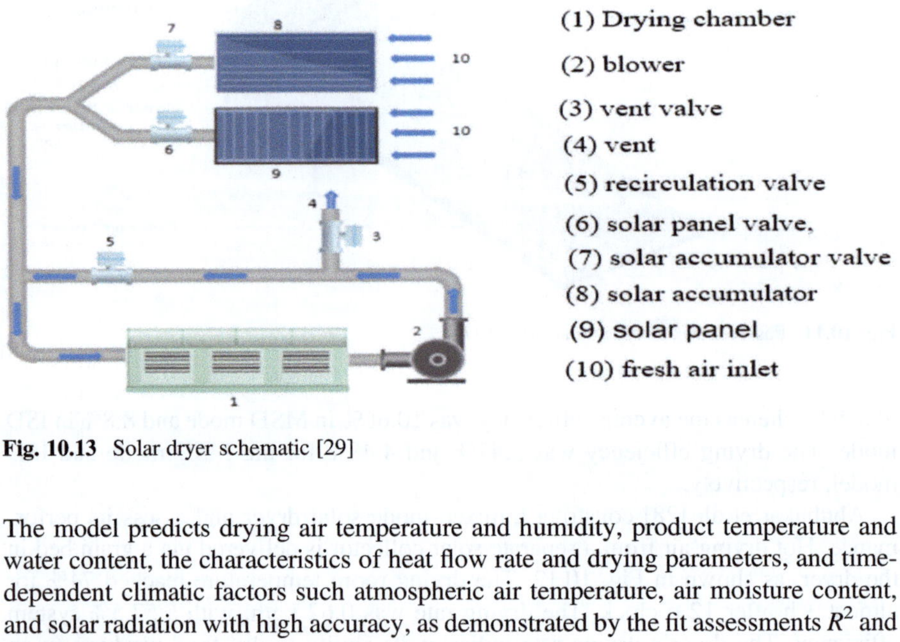

Fig. 10.13 Solar dryer schematic [29]

The model predicts drying air temperature and humidity, product temperature and water content, the characteristics of heat flow rate and drying parameters, and time-dependent climatic factors such atmospheric air temperature, air moisture content, and solar radiation with high accuracy, as demonstrated by the fit assessments R^2 and RMSE.

Akpinar [30] investigated the thin-layer drying behavior of mint leaves utilizing energy and exergy studies of the solar drying process in a solar dryer with forced convection and in the open with natural convection. The drying data was fitted with ten various theoretical models. Of the models, it was discovered that the forced solar drying and natural sun drying models developed by Wang and Singh better explained the thin-layer drying behavior of mint leaves. The drying cabinet's energy

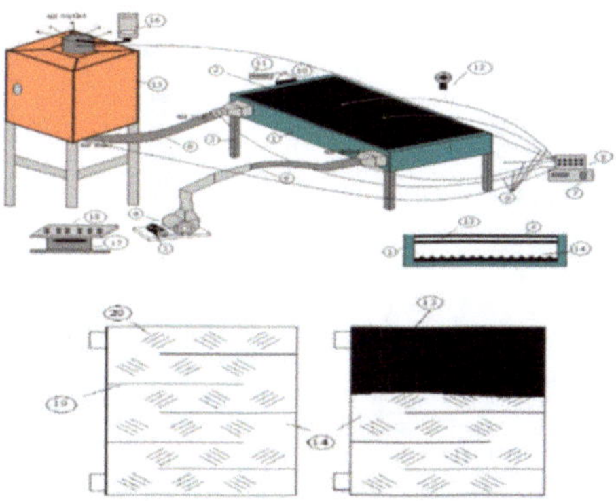

Fig. 10.14 Schematic view of experimental setup solar dryer [30]

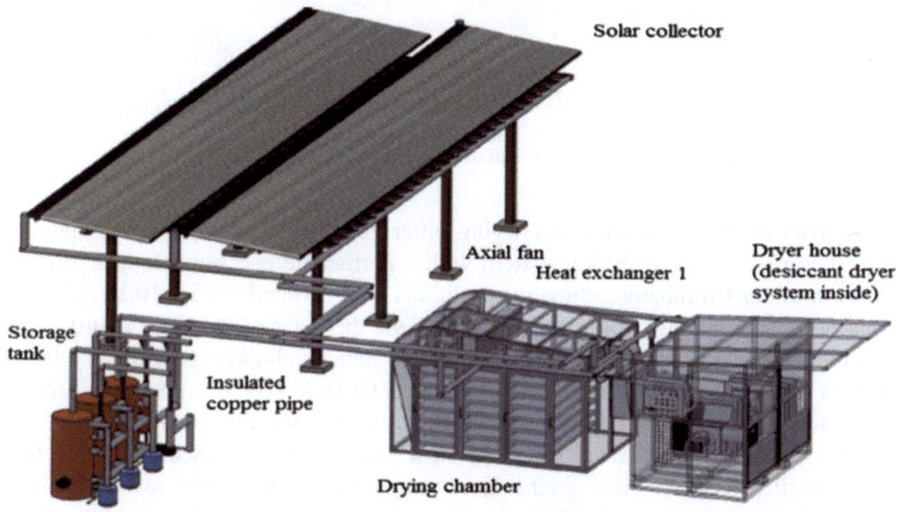

Fig. 10.15 Solar-assisted desiccant dryer [5]

utilization ratio (EUR) had values ranging from 7.826% to 46.285%. The values of energetic efficiency varied from 34.8% to 86.7%. The range of values for the improvement potential was 0 to 0.017 kJ/s (Fig. 10.14).

Crushed oil palm fronds were dried using a solar-assisted solid desiccant dryer, which was abandoned to study the drying process [5]. As shown in Fig. 10.15, a solar collector was used to heat water, and the thermal energy was transmitted to the air via two heat exchangers. When compared to open-sun drying, the first, second, and third solar dryer columns reduced drying time by roughly 64, 44, and 33 minutes, respectively. The drying rate was calculated to be 8.37 kg/h at full capacity, which was double that of open-sun drying (4.23 kg/h). The desiccant wheel was found to improve sensible and latent effectiveness by 74% and 67%, in order.

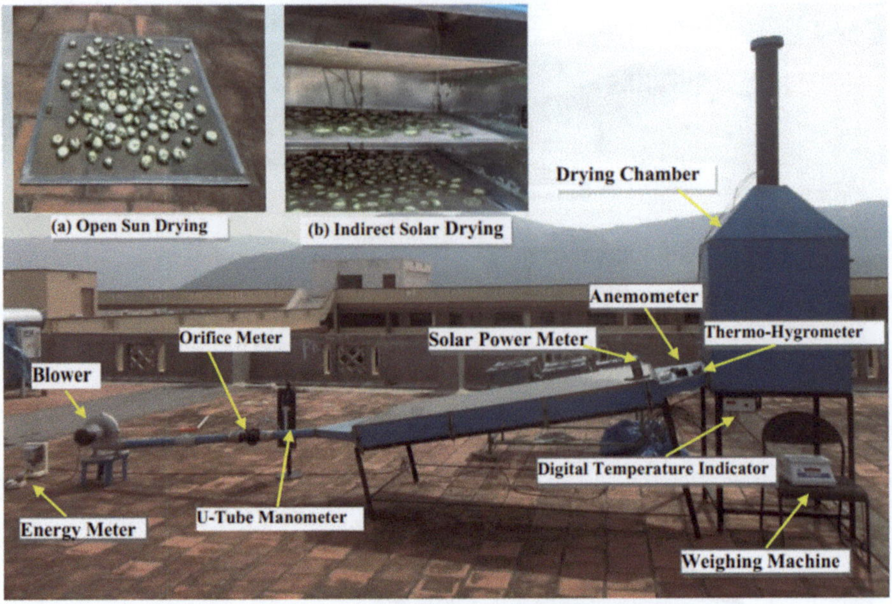

Fig. 10.16 Experimental setup for open-sun drying and indirect solar drying [31]

Vijayan et al. [31] determined drying bitter gourd slices using a cheap indirect forced convection solar drying system with a permeable bed sensible heat storage medium under Coimbatore climatic conditions as illustrated in Fig. 10.16. Using a solar dryer with a mass flow rate of 0.0636 kg/s, the mass of the bitter gourd slices was reduced by 3277 in 7 h. Average energy efficiency values range from 28.74% to 40.67% for air mass flow rates ranging from 0.0141 to 0.0872 kg/s. For different flow rates of air, the average drying air pickup efficiency ranged from 54.29% to 17.18%, and it decreased as the mass flow rate of air increased. The effective moisture diffusivity of bitter gourd slices ranged from 8.6293m^2/s to 12.9585 m^2/s, with the lowest value of 0.9568m^2/s when dried in the open sun.

Chaouch et al. [20] analyzed both direct and indirect active solar dryer with a pebble sensible heat storage medium as shown in Fig. 10.17. The sensible heat storage technology increased the solar collector's thermal efficiency by 28% for a duration of 1 h after the sunset. The drying chamber's thermal efficiency improved by 11.8%. They also conducted experiments on drying camel meat without salting in July and November under various weather conditions. The Midilli-Kucruk and logarithmic models are best for the November and July experiments, in order. Over 24 h, indirect drying efficiency averaged 18.34% in July and 15.52% in November. The average direct drying efficiency in July and November was 10.35% and 7.88%, respectively.

The effectiveness of a home solar drier with direct forced convection for banana drying is examined [21]. A parabolic-shaped solar dryer made of polycarbonate plates was constructed atop a flat plate collectors illustrated Fig. 10.18. The

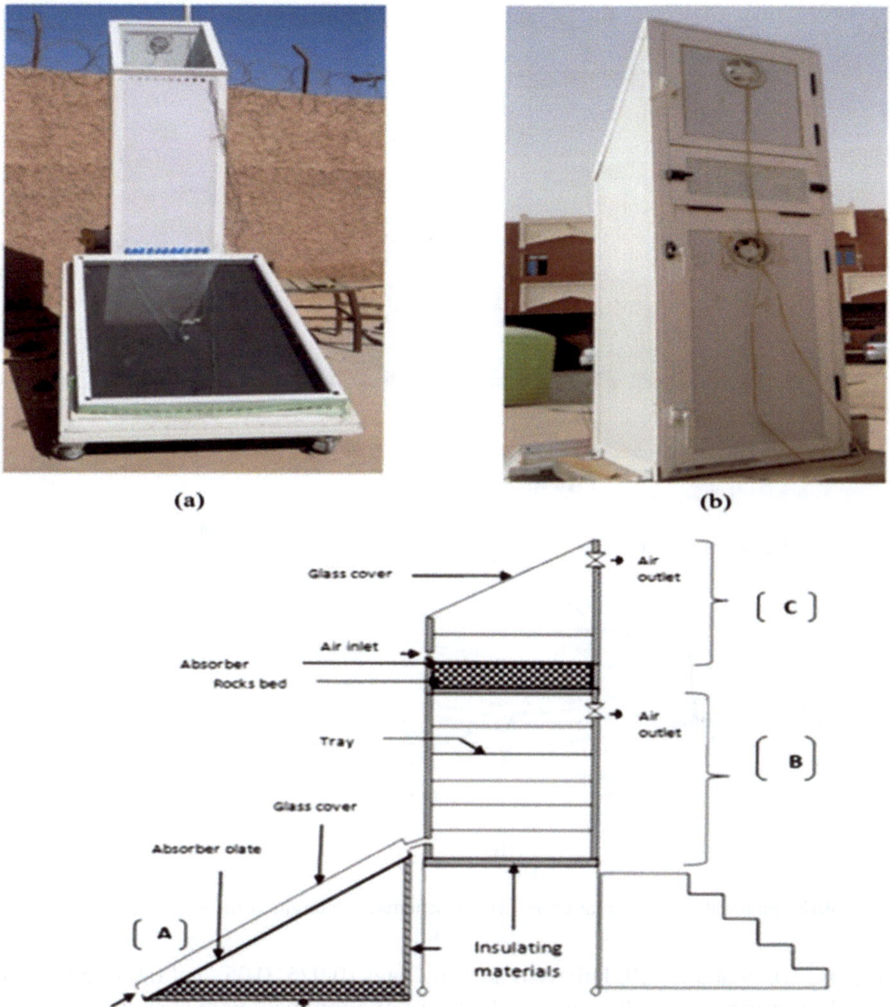

Fig. 10.17 The solar dryer' (**a**) front view, (**b**) back view, and (**c**) setup [20]

outcomes showed that from 8:00 am to 6:00 pm, the dryer's drying air had a temperature range of 35–60°C. The banana in the dryer saw a reduction in moisture content for 4 days from 72% to 28%, but the samples dried in the sun had a moisture content reduction of 40%. The solar dryer reduces drying time in half when compared to drying in the open sun. The solar-dried banana was great in terms of taste, appearance, and texture.

Iranmanesh et al. [32] investigated the potential of a thermal storage PCM and an evacuated tube solar collector (ETSC) in a solar cabinet drying system as illustrated

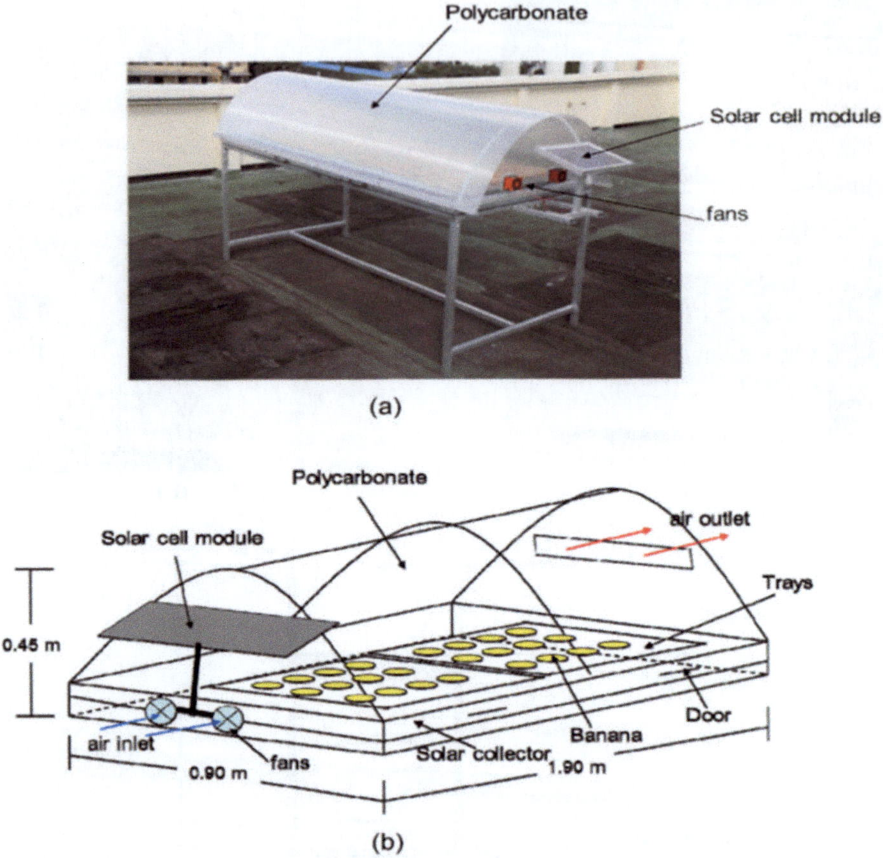

Fig. 10.18 Pictorial view of solar dryer (**a**) and schematic diagram of the system (**b**) [21]

in Figs. 10.19 and 10.20. Different air flow rates (0.025, 0.05, and 0.09 kg/s) were tested for drying apple slices with a thickness of 5 mm both with and without PCM. According to the thermal analysis finding, for airflow rates of 0.025 and 0.05 kg/s, PCM enhanced the thermal storage by 1.72% and 5.12%, respectively. The maximum total drying efficiency was achieved by the PCM-equipped system with an airflow rate of 0.025 kg/s or 39.9%.

Lakshmi et al. [33] executed an experimental analysis on drying stevia leaves using a mixed-mode forced convection solar dryer (MFSCD) and open-sun drying (OSD) as shown in Fig. 10.21. The analysis was conducted with an ambient temperature of 30°C, an average solar radiation of 567 W/m^2, and a drying air flow rate of 0.049 kg/s. After 330 min and 870 min of drying time, respectively; the safe (final) moisture content of stevia leaves of 0.053 (d.b) was reached in MFSCD and OSD. The average energy efficiency and overall dryer efficiency for the MFSCD were found to be 33.5% and 59.1%, respectively. Fresh, open-sun-dried and solar-dried stevia leaf samples were analyzed for quality. When comparing solar-dried

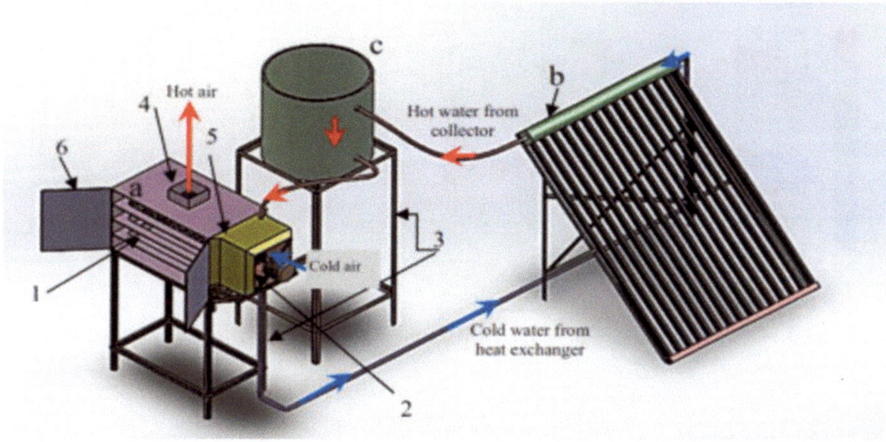

Fig. 10.19 Solar cabinet dryer including (**a**) dryer, (**b**) ETSC, and (**c**) storage tank and PCM container

samples to OSD samples, it was discovered that antioxidants and flavonoids were abundant. In comparison to OSD, color preservation in solar-dried samples is good. A sensory study of stevia leaves (flavor, fragrance, and taste) revealed that the solar-dried stevia leaves scored higher than the OSD samples.

In Karthikeyan and Murugavelh [22], turmeric (*Curcuma longa*) drying kinetics in a mixed-mode solar tunnel drier has been investigated as illustrated in Fig. 10.22. Its tilting solar air heater effectively collects solar light, which is the main advantage of the current design. In the solar tunnel drier, the turmeric was first 0.779 kg of water per kilogram of dry matter. This was changed to 0.070 kg of water per kg of dry matter by forced convection. After 12 h in the sun drier, the turmeric reached its final drying levels, whereas open-air drying took 43 h. Between 9.75% and 33.98% of energy was used on average by the solar tunnel. 49.12% was found to be the energy effectiveness of the dryer (Tables 10.4 and 10.5).

10.5 Important Parameter in Solar Drying

The most important parameters of solar dryer such as the quality of the solar dryer with its social, environmental, and economic aspects are discussed.

10.5.1 Quality Aspects

Many physical, chemical, and nutritional aspects of food are altered during drying due to mass and heat transfers. The color change and volume decrement of the dried fruits are better in solar drying with air recycling system when compared to open-sun

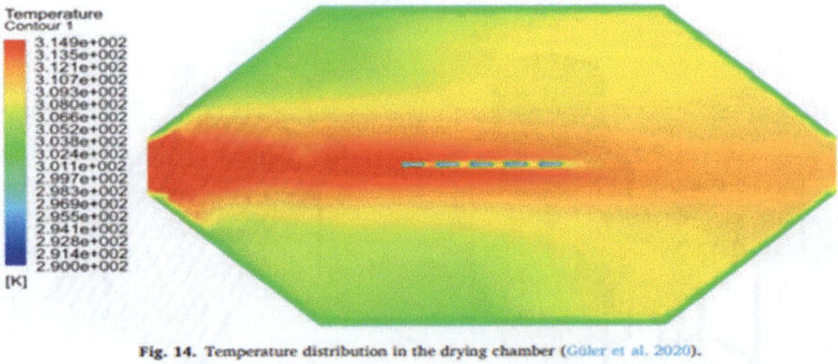

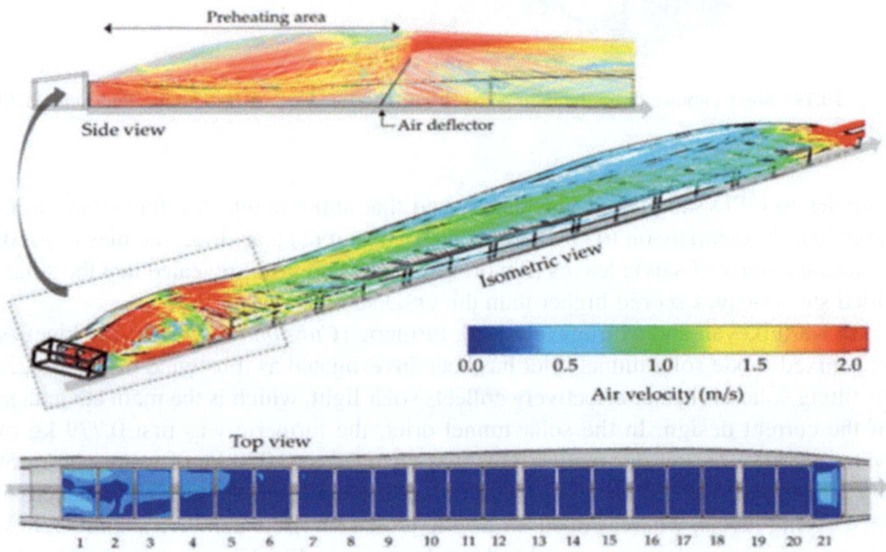

Fig. 10.20 Airflow distribution simulation for isometric, side view, and top view [32]

drying system [3]. However, there have recently been initiatives to make high-quality dried foods [39]. This is accomplished by the application of highly innovative drying technologies, the improvement and optimization of existing drying methods, and the optimization of quality parameters like as structure, color, flavor, nutrition.

10.5.2 Economic Aspect

Solar dryers help to reduce the need of traditional energy sources, such as fossil fuels. In reality, it decreases CO_2 gas emission, which is the primary cause of global warming [16]. As a result, solar dryers play great a role in environmental protection and cost used for fossil fuel purchasing [40].

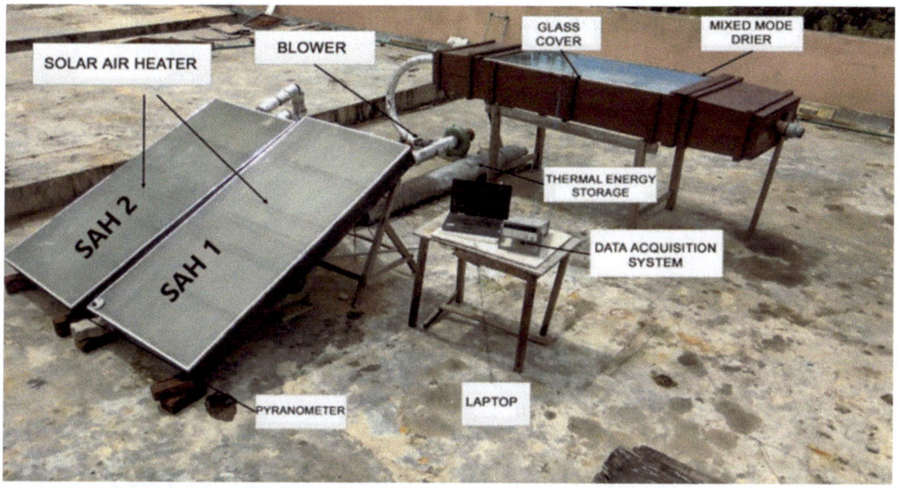

Fig. 10.21 Active solar dryer in mixed mode [33]

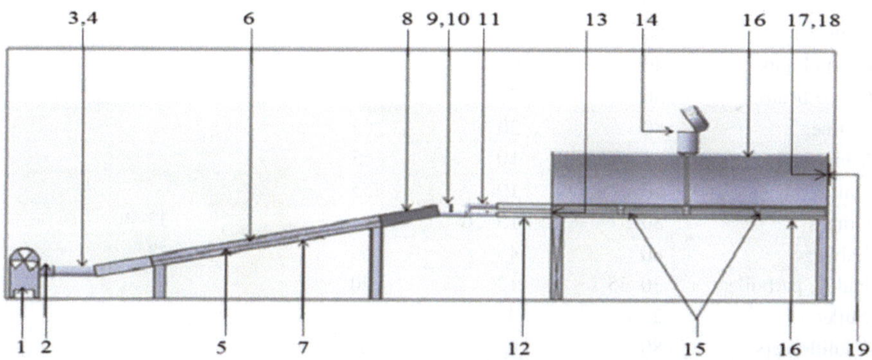

1. Blower. 2. Airflow regulating valve. 3. Solar collector inlet. 4,9,13,18. Thermocouples. 5. Aluminium sheet. 6. Transparent glass. 7,16. Insulator. 8. Pyranometer. 10. Anemometer section. 11. Airflow valve. 12. Solar dryer inlet. 14. Chimney. 15. Drying trays. 16. Transparent polycarbonate sheet. 17. Hygrometer. 19. Solar dryer exhaust.

Fig. 10.22 Dual mode solar tunnel dryer setup [22]

10.5.3 Environmental Aspects

The majority of drying systems rely on fossil fuels and electricity for energy, leading to substantial operating expenses and pollution due to high greenhouse gas emissions [16]. Therefore, food manufacturers are focusing on safe energy sources such as solar energy [41].

Table 10.4 The optimum temperature and drying time for various agricultural products [4]

Product	Moisture content (w.b) Initial %	Final %	Max. temp. (°C)	Time for drying (h)
Onions	85	6	55	48
Onions flakes	80	10	55	24
Tomatoes	95	7	60	36
Bananas	80	15	70	15
Potato chips	75	13	70	72
Apples	82	10–14	65–70	24–26
Green peas	80	5	60	8–10
Maize	35	15	60	–
Wheat	20	16	45	–
Turmeric	80	10	–	120
Pepper	71	13	–	48
Tea	80	3	–	96
Coffee	65	11	–	288
Carrot	70	5	5	–
Figs	70	20	70	32
Cotton	50	9	75	–
Ground nuts	40	9	–	–
Cocoa beans	50	7	–	–
Nutmeg	80	20	65	–
Yams	80	10	65	–
Onion rings	80	10	55	–
Grapes	80	15–20	–	32–40
Cabbage	80	4	65	48
Paddy, parboiled	30–35	13	50	–
Millet	21	4	–	–
Cauliflowers	80	6	65	–
Sweet potato	75	7	55	–
Red Lauan	90	20	–	–

10.6 Conclusion

This study examined numerous sun-dryer varieties as well as their development, performance, and application to different products. The drying of several products is the main topic of this study, and every product requires a particular drying environment. Important factors like sun radiation intensity, relative humidity, ambient temperature, beginning moisture content, and crop kinds must be taken into account when building a solar dryer. According to the moisture content and product to be dried, air velocity and temperature must be regulated to produce a better-quality

Table 10.5 A comprehensive summary of the findings on solar dryers

Types of solar dryer	Author and research finding
Direct solar dryer	Simple construction, low cost, and lower efficiency in the range of 4–7% [27]. However, a little color variation was observed which may lead to nutritional value loss [19]. A little color variation was observed may lead to nutritional value loss
Indirect solar dryer	Solar chimneys enhanced the temperature of the drying air by 7°C, and better product quality was found, and uniform drying air temperature inside indirect solar dryers are achieved [34] They claim to preserve the fruit from deterioration and save more energy [21]. Two more hours following a sunny period, the drying room temperature was considerably raised by PCM in the solar collector. With paraffin's help, the daily moisture removal was reported as 5.1% for potato slices Drying cassava root chips were modeled and experimentally investigated [35]. The model indicated the core of the chips took more drying time Numerical simulations were used to validate the effect of absorber plate shape, and the study demonstrated the success of tubular absorbers for solar heaters [36]. The thermal performance has been improved, and drying time decreased by 30% using aluminum wool inside the tubes The dryer with semicontinuous mode was tested for drying turmeric (*Curcuma longa*) and fenugreek leaves (*Trigonella foenum-graecum*) and resulted in the overall thermal efficiency of 23.6% and 34.1%, respectively [37]. The open-sun drying (OSD) thermal efficiencies were 5.7% and 5.4%, respectively
Mixed-mode solar dryer	It was examined whether a mixed-type solar dryer working in hybrid active-type dryer, MM-FC, and indirect-mode passive, IM-NC, could lower the moisture content of pears from 82.15 0.13% to 8.86 1.3% (wet basis) (Orlando, 2021). For the MM-NC, MM-FC, and IM-NC modes, the $\eta_{th,\ avg}$ of the solar air heater were 33.0%, 56.3%, and 30.9%, in order. The drying times and drying efficiencies were 5.25, 4.5, and 6.6 h and 13.6%, 11.2%, and 26.6%, respectively For the mass flow rates of air 3.72, 2.76, 1.8, and 1.08 kg/min, cluster fig dried from a moisture content (MC) of 5.67 to 0.08 kg/kg on a dry basis in 7.0, 8.0, 8.5, and 9.5 h, in order. On the other hand, open-sun drying took 14 h [38]

product. Forced convection dryers outperform natural convection dryers in drying high-moisture-content items because of their quicker drying rates. Using thermal storage, the solar dryer can be used at night and during low-intensity periods. The bulk of the researchers used paraffin wax, rock, water, metal waste and a mixture of paraffin wax and aluminum powder. The mixed-mode type solar dryer is more efficient and more suited for a variety of uses.

References

1. Singh, P., & Gaur, M. K. (2020). Review on development, recent advancement and applications of various types of solar dryers. *Energy Sources, Part A: Recovery, Utilization, and Environmental Effects*, 1–21. https://doi.org/10.1080/15567036.2020.1806951
2. Lingayat, A. B., Chandramohan, V. P., Raju, V. R. K., & Meda, V. (2020). A review on indirect type solar dryers for agricultural crops – Dryer setup, its performance, energy storage and important highlights. *Applied Energy, 258*, 114005. https://doi.org/10.1016/j.apenergy.2019.114005
3. Udomkun, P., et al. (2020). Review of solar dryers for agricultural products in Asia and Africa: An innovation landscape approach. *Journal of Environmental Management, 268*, 110730. https://doi.org/10.1016/j.jenvman.2020.110730
4. Srinivasan, G., Rabha, D. K., & Muthukumar, P. (2021). A review on solar dryers integrated with thermal energy storage units for drying agricultural and food products. *Solar Energy, 229*, 22. https://doi.org/10.1016/j.solener.2021.07.075
5. Nukulwar, M. R., & Tungikar, V. B. (2021). A review on performance evaluation of solar dryer and its material for drying agricultural products. *Materials Today Proceedings, 46*, 345–349. https://doi.org/10.1016/j.matpr.2020.08.354
6. Bal, L. M., Satya, S., & Naik, S. N. (2010). Solar dryer with thermal energy storage systems for drying agricultural food products: A review. *Renewable and Sustainable Energy Reviews, 14*(8), 2298–2314. https://doi.org/10.1016/j.rser.2010.04.014
7. Wassie, H. M., Admasu, B. T., Getie, M. Z., & Alem, M. S. (2023). Experimental investigation of parabolic solar dish concentrator-based solar dryer assisted with thermal energy storage system. In B. H. Woldegiorgis, K. Mequanint, M. Z. Getie, E. G. Mulat, & A. Alemayehu Assegie (Eds.), *Advancement of science and technology: Materials and energy* (pp. 263–283). Springer Nature Switzerland. https://doi.org/10.1007/978-3-031-33610-2_15
8. Bhavsar, H. P., & Patel, C. M. (2021). Performance investigation of natural and forced convection cabinet solar dryer for ginger drying. *Materials Today Proceedings, 47*, 6128–6133. https://doi.org/10.1016/j.matpr.2021.05.050
9. Ansarinasab, H., Hajabdollahi, H., & Fatimah, M. (2021). Life cycle assessment (LCA) of a novel geothermal-based multigeneration system using LNG cold energy- integration of Kalina cycle, stirling engine, desalination unit and magnetic refrigeration system. *Energy, 231*, 120888. https://doi.org/10.1016/j.energy.2021.120888
10. Templalexis, C., et al. (2021). Environmental conditions affecting ochratoxin a during solar drying of grapes: The case of tunnel and open air-drying. *Toxins (Basel), 13*(6), 1–18. https://doi.org/10.3390/toxins13060400
11. El Hage, H., Herez, A., Ramadan, M., Bazzi, H., & Khaled, M. (2018). An investigation on solar drying: A review with economic and environmental assessment. *Energy, 157*, 815–829. https://doi.org/10.1016/j.energy.2018.05.197
12. Wassie, H. M., Getie, M. Z., Alem, M. S., Kotu, T. B., & Salehdress, Z. M. (2022). Experimental investigation of the effect of reflectors on thermal performance of box type solar cooker. *Heliyon, 8*(12), e12324. https://doi.org/10.1016/j.heliyon.2022.e12324
13. Alvarez, A., Cabeza, O., Muñiz, M. C., & Varela, L. M. (2010). Experimental and numerical investigation of a flat-plate solar collector. *Energy, 35*(9), 3707–3716. https://doi.org/10.1016/j.energy.2010.05.016
14. Sharma, N., & Diaz, G. (2011). Performance model of a novel evacuated-tube solar collector based on minichannels. *Solar Energy, 85*(5), 881–890. https://doi.org/10.1016/j.solener.2011.02.001
15. Islam, M. T., Huda, N., Abdullah, A. B., & Saidur, R. (2018). A comprehensive review of state-of-the-art concentrating solar power (CSP) technologies: Current status and research trends. *Renewable and Sustainable Energy Reviews, 91*, 987–1018. https://doi.org/10.1016/j.rser.2018.04.097

16. Kamarulzaman, A., Hasanuzzaman, M., & Rahim, N. A. (2021). Global advancement of solar drying technologies and its future prospects: A review. *Solar Energy, 221*, 559–582. https://doi.org/10.1016/j.solener.2021.04.056
17. Chauhan, Y. B., & Rathod, P. P. (2020). A comprehensive review of the solar dryer. *International Journal of Ambient Energy, 41*(3), 348–367. https://doi.org/10.1080/01430750.2018.1456960
18. Musembi, M. N., Kiptoo, K. S., & Yuichi, N. (2016). Design and analysis of solar dryer for mid-latitude region. *Energy Procedia, 100*, 98–110. https://doi.org/10.1016/j.egypro.2016.10.145
19. Ferraresi, L., Nascimento, M., Nishioka, K., Nakamura, G., & Vieira, A. (2021). Natural and forced air convection operation in a direct solar dryer assisted by photovoltaic module for drying of green onion. *Solar Energy, 220*, 24–34. https://doi.org/10.1016/j.solener.2021.02.061
20. Braham, W., et al. (2018). Experimental investigation of an active direct and indirect solar dryer with sensible heat storage for camel meat drying in Saharan environment. *Solar Energy, 174*, 328–341. https://doi.org/10.1016/j.solener.2018.09.037
21. Nabnean, S., & Nimnuan, P. (2020). Case studies in thermal engineering experimental performance of direct forced convection household solar dryer for drying banana. *Case Studies in Thermal Engineering, 22*, 100787. https://doi.org/10.1016/j.csite.2020.100787
22. Karthikeyan, A. K., & Murugavelh, S. (2018). Thin layer drying kinetics and exergy analysis of turmeric (Curcuma longa) in a mixed mode forced convection solar tunnel dryer. *Renewable Energy, 128*, 305–312. https://doi.org/10.1016/j.renene.2018.05.061
23. Hamdi, I., Kooli, S., Elkhadraoui, A., Azaizia, Z., Abdelhamid, F., & Guizani, A. (2018). Experimental study and numerical modeling for drying grapes under solar greenhouse. *Renewable Energy, 127*, 936. https://doi.org/10.1016/j.renene.2018.05.027
24. Blanco-Cano, L., Soria-Verdugo, A., Garcia-Gutierrez, L. M., & Ruiz-Rivas, U. (2016). Modeling the thin-layer drying process of granny smith apples: Application in an indirect solar dryer. *Applied Thermal Engineering, 108*, 1086–1094. https://doi.org/10.1016/j.applthermaleng.2016.08.001
25. Lakshmi, D. V. N., Muthukumar, P., Layek, A., & Nayak, P. K. (2018). Drying kinetics and quality analysis of black turmeric (Curcuma caesia) drying in a mixed mode forced convection solar dryer integrated with thermal energy storage. *Renewable Energy, 120*, 23–34. https://doi.org/10.1016/j.renene.2017.12.053
26. Ssemwanga, M., Makule, E., & Kayondo, S. I. (2020). Performance analysis of an improved solar dryer integrated with multiple metallic solar concentrators for drying fruits. *Solar Energy, 204*, 419–428. https://doi.org/10.1016/j.solener.2020.04.065
27. Erick César, L. V., Ana Lilia, C. M., Octavio, G. V., Isaac, P. F., & Rogelio, B. O. (2020). Thermal performance of a passive, mixed-type solar dryer for tomato slices (Solanum lycopersicum). *Renewable Energy, 147*, 845–855. https://doi.org/10.1016/j.renene.2019.09.018
28. Abubakar, S., Umaru, S., Kaisan, M. U., Umar, U. A., Ashok, B., & Nanthagopal, K. (2018). Development and performance comparison of mixed-mode solar crop dryers with and without thermal storage. *Renewable Energy, 128*, 285–298. https://doi.org/10.1016/j.renene.2018.05.049
29. Vásquez, J., Reyes, A., & Pailahueque, N. (2019). Modeling, simulation and experimental validation of a solar dryer for agro-products with thermal energy storage system. *Renewable Energy, 139*, 1375–1390. https://doi.org/10.1016/j.renene.2019.02.085
30. Akpinar, E. K. (2010). Drying of mint leaves in a solar dryer and under open sun: Modelling, performance analyses. *Energy Conversion and Management, 51*(12), 2407–2418. https://doi.org/10.1016/j.enconman.2010.05.005
31. Vijayan, S., Arjunan, T. V., & Kumar, A. (2020). Exergo-environmental analysis of an indirect forced convection solar dryer for drying bitter gourd slices. *Renewable Energy, 146*, 2210–2223. https://doi.org/10.1016/j.renene.2019.08.066

32. Iranmanesh, M., Samimi, H., Saleh, M., & Jahromi, B. (2020). CFD modeling and evaluation the performance of a solar cabinet dryer equipped with evacuated tube solar collector and thermal storage system. *Renewable Energy, 145*, 1192–1213. https://doi.org/10.1016/j.renene.2019.06.038
33. Lakshmi, D. V. N., Muthukumar, P., Layek, A., & Kumar, P. (2018). Performance analyses of mixed mode forced convection solar dryer for drying of stevia leaves. *Solar Energy, 188*, 507–518. https://doi.org/10.1016/j.solener.2019.06.009
34. Benhamza, A., Boubekri, A., Atia, A., & Hadibi, T. (2021). Drying uniformity analysis of an indirect solar dryer based on computational fluid dynamics and image processing. *Sustainable Energy Technologies and Assessments, 47*, 101466. https://doi.org/10.1016/j.seta.2021.101466
35. Simo-Tagne, M., Tagne Tagne, A., Ndukwu, M. C., Bennamoun, L., Obounou Akong, M. B., El Marouani, M., & Rogaume, Y. (2021). Numerical study of the drying of cassava roots chips using an indirect solar dryer in natural convection. *AgriEngineering, 3*(1), 138–157.
36. Khanlari, A. (2021). Thermal performance improvement of an indirect solar dryer with tube-type absorber packed with aluminum wool. *Solar Energy, 217*, 328–341. https://doi.org/10.1016/j.solener.2021.02.029
37. Singh, S., Gill, R. S., Hans, V. S., & Singh, M. (2021). A novel active-mode indirect solar dryer for agricultural products: Experimental evaluation and economic feasibility. *Energy, 222*, 119956. https://doi.org/10.1016/j.energy.2021.119956
38. Poonam, J., Muthukumar, P., Bala, K., & Kumar, D. (2020, 2021). Performance studies on mixed-mode forced convection solar cabinet dryer under different air mass flow rates for drying of cluster fig. *Solar Energy*. https://doi.org/10.1016/j.solener.2021.06.086
39. Sablani, S. S. (2006). Drying of fruits and vegetables: Retention of nutritional/functional quality. *Drying Technology, 24*(2), 123–135. https://doi.org/10.1080/07373930600558904
40. Kondareddy, R., Sivakumaran, N., Radha Krishnan, K., Nayak, P. K., Sahu, F. M., & Singha, S. (2021). Performance evaluation and economic analysis of modified solar dryer with thermal energy storage for drying of blood fruit (Haematocarpus validus). *Journal of Food Processing & Preservation, 45*(9), 0–1. https://doi.org/10.1111/jfpp.15653
41. Simo-Tagne, M., & Ndi-Azese, M. (2021). Thermal, economic, and environmental analysis of a novel solar dryer for firewood in various temperate and tropical climates. *Solar Energy, 226*, 348–364. https://doi.org/10.1016/j.solener.2021.08.060

Chapter 11
Recent Developments on Design and Thermal Performance Enhancement of Parabolic Dish Solar Concentrators: A Review

Rewla E. Abreham, Muluken Z. Getie, and Yared A. Demissie

11.1 Introduction

Parabolic dish concentrators gather and focus sunlight to bring about extreme temperatures to generate hot air, which is then transformed into mechanical motion in a turbine to produce power. The four primary PDC technologies are the linear Fresnel, parabolic trough reflector, centralized tower receiver, and parabolic dish [1]. The parabolic trough received a lot of attention since it had successful commercial uses and accounted for 90% of the installed solar thermal power plants. Other technologies, including solar parabolic dish solar concentrators (PDSCs), are thus of less interest. However, among the available technologies, parabolic dish solar concentrators are the most effective, offer the highest ratios of concentration, and are ideal for producing small amounts of power in remote locations, and they have capacities in the range of 10–100 kW, making them appropriate for microscale power production in rural and small towns [2]. Due to their small size and modular construction, parabolic dish concentrators can be installed in small spaces. Most academics are drawn to them owing to their modularity and higher concentration ratio, which made them important in towns or regions where land is limited. These concentrators' modular design makes it simple to expand or reduce their size in accordance with changing energy demands. This scalability is beneficial in scenarios where energy demand fluctuates or must be gradually increased [3].

In terms of application parabolic dish concentrators are a well-known solution for many applications such as concentrated solar power (CSP), solar metallurgical processes, solar reactors for fuel production, etc. Nevertheless, this technology is

---

R. E. Abreham · M. Z. Getie (✉) · Y. A. Demissie
Faculty of Mechanical and Industrial Engineering, Bahir Dar Institute of Technology, Bahir Dar University, Bahir Dar, Ethiopia

K. Mequanint et al. (eds.), *Sustainable Development Research in Materials and Energy*, Green Energy and Technology,
https://doi.org/10.1007/978-3-031-69860-6_11

facing a tremendous challenge to become more efficient and competitive (especially within CSP field) in comparison with other technologies, and the need for advancement becomes mandatory [4].

This literature review will delve into the technical aspects of parabolic dish concentrators, including their design considerations, optical performance, efficiency, and integration with different power generation systems. Additionally, it will explore the challenges associated with parabolic dish concentrators, such as thermal losses, tracking mechanisms, and cost-effectiveness. The review will also discuss recent research advancements and potential future directions for this technology, and by synthesizing the knowledge and findings from various studies, this literature review aims to contribute to a deeper understanding of parabolic dish concentrators and provide insights that can guide further research and development in this field.

In the modern world, solar thermal technologies are widely used. For thermal processes requiring high temperatures, such as solar cooling, solar drying, solar cooking, desalination, and power generation, concentrated solar energy provides an alternate source. The thermal properties of concentrating solar collectors have been the subject of several experimental and numerical research. This paper evaluates recent progress on parabolic dish concentrator design and performance improvement. When compared to other concentrating methods, parabolic dish solar collectors have greater concentration ratios (1000–3000), according to A. Kalogirou [5]. Concentrator/collector, receiver, and tracking system are the three primary parts of a PDC system. Figure 11.1 shows the primary parts for EURODISH system design.

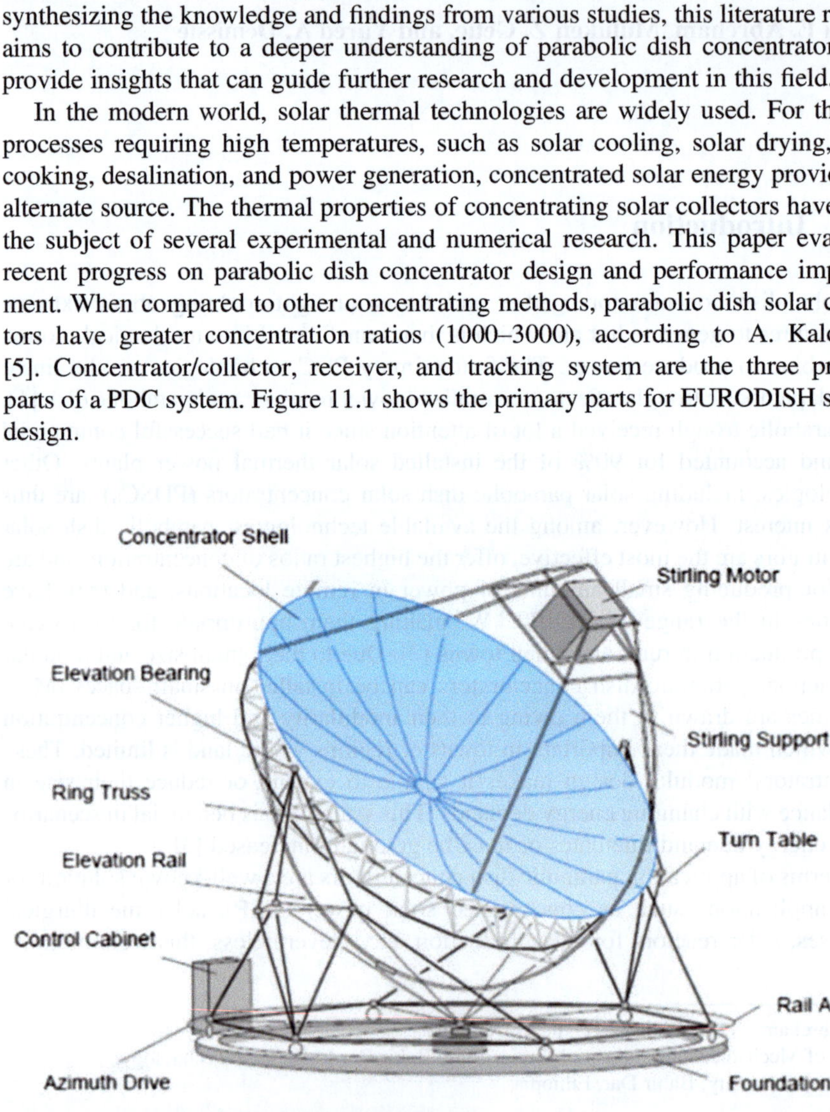

Fig. 11.1 Design diagram of EURODISH system [6]

11.2 Advancements on Parabolic Dish Solar Concentrators (PDSC)

11.2.1 Enhancement on PDSC Receivers

The main issue with concentrating solar collectors is the longevity of the receiver; as a result, improvements in the receiver's size, form, and heat transfer capacity have been highly regarded. The impact of receiver geometry on the optical performance of solar cavity receiver in PDC systems was thoroughly analyzed by Daabo et al. [7]. Using an advanced ray tracing technique, the system analyzes the optical efficiency and flux distribution of the three distinct geometries cylindrical, conical, and spherical of a cavity receiver. The conical-shaped receiver outperformed the other two shapes in terms of flux energy received, absorbed, and reflected, displaying a better optical efficiency of 75.3% at 85% surface absorptivity. The findings demonstrated a relationship between the flux distribution, the internal surfaces of the cavities, and their optical efficiency; as a result, the focus point location is influenced by the design of the cavity receiver and its absorptivity. Additionally, Idlimam et al. employed COMSOL Multiphysics as simulation software based on the finite element approach to research the impact of receiver shape on a PDC system. In terms of flux distribution and surface temperature, three receiver shapes—cubical, conical, and cylindrical—were estimated. The simulation results showed that the conical shape receiver performed better thermally than the other forms, with an internal energy of 79.04 kJ/kg and a maximum temperature of 2300° C at baseline Meteonorm 7 software data of rim angle 450, ambient temperature, and normal irradiance [8].

Three different sizes of receivers with diameters of 0.150 m, 0.228 m, and 0.304 m and thicknesses of 0.001 m were used in the experiments conducted by Vikrant Kumar et al. [9]. Each of the three is made of metal with a coating. In NIT, Kurukshetra, India, the receiver temperature reached a maximum of 3090°C. As the receiver's diameter grows, the temperature at the outer surface starts to decrease, demonstrating an inverse relationship between the two variables.

In order to determine the different factors impacting the thermal efficiency and overall heat loss of the solar cylindrical cavity receiver, Azzouzi et al. performed an experimental and analytical thermal analysis of the device. The findings showed how the downward-facing cavity receiver's total heat loss and thermal efficiency were affected by the receiver's inclination angle, water flow rate, solar concentration ratio, and report/ratio between the cavity depth and its aperture diameter L/D. Thermal efficiency falls as L/D grows, rises as inclination angle increases, and is significantly impacted by solar concentration ratio [10].

Thirunavukkarasu et al. looked into how aspect ratio affected the thermal efficiency of cavity receivers. The experiment was conducted under an average beam radiation of more than 600 W/m2, and the receivers had aspect ratios of 0.8, 1.0, and 1.2 built for a certain cone angle. The findings indicated that the receiver surface temperature decreased as the aspect ratio increased and that the thermal performance

Fig. 11.2 Coated and insulated cylindrical cavity receiver [12]

also decreased. Thus, among all the receivers tested, the one with a 0.8 aspect ratio performed the best. As a result, improving the thermal performance of the parabolic dish concentrator requires considering other factors in addition to surface temperature, such as surface area and useful optimization of the receiver's aspect ratio in the design process, as it affects thermal performance [11].

Investigations focus on the improvement of the receiver's thermal performance using alternative methods in addition to its geometrical aspect; Loni et al. [12] evaluated the influence of nano-fluid application for thermal performance improvement in hemispherical cavity receiver. In comparison to base fluid or pure oil, the results showed that using MWCNT/oil nano-fluid boosted the receiver's thermal efficiency by 12.93%. The dependence between the receiver heat gain and thermal efficiency with the operating differential temperature was modeled using group method of data handling (GMDH) artificial neural networks. By contrasting Al2O3/oil and SiO2/oil nano-fluids, Loni et al. conducted additional experimental research on the thermal performance of cylindrical cavity receiver The setup for coating the cylindrical cavity receivers for the purpose of insulation is shown in Fig. 11.2. Loni et al. [13] reported that the Al2O3/thermal oil nano-fluid was shown to have a better thermal efficiency than the other working fluids under study, according to the result. However, in several research, they used MWCNT to conduct additional experiments on wounded cylindrical cavity receivers [12–16].

An experimental investigation of a cylindrical water heater receiver was presented by Ali Reshad et al. [17]. The average and maximum energy efficiencies of the concentrator were found to be 57% and 67%, respectively, based on calculations of energy and exergy. At the Sharif University of Technology in Tehran, Iran, the water could absorb up to 1656.8 W of power and reached a maximum temperature of 540°C.

Soltani et al. examined a PDC system using a helically baffled cylindrical cavity receiver both numerically and empirically. Optimal selection of parameters as heat transfer fluid (HTF) inlet temperature, mass flow rate, and solar irradiation intensity showed 65% enhancement on the thermal performance of the system [18].

Porous media were employed in the experiments on enhancing receiver tubes (ERTs) by Zheng et al. These porous media are used to prevent hotspots in the receiver brought on by uneven flux distribution. The thermos-hydraulic performance of ERT is determined by the thermal conductivity ratio, and the results show that ERT partially filled with porous medium performs better in terms of heat transmission than ERT fully filled with porous medium [19].

In addition to those improvements made on the thermal performance and heat transfer rate on receiver, Karimi et al. developed a mathematical model for a cylindrical cavity receiver on a PDC system. This model, which was supported by experimental evidence, featured non-isothermal interior receiver walls. The effect of HTF mass flow rate and global solar irradiation intensity and some geometrical parameters of solar receiver such as receiver aperture diameter and its length on the thermal performance of the PDC were analyzed, and the results recommend that the design of receivers should consider the concentrating capacity of collectors [20].

11.2.2 Advancements on PDSC for Drying Applications

Many researches discuss parabolic dish concentrator-based solar drying methods. For the conversion of solar thermal energy, parabolic dish solar collectors are frequently employed, particularly in medium- and high-temperature applications [21]. This includes large-scale power generating and industrial heating and cooling. Parabolic dish concentrators are appropriate for the application area due to their availability at medium-high temperatures. Using drying powered by solar thermal energy, trials have been conducted for many years with the goal of preserving fruits, vegetables, and other agricultural products. This made the need for drying system modification even more urgent.

The thermal efficiency of a solar parabolic dish concentrator for remote area applications was investigated by M. Gwani et al. [22]. The solar concentrator might be made inexpensively using readily accessible materials and labor, allowing users to manually follow the sun's path. The concentrators' ability to reach temperatures between 100 and 1800°C, which are suited for both drying and cooking, was demonstrated by the results.

Manchanda and Kumar conducted an experimental investigation on solar distillation cum drying in a single-slope basin utilizing a parabolic reflector. Along with reflecting sunlight from the parabolic reflector, the heat losses from the distillation

Fig. 11.3 Photographic view of the experimental unit [23]

unit's basin are used to dry ginger. Studying the impact of water depth on distillation and ginger drying is another option. The photographic view of the experimental unit is shown in Fig. 11.3. The system produced 2 l/m^2 of distillate and 2 kg/m^2 of dried ginger, both at a 4-cm water depth, according to the data. With a rise in ambient temperature and solar radiation, evaporative and radiative heat transfer coefficients for distillation increase [23].

The design and performance study of an active indirect solar dryer integrated with PDC for drying peanuts were studied by Varghese et al. When compared to open solar drying, the parabolic dish collector (PDC) reduces moisture by 21% and produces a maximum temperature of 79°C at 12.30 p.m. when drying 1.5 kilograms of peanuts in 5 h [24].

The effectiveness of concentrating solar panels and mixed-mode solar dryers was compared by Stiling et al. [25]. According to the study, concentrating solar panel dryers may operate at temperatures up to 100°C hotter than conventional dryers. The drying time for Roma tomatoes was shortened as a result of the temperature increase. The concentrating solar panels were able to achieve the desired moisture content of 0.2 with a 27% reduction in overall drying time when compared to a standard drier.

Merlin Simo-Tagne et al. conducted a comparison analysis of various concentrator forms. In order to give a numerical model and simulation of coupled heat and mass transfer of solar dryer, a parabolic, conical, and coaxial concentrators were compared in this work. The findings showed that, in comparison to conical and coaxial concentrators, parabolic concentrators had poor drying kinetics [26].

11.2.2.1 Advancement on PDSCs with Latent Heat Storage

There have clearly been numerous demonstrations over the past few decades for drying agricultural food products using solar dryers that have solar thermal energy storage in the form of sensible heat, and mass transfer analysis has been done with heat storage drying systems to investigate the performance and proved superior to drying systems without heat storage [27]. The impact of PCMs in a solar receiver to enhance the thermal performance of parabolic dish collectors was evaluated by Senthil, R., et al. [28]. Throughout the study, experimental analysis of the heat transfer from the receiver to the heat transfer fluid with and without PCMs was conducted. The results showed that, when compared to a receiver without phase change material, the use of PCMs in the receiver's peripheral increased energy efficiency by 5.62%, exergy efficiency by 12.8%, and time to reach the boiling point of water by 20%.

For a PDSC integrated drying system, Amir et al. utilized nano-fluids as thermal storage. Experimental data were used to validate numerical simulations of the drying of mint. The outcomes demonstrate the improvement in PDSC thermal efficiency with the addition of the alumina nano-fluid [21, 29].

In order to track temperature, humidity, and moisture loss, Mandal et al. [30] created a PCM-based solar dryer that was integrated with a real-time data collecting system. The performance of the dryer for drying fresh mint and acacia leaves with high initial wetness up to 82–86% (wb) was assessed using a paraffin wax LHS material. The PCM-developed dryer demonstrated a notable difference in drying rate for both herbs while preserving the nutritional value of the herbs.

11.2.3 Advancements on PDSC Concentrators for Regeneration of Solid Desiccant Applications

Desiccants can be regenerated through mechanical adjustments, such as different drying techniques, or chemical treatment, such as an acid wash, all of which increase surface area and alter the ion exchange characteristics and adsorption capacity. The desiccant dehumidification framework has been given a lot of thought as a potential replacement for the conventional dehumidification system in recent years [31]. As the system requires medium-high temperatures, the use of PDSCs is crucial in the regeneration of desiccants.

A detailed mathematical model for an integrated system to regenerate desiccant materials was created by V. Madadi Avargani et al. [32]. The system was examined

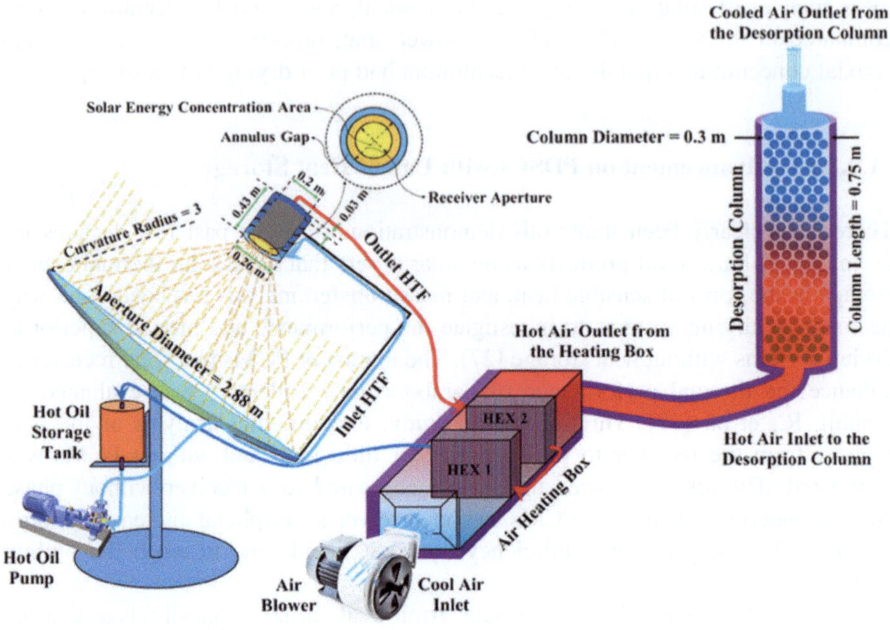

Fig. 11.4 Integrated system to regenerate desiccants [32]

both theoretically and experimentally by connecting a fixed bed of silica gel to two series finned-tube heat exchangers, a solar parabolic dish collector, and a cylindrical cavity receiver. The desiccant materials' regeneration rate and the system's typical daily thermal regeneration efficiency were both up to 0.4 kg water/h m2 and 30%, respectively. On the basis of actual solar radiation intensity and environmental factors, the system's performance was examined on a test day. Figure 11.4 explains how the integration of this system works to regenerate the desiccants.

A. Kumar et al. explored the regeneration of various desiccants using PDSC, including silica gel, activated alumina, and activated charcoal. Solar radiation, surface temperature, and desiccant starting moisture all had a significant impact on how well the regeneration process worked. As shown in Fig. 11.5, the receiver is the main component of the regeneration system. According to the findings, 1 kg of silica gel may be dried in 120 min when done on a dry basis, but it takes 560 min when done through adsorption. In comparison to alumina and activated charcoal, silica gel has a high regeneration rate and adsorption rate [33].

The regeneration of several saturated desiccant (wet basis) including silica gel, activated alumina, molecular sieve, and unique composite desiccant was the subject of experimental research by A. Malan et al. employing the Scheffler type PDSC. The findings show that molecular sieves may reach a temperature of 194.30°C and that the greatest regeneration rate for a unique composite desiccant can reach 0.07440 kg/h. The highest temperature can be reached in regeneration by Scheffler type PDSCs [34].

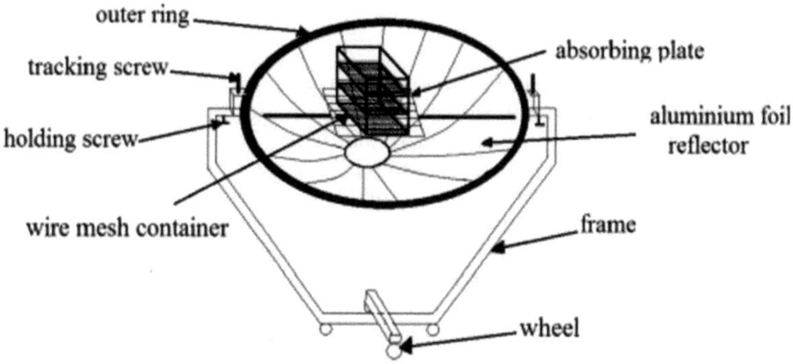

Fig. 11.5 Schematic diagram of parabolic dish collector [33]

11.3 Conclusion

In conclusion, this review has examined the recent advancements in the design of parabolic dish concentrators (PDCs), with a particular focus on receiver design, integration with phase change material (PCM)/latent heat storage, and applications for the regeneration of desiccants. The findings highlight the significant progress made in these areas and shed light on potential future research directions. Regarding receiver design, advancements have been made to enhance its efficiency and performance. The use of selective coatings and novel materials has shown promise in improving solar absorption and thermal conductivity, thereby increasing the overall energy conversion efficiency of PDCs. Integration of PCM/latent heat storage with PDCs has emerged as a promising approach for thermal energy storage. Furthermore, the review highlights the application of PDCs for the regeneration of desiccants in air dehumidification systems. By utilizing the high-temperature capabilities of PDCs, desiccants can be effectively regenerated, leading to improved energy efficiency and overall system performance. This integration of PDCs with desiccant systems has shown potential in various sectors, including agriculture, pharmaceuticals, and industrial processes.

The advancements in parabolic dish concentrator design on the receiver, integration with PCM/latent heat storage for solar drying technologies, and applications for desiccant regeneration have demonstrated the potential of this technology in enhancing energy efficiency and utilization. However, further research and development are necessary to enable the successful implementation of PDC systems in various sectors, contributing to sustainable energy solutions and addressing future energy challenges.

Acknowledgments The research has been funded by the faculty of Mechanical and Industrial Engineering, Bahir Dar Institute of Technology, Bahir Dar University, Ethiopia.

References

1. Islam, M. T., Huda, N., & Saidur, R. (2019). Current energy mix and techno-economic analysis of concentrating solar power (CSP) technologies in Malaysia. *Renewable Energy, 140*, 789–806. https://doi.org/10.1016/j.renene.2019.03.107
2. Prinsloo, G., & Dobson, R. (2015). Combined solar heat and power with microgrid storage and layered smartgrid control toward supplying off-grid rural villages. *Energy Science & Engineering, 3*(2), 135–144. https://doi.org/10.1002/ese3.57
3. Sahu, S. K., Arjun Singh, K., & Natarajan, S. K. (2021). Design and development of a low-cost solar parabolic dish concentrator system with manual dual-axis tracking. *International Journal of Energy Research, 45*(4), 6446–6456. https://doi.org/10.1002/er.6164
4. Canavarro, D., Chaves, J., & Collares-Pereira, M. (2019). Simultaneous multiple surface method for the design of new parabolic dish-type concentrator using a Cassegranian approach. In *AIP conference proceedings*. American Institute of Physics. https://doi.org/10.1063/1.5117584
5. Wang, R., & Zhai, X. (2018). Handbook of Energy Systems in Green Buildings.
6. Hafez, A. Z., Soliman, A., El-Metwally, K. A., & Ismail, I. M. (2017). Design analysis factors and specifications of solar dish technologies for different systems and applications. *Renewable and Sustainable Energy Reviews, 67*, 1019–1036. https://doi.org/10.1016/j.rser.2016.09.077
7. Daabo, A. M., Mahmoud, S., & Al-Dadah, R. K. (2016). The optical efficiency of three different geometries of a small scale cavity receiver for concentrated solar applications. *Applied Energy, 179*, 1081–1096. https://doi.org/10.1016/j.apenergy.2016.07.064
8. Idlimam, R., Bah, A., Asbik, M., Malha, M., & Kazdaba, H. (2019). Effect of receiver shape on a parabolic solar concentrator system thermal behavior. In *AIP conference proceedings*. American Institute of Physics. https://doi.org/10.1063/1.5116975
9. Kumar, V., & Yadav, A. (2020). Experimental investigation of parabolic dish concentrator with various sizes of receiver. *Materials Today: Proceedings, 44*, 4966–4971. https://doi.org/10.1016/j.matpr.2020.12.883
10. Azzouzi, D., Boumeddane, B., & Abene, A. (2017). Experimental and analytical thermal analysis of cylindrical cavity receiver for solar dish. *Renewable Energy, 106*, 111–121. https://doi.org/10.1016/j.renene.2016.12.102
11. Venkatachalam, T., & Cheralathan, M. (2019). Effect of aspect ratio on thermal performance of cavity receiver for solar parabolic dish concentrator: An experimental study. *Renewable Energy, 139*, 573–581. https://doi.org/10.1016/j.renene.2019.02.102
12. Loni, R., Asli-Ardeh, E. A., Ghobadian, B., & Kasaeian, A. (2018). Experimental study of carbon nano tube/oil nanofluid in dish concentrator using a cylindrical cavity receiver: Outdoor tests. *Energy Conversion and Management, 165*, 593–601. https://doi.org/10.1016/j.enconman.2018.03.079
13. Loni, R., Pavlovic, S., Bellos, E., Tzivanidis, C., & Asli-Ardeh, E. A. (2018). Thermal and exergy performance of a nanofluid-based solar dish collector with spiral cavity receiver. *Applied Thermal Engineering, 135*, 206–217. https://doi.org/10.1016/j.applthermaleng.2018.02.070
14. Loni, R., Asli-Ardeh, E. A., Ghobadian, B., Ahmadi, M. H., & Bellos, E. (2018). GMDH modeling and experimental investigation of thermal performance enhancement of hemispherical cavity receiver using MWCNT/oil nanofluid. *Solar Energy, 171*, 790–803. https://doi.org/10.1016/j.solener.2018.07.003
15. Loni, R., Askari Asli-Ardeh, E., Ghobadian, B., Bellos, E., & Le Roux, W. G. (2018). Numerical comparison of a solar dish concentrator with different cavity receivers and working fluids. *Journal of Cleaner Production, 198*, 1013–1030. https://doi.org/10.1016/j.jclepro.2018.07.075
16. Loni, R., Askari Asli-Ardeh, E., Ghobadian, B., Kasaeian, A. B., & Bellos, E. (2018). Thermal performance comparison between Al2O3/oil and SiO2/oil nanofluids in cylindrical cavity receiver based on experimental study. *Renewable Energy, 129*, 652–665. https://doi.org/10.1016/j.renene.2018.06.029

17. Reshad, A., & Boroushaki, M. (2023). Fabrication and assessment of a parabolic dish concentrator using a water heater cylindrical receiver. *Energy Equipment and Systems, 11*(1), 55–68.
18. Soltani, S., Bonyadi, M., & Madadi Avargani, V. (2019). A novel optical-thermal modeling of a parabolic dish collector with a helically baffled cylindrical cavity receiver. *Energy, 168*, 88–98. https://doi.org/10.1016/j.energy.2018.11.097
19. Zheng, Z. J., Li, M. J., & He, Y. L. (2017). Thermal analysis of solar central receiver tube with porous inserts and non-uniform heat flux. *Applied Energy, 185*, 1152–1161. https://doi.org/10.1016/j.apenergy.2015.11.039
20. Karimi, R., Gheinani, T. T., & Madadi Avargani, V. (2018). A detailed mathematical model for thermal performance analysis of a cylindrical cavity receiver in a solar parabolic dish collector system. *Renewable Energy, 125*, 768–782. https://doi.org/10.1016/j.renene.2018.03.015
21. Arkian, A. H., Najafi, G., Gorjian, S., Loni, R., Bellos, E., & Yusaf, T. (2019). Performance assessment of a solar dryer system using small parabolic dish and alumina/oil nanofluid: Simulation and experimental study. *Energies (Basel), 12*(24). https://doi.org/10.3390/en12244747
22. Gwani, M., Abubakar, G. A., Abbas, M., Allah, M. N., & Danyaro, J. (2015). Design, fabrication and experimental study of solar parabolic dish concentrator for remote area application. *International Journal of Sciences: Basic and Applied Research (IJSBAR), 23*(1), 230–241.
23. Manchanda, H., & Kumar, M. (2017). Performance analysis of single basin solar distillation cum drying unit with parabolic reflector. *Desalination, 416*, 1–9. https://doi.org/10.1016/j.desal.2017.04.020
24. Varghese, J., Rupesh, S., Augustine, J., & Nair, A. (2021, April). Design and analysis of a solar drier with a parabolic shaped dish type collector for drying peanut. *IOP Conference Series: Materials Science and Engineering, 1132*(1), 012046. https://doi.org/10.1088/1757-899x/1132/1/012046
25. Stiling, J., et al. (2012). Performance evaluation of an enhanced fruit solar dryer using concentrating panels. *Energy for Sustainable Development, 16*(2), 224–230. https://doi.org/10.1016/j.esd.2012.01.002
26. Simo-Tagne, M., & Ndukwu, M. C. (2021). Study on the effect of conical and parabolic solar concentrator designs on hybrid solar dryers for apricots under variable conditions: A numerical simulation approach. *International Journal of Green Energy, 18*(15), 1613–1631. https://doi.org/10.1080/15435075.2021.1914632
27. Srivastava, A. K., & Shukla, S. K. (2017). Thermal modeling of indirect solar drying system: An experimental validation. *Distributed Generation and Alternative Energy Journal, 32*(3), 19–51. https://doi.org/10.1080/21563306.2017.11878944
28. Senthil, R., & Cheralathan, M. (2017). Simultaneous testing of a parabolic dish concentrated PCM and non-PCM solar receiver. *International Journal of Mechanical and Production Engineering Research and Development, 7*(6), 79–85.
29. Thirunavukkarasu, V., & Cheralathan, M. (2020). An experimental study on energy and exergy performance of a spiral tube receiver for solar parabolic dish concentrator. *Energy, 192*. https://doi.org/10.1016/j.energy.2019.116635
30. Mandal, S., Sharma, P. K., Mani, I., Kushwaha, H. L., TV, A. K., & Sarkar, S. (2020). Design and development of phase change material (PCM) based hybrid solar dryer for herbs and spices. *Indian Journal of Agricultural Sciences, 90*(11), 2217–2224.
31. Malan, A. (2017). *Regeneration of solid and composite desiccant using scheffler solar concentrator*. Doctoral dissertation. National Institute of Technology Kurukshetra.
32. Avargani, V. M., Karimi, R., & Gheinani, T. T. (2019). Mathematical modeling of an integrated system for regeneration of solid desiccants using a solar parabolic dish concentrator. *International Journal of Heat and Mass Transfer, 142*. https://doi.org/10.1016/j.ijheatmasstransfer.2019.118479

33. A. Kumar, A. Chaudhary, and A. Yadav, "The regeneration of various solid desiccants by using a parabolic dish collector and adsorption rate: An experimental investigation," International Journal of Green Energy, vol. 11, no. 9, pp. 936–953, Oct. 2014, doi: https://doi.org/10.1080/15435075.2013.833514.
34. Malan, A., Kamboj, V., Sharma, A. K., & Yadav, A. (2020). The regeneration of various saturated solid and novel composite desiccant using Scheffler solar concentrator: An experimental investigation. *International Journal of Ambient Energy, 41*(2), 224–236. https://doi.org/10.1080/01430750.2018.1456965

Chapter 12
Parameter Identification of Single-Diode Solar Photovoltaic Model Using Particle Swarm Optimization Hybrid with Newton-Raphson Method

Edemialem Gedefaye Mehretie, Samuel Lakeou, Tassew Tadiwose Zewdie, and Tefera Terefe Yitayew

12.1 Background

Fossil fuels are one of the energy sources that can best meet human needs. Fossil fuels have grown in popularity due to their low cost, high efficiency, and ease of transportation [1]. However, due to worsening industrial pollution, environmental toxins, global warming, and energy source depletion [2], there has been a significant effort to develop renewable energy sources. As a result, producing energy from renewable sources rather than fossil fuels is becoming more widespread.

Solar energy sources emerge as a viable option. It contains a high amount of energy and has the potential to meet all the energy needs of current human activities [3], in addition to having excellent environmental, technical, and economic properties. Photovoltaic (PV) panels can produce electricity from solar energy, which reaches the Earth daily [2–14]. However, the PV output depends on radiation, temperature, and the physical properties of the cell. PV cell materials' nonlinear, multivariable, and tightly coupled characteristics lead to a multi-peaked problem with lots of local maxima. This property distorts the sinusoidal waveforms of current

E. G. Mehretie (✉) · T. T. Zewdie
Faculty of Electrical and Computer Engineering, Bahir Dar Institute of Technology, Bahir Dar University, Bahir Dar, Ethiopia

S. Lakeou
Department of Electrical and Computer Engineering, University of the District of Columbia, Washington, DC, USA
e-mail: slakeou@udc.edu

T. T. Yitayew
Faculty of Electrical and Computer Engineering, Adama Science and Technology University, Adama, Ethiopia

K. Mequanint et al. (eds.), *Sustainable Development Research in Materials and Energy*, Green Energy and Technology,
https://doi.org/10.1007/978-3-031-69860-6_12

and voltage, resulting in harmonic components in the system. Similarly, PV cell performance and control quality need accurate parameter estimation to improve. Therefore, developing effective models and obtaining model parameters are crucial for monitoring and controlling the actual operation of simulating PV modules and predicting their power generation. Thus, the current-voltage and voltage-power relationships of a solar PV module model involve mathematical modeling and model parameter identification [1–3].

There are numerous models and methods for demonstrating the behavior of PV cells and accurately estimating their parameters. The single-diode (SDM), double-diode (DDM), and triple-diode (TDM) models are known to accurately estimate five, seven, and ten parameters, respectively. Furthermore, the most common methods for extracting solar PV cell parameters are analytical, numerical, and metaheuristic algorithms. The analytical method primarily employs open circuit voltage, short circuit current, maximum power point current, and voltage [4]. It establishes the characteristic curves using a streamlined and approximative method [5]. Thus, analytical methods are unreliable and provide unsatisfactory results in most cases when many local minimum points exist.

PV cell model parameters are estimated using numerical methods by fitting points on PV characteristic curves [6–9]. Compared to analytical methods, numerical methods can produce relatively accurate estimates. However, numerical methods involve complex calculations, and their estimation accuracy and convergence are overly dependent on the initial value, which remains a difficult task.

Metaheuristic algorithms are other techniques for determining the best unknown parameters [9–12]. Based on the global optimization population algorithm, there is no restriction on the continuity of the objective function; the implementation is simple; the robustness is strong, and it can deal with multimodal optimization problems more simply. However, these algorithms generate arbitrary initial values and then use a repeated procedure to determine the variation in the fitness function of the parameter problem. Moreover, it takes a long computational time, needs programming knowledge, is unevenly balanced between exploration and development, and has slow convergence. Furthermore, some algorithms perform poorly in global optimization and are prone to entering local optimization [12].

In light of this, this study suggests using the Newton-Raphson method. This method avoids mathematical simplifications and physical assumptions while still providing an accurate estimate of the roots of the nonlinear characteristics of PV cells via numerical analysis.

The main contributions of this study are as follows:

- It describes the mathematical models for SDM, DDM, and PV module parameter extraction.
- The statistical outcomes of the proposed technique are compared using RMSE, R^2, and algorithmic settings.

- The proposed technique evaluated the effects of temperature, irradiance, and internal resistance variation on the output parameters of a PV system.
- Future work focus options are analyzed and provided, along with current challenges.

12.2 Problem Formulation

12.2.1 Solar Cell Models

In the literature, several models are available to describe the I–V characteristics of solar cells. In practice, two commonly used models are single-diode and double-diode models [12]. In addition, this subsection provides a brief overview of these two models.

12.2.1.1 Single Diode Mode

Due to its simplicity and accuracy, the single-diode model is well-regarded in Fig. 12.1.

Equation (12.1) calculates the output current and the current that passes through the shunt resistance using the single-diode model.

$$I_s = I_L - I_D - I_{sh} \tag{12.1}$$

where I_L, I_D, and I_{sh} represent the current generated by the incidence of light, the equivalent diode current, and the shunt resistance current, respectively. Eq. (12.2) represents the photocurrent equation.

$$I_L = [I_{sc} + 0.0017(T - 298)]\frac{G}{1000} \tag{12.2}$$

Fig. 12.1 The equivalent PV cell circuit model

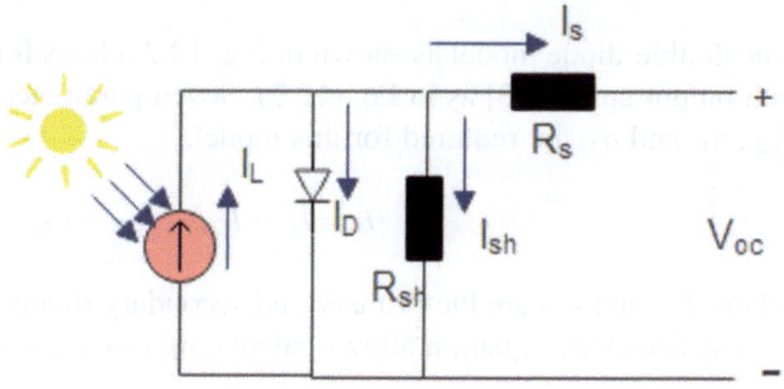

where I_{sc}, T, and G represent short circuit current; T is the temperature in C; and G is the irradiance in W/m^2, respectively. Calculate the diode currents (I_D) using the Shockley equation (12.3):

$$I_D = I_o \left[\exp\left(\frac{q(V_o + I_s R_s)}{akT_o} \right) - 1 \right] \quad (12.3)$$

where I_o is the equivalent diode reverse saturation current, a is the diode ideality constant, q is the electron charge (1.60210^{-19} C), k is the Boltzmann constant (1.380650310^{-23} J/K), and T_o is the temperature of the p-n junction in Kelvin.

Equation (12.4) determines the shunt resistor current (I_{sh}).

$$I_{sh} = \frac{(V_o + I_s R_s)}{R_{sh}} \quad (12.4)$$

PV cell p-n junction ideality factor "a" is mainly assumed to be constant and temperature-independent. However, the ideality factor varies with temperature for most semiconductor materials according to the general expression given in Eq. (12.5).

$$a = a_R = \frac{\alpha_{A T_o^2}}{T_o + \beta_o} \quad (12.5)$$

where a_R is the ideality factor at absolute zero temperature, α_A is the ideality factor temperature coefficient, and β_D is Debye's temperature.

Equation (12.6) calculates the output current of the PV cell (I_s) by substituting the diode current and shunt current from Eq. (12.1) for series-parallel PV cells.

$$I_s = I_L - I_o \left[\exp\left(\frac{q(V_o + I_s R_s)}{\mathrm{ak}T_o} \right) - 1 \right] - \frac{(V_o + I_s R_s)}{R_{sh}} \quad (12.6)$$

The single-diode model requires five parameters: I_L, I_o, R_s, R_{sh}, and a.

12.2.1.2 Double-Diode Model

The double-diode model as shown in Fig. 12.2 allows for the expression of the solar cell output current [3] as in Eq. (12.7). Seven parameters, including I_L, I_{D1}, I_{D2}, R_s, R_{sh}, a_1, and a_2, are required for this model.

$$I_s = I_L - I_{D1} - I_{D2} - I_{sh} \quad (12.7)$$

where I_{D1} and I_{D2} are the primary and secondary diode currents, respectively.

The Shockley equation allows calculating diode currents I_{d1} and I_{d2} as Eq. 12.8.

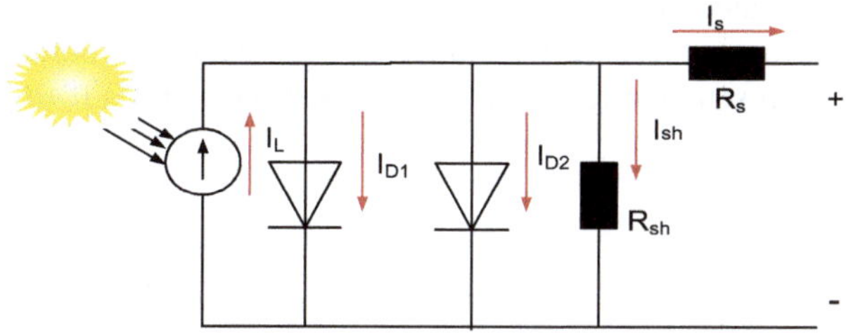

Fig. 12.2 Equivalent circuit of double-diode PV cell model

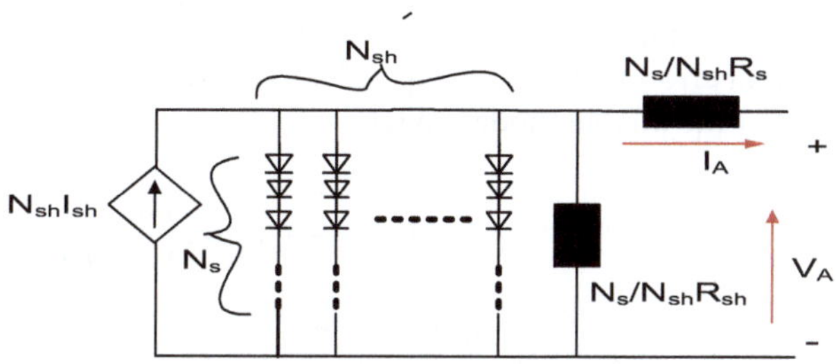

Fig. 12.3 A PV module's equivalent circuit

$$I_s = I_L - I_{o1}\left[\exp\left(\frac{q(V_o + I_s R_s)}{a_1 k T_o}\right) - 1\right] - I_{o2}\left[\exp\left(\frac{q(V_o + I_s R_s)}{a_2 k T_o}\right) - 1\right] - \frac{(V_o + I_s R_s)}{R_{sh}} \tag{12.8}$$

12.2.2 Photovoltaic Module

As shown in Fig. 12.3, this single-cell equivalent circuit is scalable to support any series or parallel PV module configuration. These modules are connected both in series and parallel to create PV arrays.

The current-voltage relationship described in Eq. (12.9) forms the mathematical model that predicts the power output of the PV generator.

$$I_s = N_p I_L - N_p I_o\left[exp\left(\frac{q}{akT_o}\left(\frac{V_o}{N_s} - \frac{I_s R_s}{N_p}\right)\right) - 1\right] - \frac{N_p}{R_{sh}}\left(\frac{V_o}{N_s} + \frac{I_s R_s}{N_p}\right) \tag{12.9}$$

12.2.3 Objective Function

The objective function defines extracting the parameters of various solar cell models from the I–V data using optimization techniques. And to improve the fit between the simulated and experimental data, the result of the objective function declines. The narrower the objective functions, the better the solutions are. Therefore, the root mean square error (RMSE), as described in Eq. (12.10), is an objective function of this study.

$$\text{Minimize } F(x) = \sqrt{\frac{1}{N}\sum_{i=1}^{N} f_j(V_{o,} \quad I_S, x)^2} \tag{12.10}$$

where x is the solution vector, N is the actual data amount, and $f(V, I, x)$ computes the current error as shown in Eqs. 12.11 and 12.12.

For the double-diode model,

$$f(V_o, I_s, x) = I_L - I_{o1}\left[exp\left(\frac{q(V_o + I_s R_s)}{a_1 k T_o}\right) - 1\right]$$
$$- I_{o2}\left[\exp\left(\frac{q(V_o + I_s R_s)}{a_2 k T_o}\right) - 1\right] - \frac{(V_o + I_s R_s)}{R_{sh}} - I_s \tag{12.11}$$

$$x = \{I_L, I_{o1}, I_{o2}, R_s, R_{sh}, a_1, a_2\}$$

For the single-diode model,

$$f(V_o, I_s, x) = I_L - I_o\left[\exp\left(\frac{q(V_o + I_s R_s)}{akT_o}\right) - 1\right] - \frac{(V_o + I_s R_s)}{R_{sh}} - I_s \tag{12.12}$$

$$x = \{I_L, I_o, R_s, R_{sh}, a\}$$

12.2.4 Newton-Raphson Method and Curve Fitting Model

The Newton-Raphson method is a simple and accurate method that requires the measured I–V characteristics of PV panels for a given environmental condition as an input. The data cover the whole range of PV panels' operation between short and open circuit. These I–V curves are fit to generate the five SDM parameter equations and the seven DDM parameter equations. The measured I–V curves should satisfy the model in the summation form given in Eqs. (12.13) and (12.14).

$$0 = \sum_{i=1}^{N} \left[I_i - I_L + I_o \left(e^{\frac{q(V_i + I_i R_s)}{akT_o}} - 1 \right) + \frac{V_i + I_i R_s}{R_{sh}} \right] \tag{12.13}$$

$$0 = \sum_{i=1}^{N} \left[I_i - I_L + I_{o1} \left(e^{\frac{q(V_i + I_i R_s)}{a_1 kT_o}} - 1 \right) + I_{o2} \left(e^{\frac{q(V_i + I_i R_s)}{a_2 kT_o}} - 1 \right) + \frac{V_i + I_i R_s}{R_{sh}} \right] \tag{12.14}$$

where N is the number of the measured points of the I–V curve.

It is possible to solve the equation and determine the exact values of the parameters if the error term is zero or at a reasonable minimum value. The sum of the error squares (E) is given by Eqs. 12.15 and 12.16.

$$E = \sum_{i=1}^{N} \left[I_i - I_L + I_o \left(e^{\frac{q(V_i + I_i R_s)}{akT_o}} - 1 \right) + \frac{V_i + I_i R_s}{R_{sh}} \right]^2 \tag{12.15}$$

$$E = \sum_{i=1}^{N} \left[I_i - I_L + I_{o1} \left(e^{\frac{q(V_i + I_i R_s)}{a_1 kT_o}} - 1 \right) + I_{o2} \left(e^{\frac{q(V_i + I_i R_s)}{a_2 kT_o}} - 1 \right) + \frac{V_i + I_i R_s}{R_{sh}} \right]^2 \tag{12.16}$$

Performing partial differentiation on I_{ph}, I_o, R_{sh}, R_s, and a, and equating to zero yields the following expressions:

$$\frac{\partial E}{\partial x} = 0 \rightarrow f_x = \pm 2 \sum_{i=1}^{N} \left[I_i - I_L + I_o \left(e^{\frac{q(V_i + I_i R_s)}{akT_o}} - 1 \right) + \frac{V_i + I_i R_s}{R_{sh}} \right]^2 \tag{12.17}$$

$$\frac{\partial E}{\partial x} = 0 \rightarrow f_x = \pm 2 \sum_{i=1}^{N} \left[\begin{array}{c} I_i - I_L + I_{o1} \left(e^{\frac{q(V_i + I_i R_s)}{a_1 kT_o}} - 1 \right) \\ + I_{o2} \left(e^{\frac{q(V_i + I_i R_s)}{a_2 kT_o}} - 1 \right) + \frac{V_i + I_i R_s}{R_{sh}} \end{array} \right]^2 \tag{12.18}$$

The Newton-Raphson method has fast convergence and robust performance in nonlinear equations. Equations 12.17, 12.18, and 12.19 express five and seven unknown nonlinear equations for SDM and DDM, respectively. The Newton-Raphson method has the following general expression:

$$x^{(t+1)} = x^t - \left[J\left(X^{(t)} \right) \right]^{-1} [F(X^t)], \tag{12.19}$$

where t is the index for the iteration number. Both $x(t + 1)$ and $x(t)$ are five-element vectors, and J ($X(t)$) is the Jacobian matrix that contains the partial derivatives for each equation corresponding to each of the five parameters; therefore, the matrices

Table 12.1 The boundary and parameter range of the PV modules model

Parameter	R.T.C France boundary		STC specification	
	LB	UB	Parameter	Value
I_L(A)	0	1	Voltage at P_{max} (V_{ppm}(V))	0.45
I_D, I_{D1},I_{D2} (μA)	0	1	Current at P_{max} (I_{ppm}(A))	0.69119
R_s (Ω)	0	0.5	Open circuit voltage (V_{oc} (V))	0.5728
R_{sh} (Ω)	0	100	Short circuit current (I_{sc} (A))	0.760
A, a_1, a_2	1	2	Maximum power (P_{max} (W))	50

computed for the current value of the five parameters, $F\ (X(t))$, contain the five partial differential equations. The following is an expression for the created system of nonlinear equations:

$$\begin{bmatrix} I_L^{t+1} \\ I_o^{t+1} \\ R_{sh}^{t+1} \\ R_s^{t+1} \\ a^{t+1} \end{bmatrix} = \begin{bmatrix} I_L^{t} \\ I_o^{t} \\ R_{sh}^{t} \\ R_s^{t} \\ a^{t} \end{bmatrix} - \begin{bmatrix} \frac{\partial f_1}{\partial I_L} & \frac{\partial f_1}{\partial I_o} & \frac{\partial f_1}{\partial R_{sh}} & \frac{\partial f_1}{\partial R_s} & \frac{\partial f_1}{\partial a} \\ \frac{\partial f_2}{\partial I_L} & \frac{\partial f_2}{\partial I_o} & \frac{\partial f_2}{\partial R_{sh}} & \frac{\partial f_2}{\partial R_s} & \frac{\partial f_2}{\partial a} \\ \frac{\partial f_3}{\partial I_L} & \frac{\partial f_3}{\partial I_o} & \frac{\partial f_3}{\partial R_{sh}} & \frac{\partial f_3}{\partial R_s} & \frac{\partial f_3}{\partial a} \\ \frac{\partial f_4}{\partial I_L} & \frac{\partial f_4}{\partial I_o} & \frac{\partial f_4}{\partial R_{sh}} & \frac{\partial f_4}{\partial R_s} & \frac{\partial f_4}{\partial a} \\ \frac{\partial f_5}{\partial I_L} & \frac{\partial f_5}{\partial I_o} & \frac{\partial f_5}{\partial R_{sh}} & \frac{\partial f_5}{\partial R_s} & \frac{\partial f_5}{\partial a} \end{bmatrix}^{-1} \times \begin{bmatrix} f_1\left(I_L^t, I_o^t, R_{sh}^t, R_s^t, a^t\right) \\ f_2\left(I_L^t, I_o^t, R_{sh}^t, R_s^t, a^t\right) \\ f_3\left(I_L^t, I_o^t, R_{sh}^t, R_s^t, a^t\right) \\ f_4\left(I_L^t, I_o^t, R_{sh}^t, R_s^t, a^t\right) \\ f_5\left(I_L^t, I_o^t, R_{sh}^t, R_s^t, a^t\right) \end{bmatrix}$$

The decision vector, or x, contains the extracted parameters. Table 12.1 displays the lower and upper bounds for each PV cell and PV module parameters for the double- and single-diode models. In addition, Fig. 12.4 displays the overall flowchart of the proposed method, and Table 12.2 expresses the modules' specification at the standard test condition.

12.2.5 Particle Swarm Optimization Algorithm

Particle swarm optimization is a population-based evolution algorithm search method based on the behavior of bird flocks [14]. This method maintains a swarm of individuals (referred to as particles), each of which represents a candidate solution. The particle position (x_i) is changed based on Eq. (12.20) [14].

$$x_i^{k+1} = x_i^k + v_i^{k+1} \tag{12.20}$$

Fig. 12.4 Newton-Raphson method flowchart

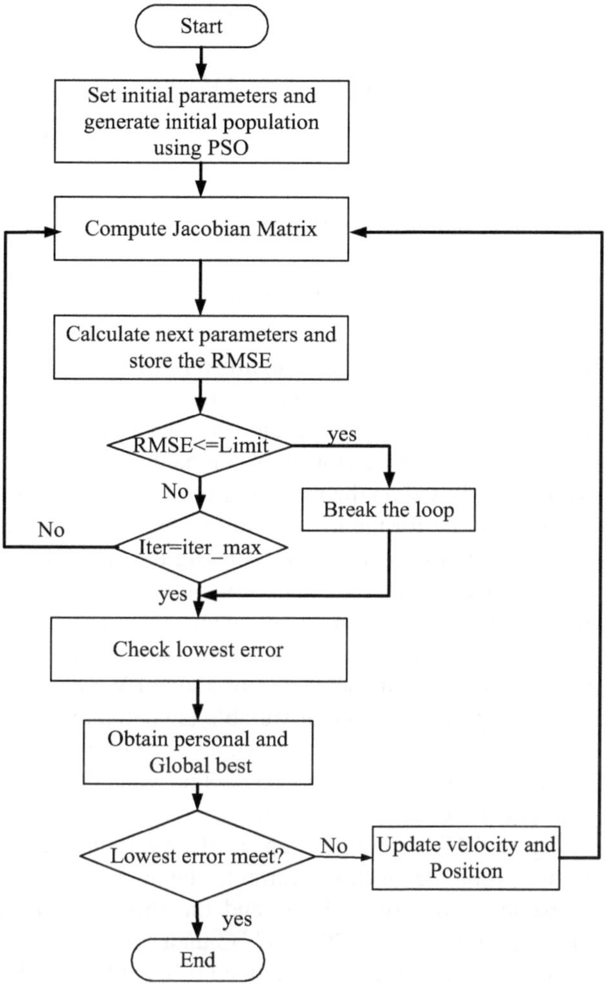

Table 12.2 Comparison of R.T.C France extracted parameters

Technique	I_{ph} (A)	I_D (μA)	R_{sh} (Ω)	R_s (mΩ)	a	RMSE ($\times 10^{-4}$)
APLO [1]	0.7607755	0.3230208	53.71852400	0.0363771	1.4811855	9.8602188
GA [3]	0.7608	0.3231	53.7378	0.0364	1.4812	9.8602
GSK [4]	0.7608	0.3231	53.7227	0.0364	1.4812	9.8248
BO [12]	0.7607755	0.3230	53.71852270	0.0363771	1.4811836	9.8519
FPSO [8]	0.7607755	0.32302	53.71852	0.03637	1.4811082	9.8602
Proposed method	0.76091	0.32157	53.71232	0.036324	1.4811083	9.6561

And then, the velocity of the particle (i) is updated in Eq. (12.21):

$$v_i^{k+1} = wv_i^k + c_1r_1\{P_{\text{best}} - x_i^k\} + c_2r_2\{G_{\text{best}} - x_i^k\} \tag{12.21}$$

12.3 Result and Simulation

12.3.1 MATLAB/Simulink Model

MATLAB/Simulink plays a significant role in the modeling and analysis of linear and nonlinear systems. It reduces the need for prototypes. Figure 12.5 shows the boundary diagram related to the PV solar cell. The study used the PSO algorithm combined with the Newton-Raphson method to optimize the extraction of solar cell parameters. Based on the measured I–V curve data, the Newton-Raphson method is best for solving for the ideal values of the solar cell parameters iteratively, and the PSO algorithm is very suited to finding the optimal initial values of the parameters that can use as input for the Newton-Raphson method. This approach can help improve the precision and convergence rate of the parameter extraction process.

Temperature and irradiation are unquestionably constraints that incorrectly influence the electrical characteristics of a solar PV module. Figure 12.6 depicts the direct relationship between P-V at variable temperatures (37, 33, and 29 °C) within fixed irradiation. And the irradiation varies greatly, ranging from 200 to 1000 W/m^2. It indicates that the nominal power decline when the radiation decreases. However, the open circuit voltage is constant.

Figure 12.7 accurately depicts the profound effect of mild temperatures and irradiation on the electrical current. The desired voltage dramatically decreases, the temperature slightly declines, and the short circuit current stays constant as the temperature rises from 29 to 37 °C under fixed illumination. Similarly, the current decreases as the irradiation decline (1000, 800, 600, 400, and 200 W/m^2).

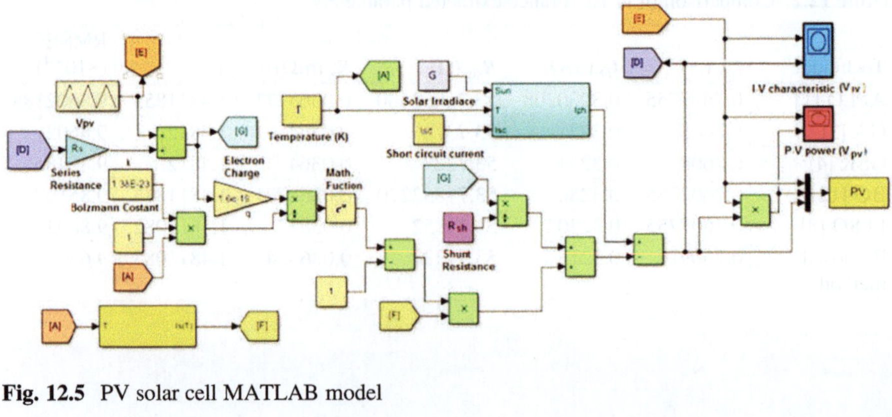

Fig. 12.5 PV solar cell MATLAB model

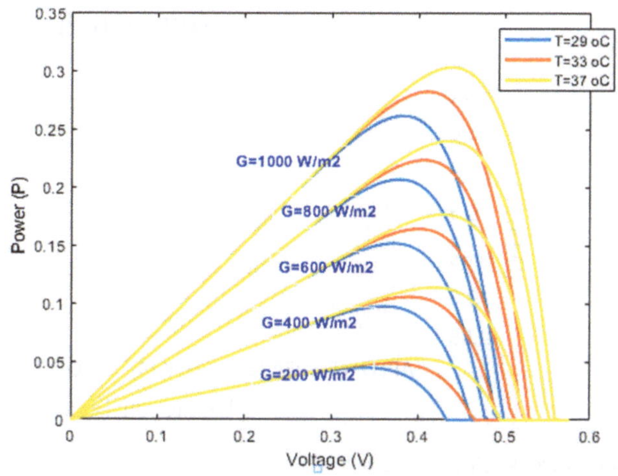

Fig. 12.6 P-V characteristics with the variation of irradiation and temperature

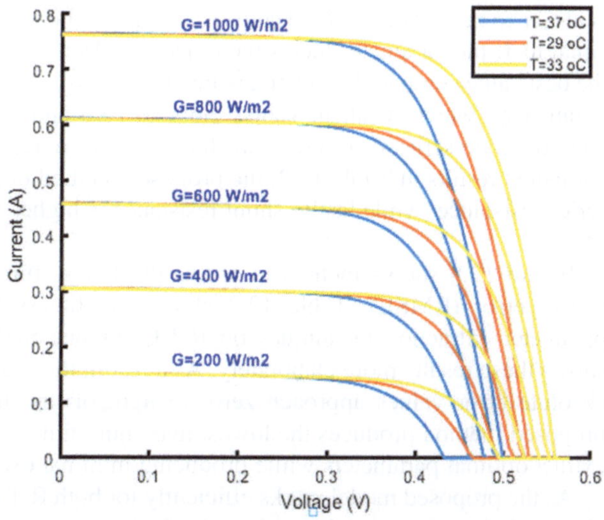

Fig. 12.7 I–V characteristics with the variation in the temperature and irradiation

Furthermore, the desired voltage and current values decrease slightly when the temperature and irradiation remain constant at STC conditions and the solar parameters change. As a result of the desired outcome, power output gradually decreases.

Table 12.2 shows an extracted parameter comparison of various techniques used by R.T.C. France solar cells. The result depicts five unknown variables. The finding described the first parameter, I_{ph}. This outcome demonstrates that the BO technique computed the lowest values, 0.7607755 A and 0.76091; the proposed method obtained the highest, 0.76091 amperes.

The second parameter, the diode saturation current I_D, is expected to be very low. The results from the table show that there is the highest result for the FPSO technique and the least for the proposed technique. The proposed method output is best since it

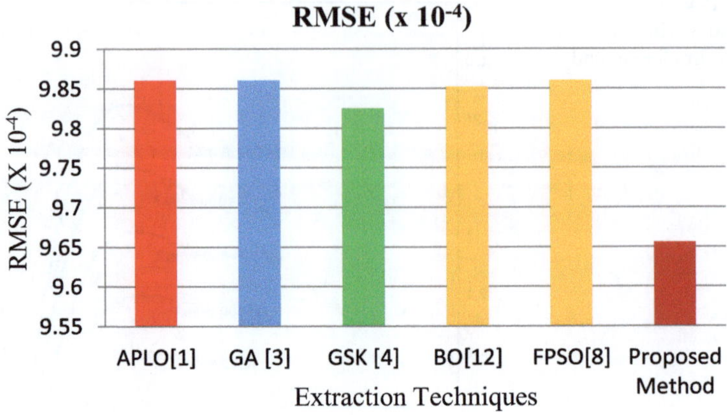

Fig. 12.8 RMSE result graph

is closer to zero. The diode ideality parameter is the third parameter. If the result is equal to 1, the parameter meets the objective. As a result, method APLO results in the best value compared to others, which is close to 1. In addition, the fourth and fifth parameters, which result in values close to zero and infinity for the standard test condition, are series resistance and shunt resistance, respectively. As a result of the computed results in Table 12.2, the proposed method outperforms the least value in series resistance, while lastly, shunt resistance is higher in the GSK method than in others.

Figure 12.8 shows metrics to assess the fitness function, including root mean square error (RMSE). Table 12.2 shows the fitness function result of various parameter extraction techniques on R.T.C. France SDM solar cells. Figure 12.8 shows these results more elaborately. As a result, the root mean error and individual absolute error values approach zero. In light of the findings, it is clear that the proposed method produces the lowest maximum fitness function. As a result, it can extract optimal parameters while producing minimal oscillations.

As the proposed model works efficiently for both R.T.C. France solar cells, it will also work for other modules and can be further experimented with in future studies. However, these types of solar cells were chosen based on the availability of experimental data for validation purposes only, which is promising for commercial modules that lack experimental datasets.

12.4 Conclusions

This study used the PSO hybrid with Newton-Raphson method to describe PV cell parameters and forecast how PV modules would perform in different weather scenarios and with various internal material properties. The proposed method was validated using measured data from the R.T.C. solar cell experimental model. The study comes to the following conclusions: (a) Temperature and irradiation are unquestionably constraints that incorrectly influence the electrical characteristics of a solar PV module; (b) the PV cells and modules model problem formulations have been clarified; (c) the obtained results were comparable to or, in some cases, better than previous studies; and (d) the result shows that the increase in series resistance increases the fill factor and vice versa in shunt resistance. As a result, it applies to all types of photovoltaic systems and produces accurate results comparable to modeling with multiple measurement data points. There will also be research on solar PV-based microgrid management and tracking the maximum power point.

References

1. Ebrahimi, S. M., Salahshour, E., Malekzadeh, M., & Gordillo, F. (2019). Parameters identification of PV solar cells and modules using flexible particle swarm optimization algorithm. *Energy, 179*, 358–372.
2. Patil, S. R., Burade, P. G., & Kadam, D. P. (2023). Five parameters extraction of single diode PV model by metaheuristic optimization method by identified built-up data. *Bulletin of Electrical Engineering and Informatics, 12*(3), 1371–1387.
3. Wang, R. (2021). Parameter identification of photovoltaic cell model based on enhanced particle swarm optimization. *Sustainability, 13*(2), 840.
4. Gu, Z., Xiong, G., & Fu, X. (2023). Parameter extraction of solar photovoltaic cell and module models with metaheuristic algorithms: A review. *Sustainability, 15*(4), 3312.
5. Tajjour, S., Chandel, S. S., Malik, H., Alotaibi, M. A., & Ustun, T. S. (2022). A novel metaheuristic approach for solar photovoltaic parameter extraction using manufacturer data. *InPhotonics, 9*(11), 858. MDPI.
6. Xiong, G., Li, L., Mohamed, A. W., Yuan, X., & Zhang, J. (2021). A new method for parameter extraction of solar photovoltaic models using gaining–sharing knowledge-based algorithm. *Energy Reports, 7*, 3286–3301.
7. Adak, S., Cangi, H., Yilmaz, A. S., & Arifoglu, U. (2023). Development software program for extraction of photovoltaic cell equivalent circuit model parameters based on the Newton–Raphson method. *Journal of Computational Electronics, 22*(1), 413–422.
8. Alanazi, M., Alanazi, A., Almadhor, A., & Rauf, H. T. (2022). Photovoltaic models' parameter extraction using new artificial parameter less optimization algorithm. *Mathematics, 10*(23), 4617.
9. Gong, W., & Cai, Z. (2013). Parameter extraction of solar cell models using repaired adaptive differential evolution. *Solar Energy, 94*, 209–220.
10. Abd El-Mageed, A. A., Abohany, A. A., Saad, H. M., & Sallam, K. M. (2023). Parameter extraction of solar photovoltaic models using queuing search optimization and differential evolution. *Applied Soft Computing, 134*, 110032.
11. Gu, Z., Xiong, G., Fu, X., Mohamed, A. W., Al-Betar, M. A., Chen, H., & Chen, J. (2023). Extracting accurate parameters of photovoltaic cell models via elite learning adaptive differential evolution. *Energy Conversion and Management, 285*, 116994.
12. Premkumar, M., Jangir, P., & Sowmya, R. (2023). Parameter extraction of the three-diode solar photovoltaic model using a new metaheuristic resistance–capacitance optimization algorithm and improved Newton–Raphson method. *Journal of Computational Electronics, 22*(1), 439–470.

13. Abdulrazzaq, A. K., Bognár, G., & Plesz, B. (2022). Accurate method for PV solar cells and modules parameters extraction using I–V curves. *Journal of King Saud University-Engineering Sciences, 34*(1), 46–56.
14. Kiani, A. T., Nadeem, M. F., Ahmed, A., Sajjad, I. A., Haris, M. S., & Martirano, L. (2020). Optimal parameter estimation of solar cell using simulated annealing inertia weight particle swarm optimization (SAIW-PSO). In *2020 IEEE international conference on environment and electrical engineering and 2020 IEEE industrial and commercial power Systems Europe (EEEIC/I&CPS Europe)* (pp. 1–6). IEEE.

Chapter 13
Thermal Modeling and Fuzzy Logic Control Design for Forced Conviction Solar Dryer

Fiseha Kedir Jemal and Betelihem Zemedkun Lodamo

Nomenclature

$\dot{Q}_{c1}$	Heat flow due to conviction between drying air temperature
$\dot{Q}_{c2}$	Heat flow due to conviction between product and chamber air, respectively
$\dot{Q}_{evap}$	Heat flow due to evaporation between product and chamber moist air
$\dot{Q}_{rad}$	Heat flow due to radiation between product and chamber moist air
I_{ba}	Energy due to solar radiation, which is, absorbed moisture air
$\dot{Q}_{cd}$	Conduction heat transfer between the chamber air and the external environment
C_{bs} (J/ °C)	Accumulation of heat capacity on the surface of banana
hc_1	Heat transfer coefficient due to convection between drying air temperature and product temperature, (w/m^2 °C)
T_{hs}	Drying air temperature from heater system of dryer (°C)
T_{bs}	Product surface temperature (°C)
A_{bs}	The area of GBS (m^2)
A_{cr}	Cros-sectional area (m^2)
ρ	Density of air (kg/^{m}3)

F. K. Jemal (✉)
Department of Electrical and computer Engineering, Arba Minch Institute of Technology, Arba Minch University, Arba Minch, Ethiopia

Department of Mechanical Engineering, Arba Minch Institute of Technology, Arba Minch University, Arba Minch, Ethiopia

B. Z. Lodamo
Department of Mechanical Engineering, Arba Minch Institute of Technology, Arba Minch University, Arba Minch, Ethiopia

K. Mequanint et al. (eds.), *Sustainable Development Research in Materials and Energy*, Green Energy and Technology,
https://doi.org/10.1007/978-3-031-69860-6_13

hc_2	Convective heat-transfer coefficient between the product and the moist air (w/m^2 °C)
hc_{evp}	Evaporative heat transfer coefficient (w/m^2 °C)
hc_{rad}	Radiation heat transfer coefficient from the product to the moist air (w/m^2 °C)
α_{ba}	Absorption constant for banana
G	Incident solar radiation rate on the GBS (W/m^2)
C_{ch}	Accumulation of heat capacity in the chamber (J/°C)
C_{pa}	Specific heat of air (J/kg°C)

13.1 Introduction

Drying is the process of vigorous mass and heat transfer that eliminates moisture from the product that is being dried. The banana drying system's coupling effect can alter the temperature or humidity in the chamber.

The highly nonlinear, time-varying, and overall complexity of the problem cannot be resolved through conventional control performance because it is challenging to predict the system behavior of the thermal model. For drying green bananas, intelligent control systems based on fuzzy logic control design were used. The reduction in moisture content serves as the basis for a qualitative analysis of the green banana drying process. The green banana's initial moisture content dropped from 0.7 to 0.2 (wet base) toward the end [1]. The most crucial effective aspect during drying procedures is the hot air speed and temperature. Based on the thin layer-drying model's drying kinematics, the characteristics of banana drying were investigated.

Banana slices were dried at 40, 60, and 80 °C as [2] studied. When the samples had 20% moisture, drying was finished. The ideal temperature for banana drying was 80 °C. The thin-layer drying kinetics were investigated using hot air and sun drying at temperatures of 50, 60, and 70 °C to determine how bananas behaved during drying.

By changing the hot air speed and drying temperature, [3–5] analyze system modeling using the thermal model. On tomatoes, some drying tests were carried out, and the resulting model's validity was validated by comparing the experimental and theoretical findings. The only form of transfer represented by the pseudo-bond graph model investigated by [5] is heat transfer, although during the drying process, drying kinetics is mostly used to assess the mass transfer events.

A solar dryer's control system was developed integrating fuzzy logic using a low-cost model-based sensor to perform "human-operator-like" control [6]. For the meat drying process, it is necessary to create a system model and an intelligent controller that can offer more control accuracy than the proportional-integral-derivative (PID) control system [7]. Setting the dryer's ideal drying temperature and removing unnecessary moisture from the chamber are the control objectives for the earlier efforts.

The majority of the studies previously stated only considered the thermal model of product drying processes. A general analytical model with design assumptions was investigated by the control objective in the present paper. The control goal is to lower the chamber moisture at the desired drying temperature for banana drying procedures. The banana drying kinematics model is a mathematical simulation of the heat and mass balance. The mass and heat balance of the thermal model is used to build the thermal model. Therefore, the major objective of this study is to implement fuzzy logic control for the forced conviction banana dryer.

13.2 Modeling Design and Methodology

13.2.1 Methodology Summary

Dynamic thermal model is based on mass and heat transfer in the drying chamber. This model method is analyzed based on bond graph approach. Design proposed dryer system: this dryer system is designed to dry 100 kg of green or unripe banana in the chamber. Since it has two heating such as solar flat plate collector and auxiliary air hater, it works for every season under control objectives. The thermal simulation is studied by the parameter variation of DAT and HAS. Simulation the control design: This is performed based on the MATLAB Simulink, used to simulate control design of chamber temperature. The banana drying curve was studied to verified the thermal model.

The dryer's heater system and dryer chamber constitute its main parts. Auxiliary heating systems and solar air heaters have supplied the heater systems. The schematic diagram of the suggested dryer system and the structure dryer model, respectively, are shown in Figs. 13.1 and 13.2.

This dryer is a forced conviction solar dryer type, and it is assumed to design the following consideration in Table 13.1.

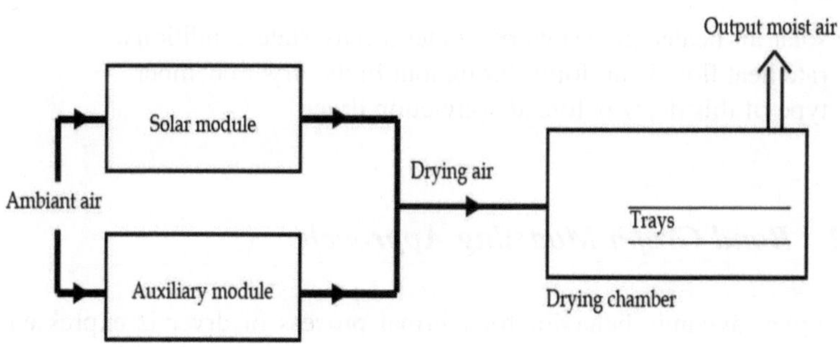

Fig. 13.1 The overflow system model design [8]

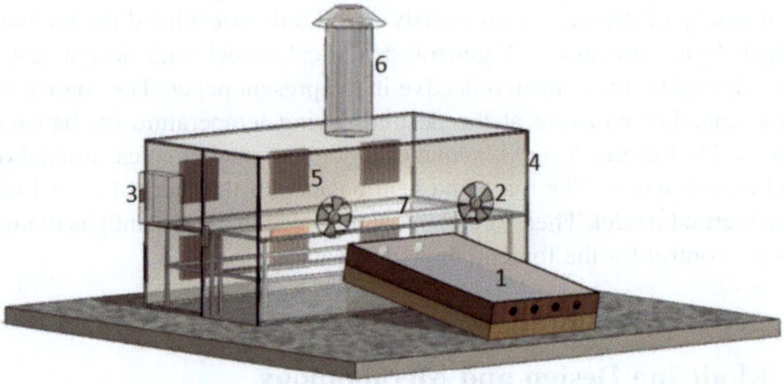

Fig. 13.2 The 3D drawing of the dryer system: (1) solar air heater, (2) exhaust fan, (3) control panel board, (4) drying chamber, (5) auxiliary heater, (6) chimney, and (7) tray

Table 13.1 Design assumption

Items	Optimum values and assumptions
Location and latitude	Arba Minch, Ethiopia (6°2′N 37°33′E)
Fruit	Unrip green banana slice
Design capacity	100 kg/batch
Bulk density	514.76 kg/m³ [9]
Allowable drying temperature range	40–60 °C [10]
Initial and final moisture content	0.7 to 0.212 (wet basis) [1]
Average unripe banana slice thickness	3–4 mm [5]
Drying time per day (sunshine hours)	7 h
Drying period	Works every season
Forced convection velocity of hot air speed	(0.5–5) m/s (design)
Annual average ambient temperature	23.8 °C or 297.3K (meteorology)
Collector type and tilt angle	Flat plate and 36.7° and 45° [4]
Collector efficiency	30–50% [3]

The proposed dryer model is based on the following consideration.

- The solar air heater (SAH) works under steady-state conditions.
- The rate heat flow is uniform throughout in the dryer chamber.
- The type of this dryer is forced conviction dryer.

13.2.2 Bond Graph Modeling Approach

The complex dynamic behavior for thermal process of dryer is expressed by the nonlinear differential equation. Since their resolution and formulation by the classic numerical methods are difficult to solve [11], it is required to solve based on the equations which are really associated with the physical phenomena such as storage and energy dissipation. So that bond graph approach method is introduced by their graphical description to show these energy exchanges in the system. For the drying of green banana process, the modeling for the system is based on the following assumption.

- DAT, T_{hs}, is the temperature of heating system, which results from the average value of solar air heater, T_{co}, and auxiliary heater, T_{au}.
- The output collector temperature, T_{co}, results from the dynamic model of SAH assumed to be taken as the average value.
- The type of this dryer is forced conviction dryer.
- The output collector temperature, T_{co}, results from the dynamic model of solar air heater SAH assumed to be taken as the average value.
- The selected mathematical model for the dryer is used to control the chamber air temperature and the chamber relative humidity.
- The thermal model for green banana slice product is validate based on the thin-layer model.
- The system has evenly uniform distribution of product and chamber temperature with lumped parameter of steady-state analysis.
- The system has evenly uniform distribution of product and chamber temperature with lumped parameter of steady-state analysis.
- The surface contact temperature of green banana slice implies the product temperature.
- The inlet air mass flow rate is assumed equivalent to outlet air mass flow rate.

The first step is to divide the general system into subsystems that is called the word pseudo-bond graph approach method, and as a result there will be exchange of power among them as shown in Fig. 13.3. The effort and flow variables are marked at the input and output of each subsystem [12].

Based on above physical phenomena of the drying process, the word pseudo-bond graph model approach is replaced by a symbol, and the model becomes complete as shown in Figs. 13.3 and 13.4.

A forced convection solar-powered dryer was created, with modeling based on a pseudo-bond graph technique generated from the system's energy balances. The pseudo-bond graph model has six components: 0 junctions, 1 junctions, effort sources (Se), flow sources (Sf), modulated or controlled flow sources (MSf), and capacitive elements (C, R).

The governing model equation for plant has been developed by junction (0, 1). Then, this model was applied by bond graph approach method, and using heat and mass balance, the thermal modeling has been developed.

- Energy balance of banana product surface
- Energy balance of chamber air
- Mass balance for air moisture or humidity and for product moisture

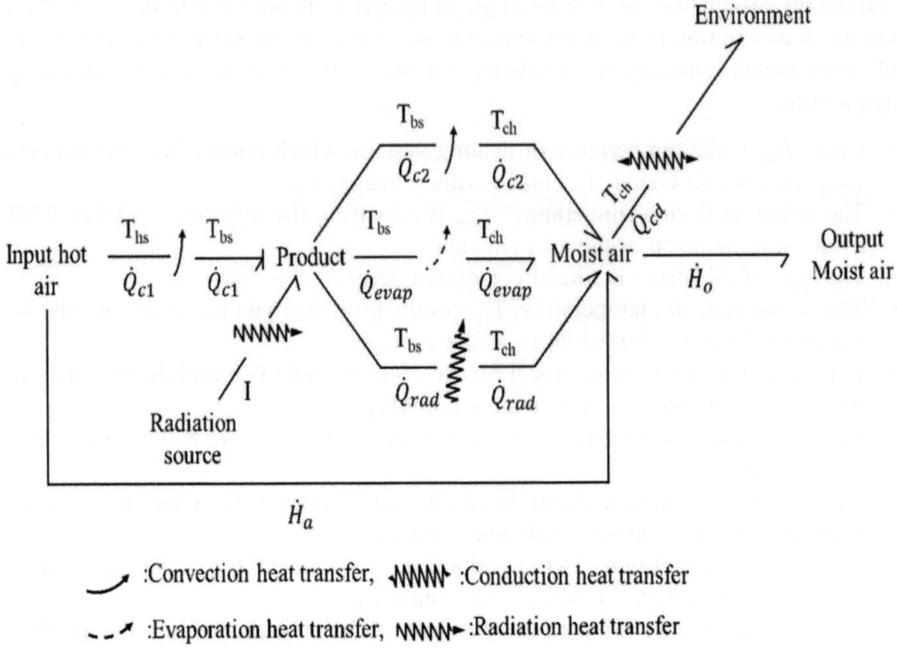

Fig. 13.3 Word pseudo-bond graph model

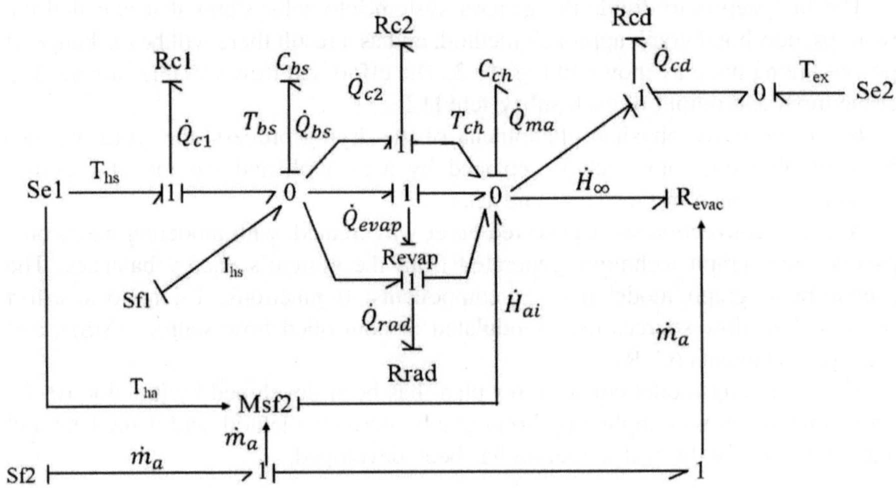

Fig. 13.4 Pseudo-bond graph model

The energy balance for banana product surface temperature and chamber air are

$$\dot{Q}_{bs} = \dot{Q}_{c1} - \dot{Q}_{c2} - \dot{Q}_{evap} - \dot{Q}_{rad} + I_{ba} \tag{13.1}$$

$$\dot{Q}_{ch} = \dot{H}_{ai} - \dot{H}_{ao} + \dot{Q}_{c2} + \dot{Q}_{rad} + \dot{Q}_{evap} - \dot{Q}_{cd} \tag{13.2}$$

Mass balance for product moisture is the amount of water changes in the banana product. $m_{pr}X$ is equivalent to the opposite value of evaporated water from the banana product m_v. So the last model equation is generating, and the amount of water changes inside chamber, $m_a\frac{dY}{dt}$, is expressed as

$$m_a\frac{dY}{dt} = m_aY_{in} - Y + m_v \tag{13.3}$$

where

$$m_{pr}X/\mathrm{dt} = -m_v \tag{13.4}$$

When the nonlinear dynamic differential equation is applied to a control system, the system model can be expressed as state space form

$$\dot{x} = f\ (t, x, u, v) \tag{13.5}$$

where x is state variables such as product temperature, room chamber air temperature, and chamber air moisture or humid air; u is the control output; v is external disturbances such as solar radiation, outdoor temperature, and humidity; t denotes time; and f (. . .) is a nonlinear function.

In this research the thermal model is used to develop the system dryer model, and the dynamic system model is represented by the differential numerical approaches. Using the correlation from Eqs. (13.1, 13.2, 13.3, and 13.4), the thermal model is analyzed.

$$C_{bs}\frac{dT_{bs}}{dt} = [a_{11}]T_{bs} + [a_{12}]T_{ch} + [b_{11}]T_{hs} + [b_{13}]\mathrm{G} \tag{13.6}$$

$$C_{ch}\frac{dT_{ch}}{dt} = [a_{21}]T_{bs} + [a_{22}]T_{ch} + [b_{21}]T_{hs}u_{air} - [b_{24}]T_{am} \tag{13.7}$$

$$m_a\frac{dY}{dt} = [a_{33}]\mathrm{Y} + [b_{25}]Y_{in} - \frac{[m_{pr}X]}{dt} \tag{13.8}$$

where

$$[a_{11}] = -(hc_2 + hc_{evp} + hc_{rad} - hc_1)A_{bs}$$

$$[a_{12}] = [a_{21}] = (hc_2 + hc_{evp} + hc_{rad})A_{bs}$$

$$[a_{22}] = -C_{pa}\,\rho A_{cr} u_{air} - hc_2 A_{bs} - hc_{evp} A_{bs} - hc_{rad} A_{bs} - hc_{cd} A_{aw}$$

$$[a_{33}] = -\rho A_{cr} u_{air}$$

$$[b_{11}] = hc_1 A_{bs}$$

$$[b_{13}] = \alpha_{ba} A_{bs}$$

$$[b_{21}] = C_{pa} \rho A_{cr}$$

$$[b_{24}] = hc_{cd} A_{aw}$$

$$[b_{25}] = \rho A_{cr} u_{air}$$

The product surface temperature, interior chamber air temperature, and room air moisture or humid air can be expressed as the dynamic state variables $x1$, $x2$, and $x3$, respectively, in the state space form. The output of the system is regarded as the state variable, along with the drying temperature from the heater, forced air speed, solar irradiation rate, ambient temperature, and outside humidity expressed as $u1$, $u2$, $u3$, $u4$, and $u5$, respectively. The state space form of the nonlinear third-order differential plant system is represented using Eqs. 13.6, 13.7, and 13.8 as it has five inputs and three outputs.

$$C_{bs}\dot{x}_1 = -[a_{11}]x_1 + [a_{12}]x_2 + [b_{11}]u_1 + [b_{13}]u_3 \tag{13.9}$$

$$C_{ch}\dot{x}_2 = [a_{21}]x_1 + [a_{22}]x_2 + [b_{21}]u_1 u_2 - [b_{24}]u_4 \tag{13.10}$$

$$m_a \dot{x}_3 = [a_{33}]x_3 + [b_{25}]u_5 - \frac{m_{pr}X}{dt} \tag{13.11}$$

13.3 Control Design

Based on the operator's experience and knowledge, which are qualitatively characterized using the fuzzy rule inference system approach, fuzzy logic control design is made for the system model. The challenge of a highly nonlinear thermal dynamic model is dealt with by an relatively simple application of fuzzy logic control. For the FL controller design as stated in [6, 13], an assumption is made.

The temperature of the chamber air is controlled through fuzzy logic controller. The coupling effect is a problem for controlling moisture in the chamber which is, regulating the air temperature in the chamber to the ideal drying air temperature while lowering the coupling signal of the chamber moisture. However, an actuation signal was used to adjust the temperature of the chamber air in order to counteract the coupling effect of moisture. The forced conviction heated air speed is controlled by a variable voltage acting as an actuation signal. This signal's voltage level is changeable and ranges from 1 to 5 volts. Quantized level for input variable is described.

Table 13.2 Quantized level description for temperature control

Linguistic variable	Quantized level description
Error is the first input	Neg: If desired temperature is less than negative Optimum: If chamber temperature is optimum at desired temperature, then error is optimum Pos: If desired temperature is greater than chamber temperature is greater than chamber temperature then error is positive
Rate of error is the second input	N: If desired temperature is less than chamber temperature, then rate of error is negative Zero: If chamber temperature is optimum at desired temperature, then rate of error is zero P: If desired temperature is greater than chamber temperature then rate of error is positive

The input of chamber air temperature (CAT) is identified as error and rate of error. In Table 13.2 the linguistic variable and quantized level for the error and rate of error of input variable are presented.

13.3.1 Quantized Level for Input Variable Description

The input of chamber air temperature (CAT) is identified as error and rate of error. In Table 13.2. the linguistic variable and quantized level for the error and rate of error of input variable is presented.

The objective of the control design is to set a desired chamber temperature based on the characteristic of drying process of banana to be dried on the optimal temperature conditions. The membership function for each descriptor of the first and second input variable design is shown in Figs. 13.5 and 13.6, respectively.

13.3.2 Quantized Level for Output Variable Description

The quantized level for forced conviction of hot air speed and the drying air temperature DAT, as output variable, is described as shown in Table 13.3. The membership function for the output is shown in Fig. 13.7.

13.3.3 Rule Inference Fuzzy Logic System

The rule is extracted from the thermal model behavior, and the if-then rule is applied to meet the control objective. As the thermal model behavior is studied, the if-then rule is presented in Table 13.4.

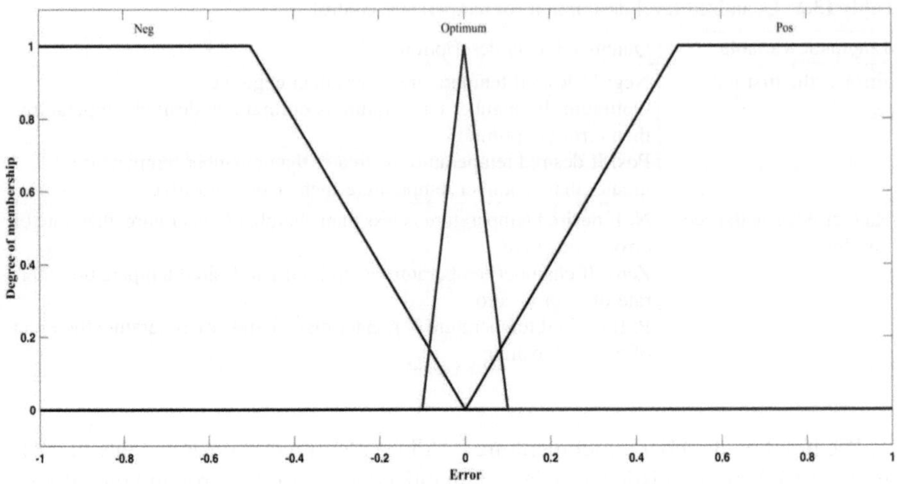

Fig. 13.5 Membership functions for first input variables

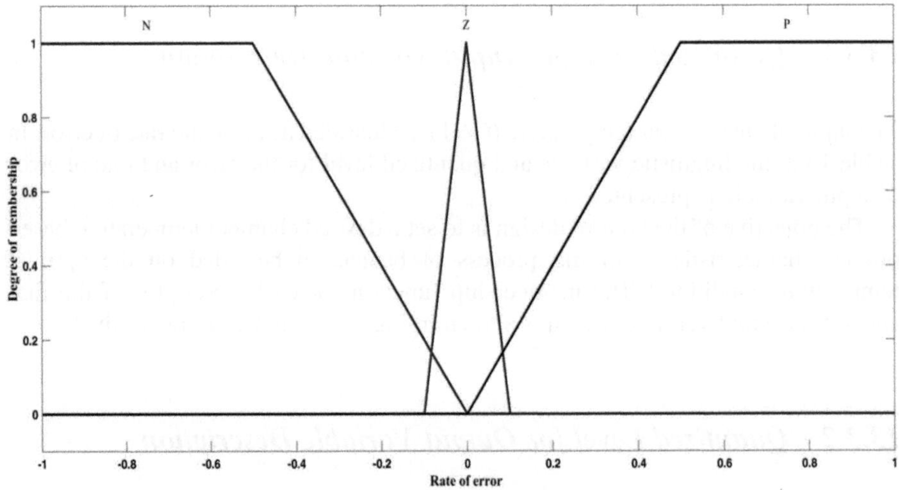

Fig. 13.6 Membership functions for second input variable

Then the if-then rule is performed to track the desire input as follows:

1. If (Error is Neg), then (DAT is VLD) (1).
2. If (Error is Pos), then (DAT is MMD) (1).
3. If (Error is optimum) and (Rate of error is N), then (DAT is LD) (1).
4. If (Error is optimum) and (Rate of error is Z), then (DAT is OD) (1).
5. If (Error is optimum) and (Rate of error is N), then (DAT is LD) (1).
6. If (Error is optimum) and (Rate of error is P), then (DAT is MD) (1).

Table 13.3 Quantized level for output description for temperature control

Linguistic variable	Quantized level description
Drying air temperature (DAT)	VLD: If the heater temperature is at very low drying temperature, then the DAT is negative large LD: If the heater temperature is at low drying temperature, then the DAT is low drying OD: If the heater temperature is at optimum drying temperature, then the DAT is at optimum drying MD: If the heater temperature is at high drying temperature, then the DAT is at high drying MMD: If the heater temperature is at maximum drying temperature, then the DAT is at maximum drying

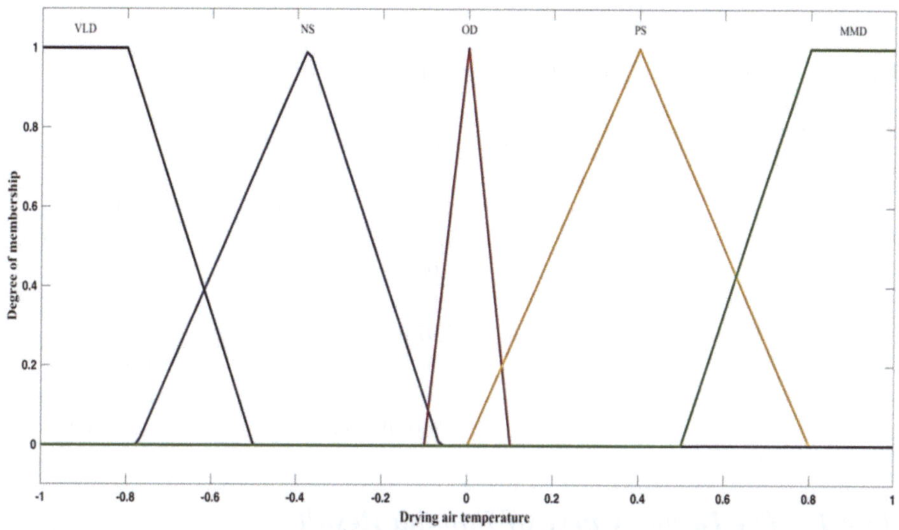

Fig. 13.7 Membership function for the output variable

Table 13.4 Rule inference

		Rate of error		
DAT		N	Z	P
Error	Neg	VLD	LD	MMD
	Optimum	LD	OD	MD
	Pos	MMD	MD	OD

13.4 Result and Discussion

The thermal model with parameter modification and control design was built using MATLAB Simulink, as illustrated in Fig. 13.8, and the simulation result was reviewed. Table 13.5 shows the thermal model parameter values. The performance of tracking and the robustness of fuzzy logic controller with and without disturbance signal effect is tested. The model of thermal is verified by the banana drying kinematics.

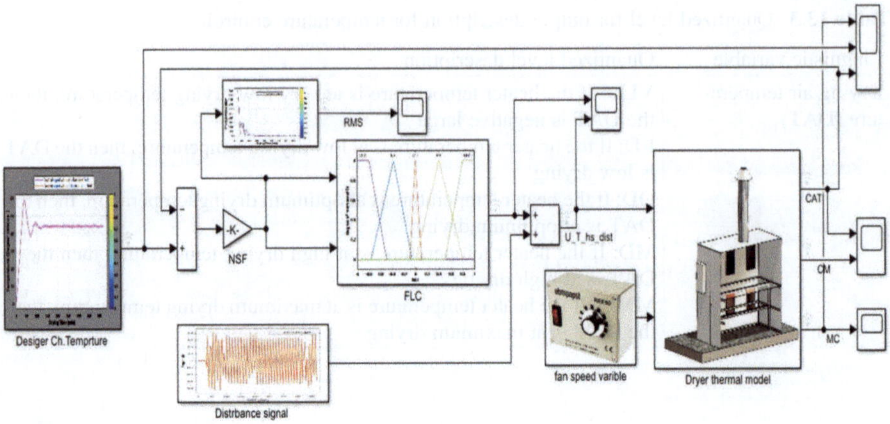

Fig. 13.8 Simulink realization of FLC for thermal model

Table 13.5 Thermal model parameter value

Parameter	Value	Parameter	Value
Abs	16 m^2	^{x}in	0.7 *wb*
^{A}wa	4 m^2	$^{\rho}ma$	1.16 kg/m^3
^{a}w	0.706	$^{\lambda}a$	0.0262 W/m °C
D	0.005 m	$^{\gamma}ma$	0.86 *dec*
L	0.01 m	$^{\alpha}pr$	0.57
^{d}wa	0.07 m	ε	0.9
C_d	0.6	*L*	1 m
Cp,ch	°	n	2×10^{-5} m/s^2
Cp,bs	1760 J/kg °C	$^{\lambda}a^{w}$	0.193 W/m °C

13.4.1 The Drying Curve of Banana Result

The optimal value of drying air temperature is determined and selected by the banana moisture content of drying curve. The initial moisture content of banana (0.7 WB) has been dried to the final (0.2 WB) moisture content of banana [1]. A number of simulations was conducted to fit the drying curve of banana. The mathematical model is verified by the drying curve of banana as it was discussed in the previous chapter.

A vital effect was observed on the variation moisture content (MC) by drying air temperature (DAT) and hot air speed (HAS) during a number of serious simulations as it is shown in Figs. 13.9 and 13.10, respectively. Comparatively, the effect of drying air temperature is higher than the hot air speed. The design parameter value used for moisture content at drying air temperature variation is explained in Table 13.6.

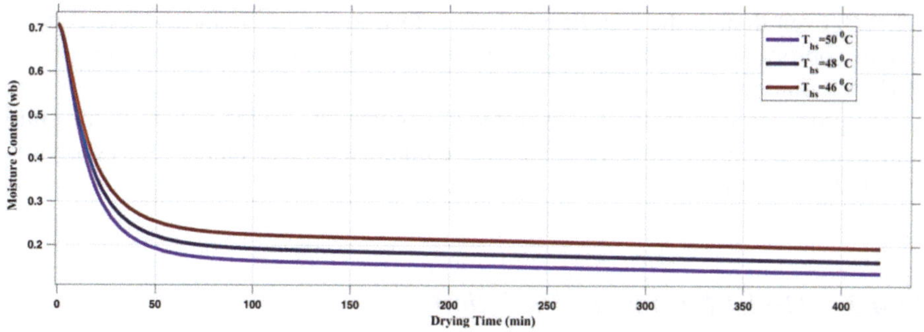

Fig. 13.9 Drying curve on the DAT variation

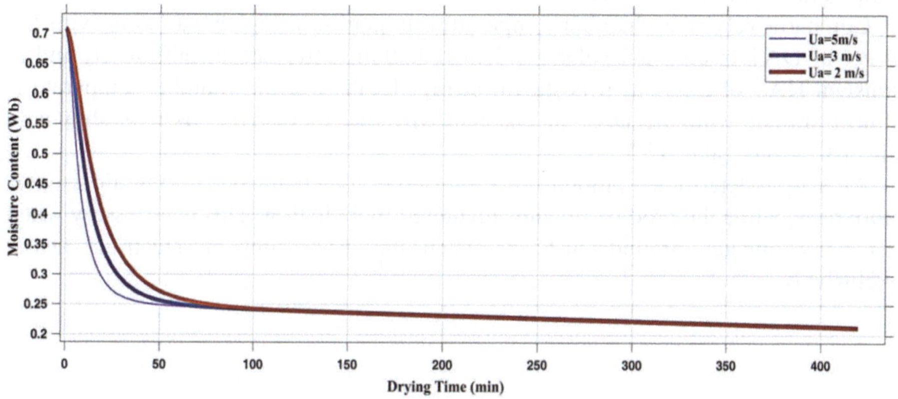

Fig. 13.10 Drying curve on the HAS variation

Table 13.6 Design parameter value for MC at DAT variation

HAS (°C)	DAT (°C)	MC (wb)
2	50	0.212
	48	0.201
	46	0.211

The response of moisture content simulation at final drying time is more appropriate to make decision on the selection of optimum drying temperature and hot air speed.

Then optimal value of drying air temperature (DAT) is determined and selected at 48 °C since it has similar value to the standard of final moisture content of banana as it is shown in Fig. 13.9. The appropriate designed value for the dryer system is good at the optimum range of DAT between 46 and 50 °C.

The effect of HAS on the variation of moisture content has not shown significant changes as it is shown in Fig. 13.10. The simulation is conducted to determine the optimum value of HAS as to fit the standard of final moisture content of banana of

0.2 WB, even though the significant effect is low on the variation moisture content and the appropriate designed value for the dryer system is good at the optimum range of HAS variation.

The proposed system dryer is expected to simulate the drying of banana moisture content and moisture ratio as reference and comparison based on [1, 10] as shown in Figs 13.9 and 13.10.

13.4.2 Model Parameter Variation

The simulation result at Fig. 13.11 shows the variation of DAT on the chamber temperature variation at constant of (HAS) and solar incident radiation rate (SIR).

Comparatively, the chamber temperature is getting a higher variation with the effect of DAT at constant solar incident radiation (SIR) rate of 500 W/m^2 and at constant HAS of 2 m/s as it is shown in Fig. 13.11. But the chamber temperate is almost similar with the DAT at constant SIR rate of 0 W/m^2 as it is shown in Fig. 13.12 and Table 13.7.

After a predetermined amount of drying time, the chamber's drying air temperature rises, while the product remains unaltered by radiation. This is necessary when it's rainy outside, around dusk, or when the dryer uses an electrical heater exclusively.

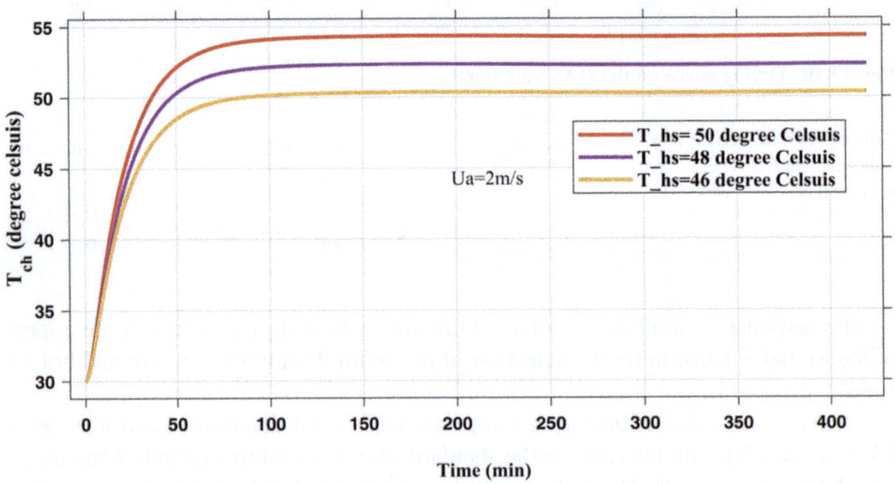

Fig. 13.11 Chamber temperature at SIR of 500 W/m^2

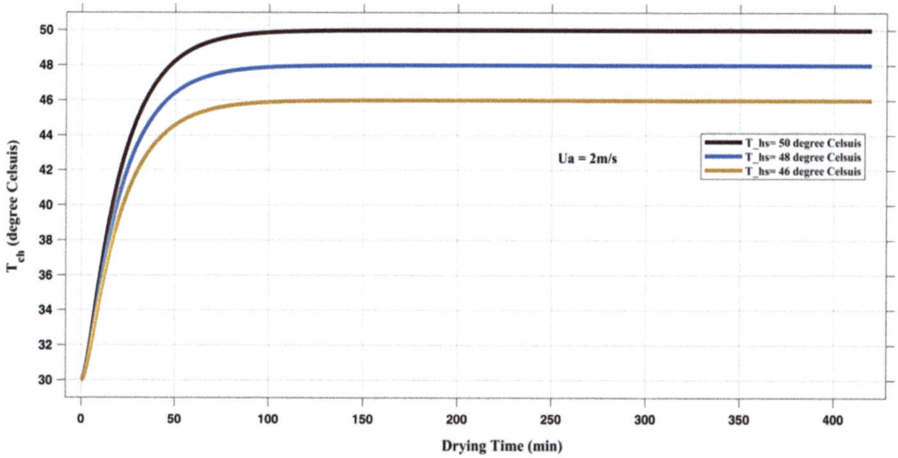

Fig. 13.12 Chamber temperature at SIR of 0 W/m^2

Table 13.7 The design parameter value at constant of HAS of 2 m/s

G (W/m^2)	T_{hs} (°C)	T_{ch} (°C)	T_{bs} (°C)
0	50	50.001	50.001
	48	48.002	48.002
	46	46.002	46.002
500	50	54.399	54.401
	48	52.399	52.400
	46	50.399	50.400

13.4.3 Chamber Moisture

The chamber's heat transmission is what causes humid air to reside there. If there is a lot of moisture in the chamber, it could hinder the drying process. Figures 13.13 and 13.14 illustrate the relationship between the amount of moisture in the chamber and changes in drying temperature and hot air speed, respectively. The amount of moisture increases as the drying air temperature rises, whereas the moisture in the chamber decreases when hot air speed (HAS) rises.

The optimum chamber moisture is achieved at a HAS of 5 m/s and a DAT of 50 ° C; however, modeling results show that DAT variation has less of an impact than HAS variation.

13.4.4 Result AI of FL Control

Setting the ideal range of chamber air temperature is the control goal. It is founded on DAT's control output. The outcome was made possible by the intelligent control design of the fuzzy logic (FL) controller covered in the preceding chapter.

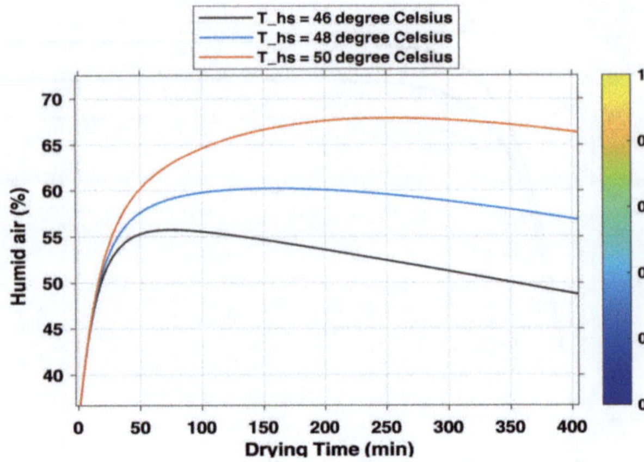

Fig. 13.13 Chamber moisture at the DAT variation

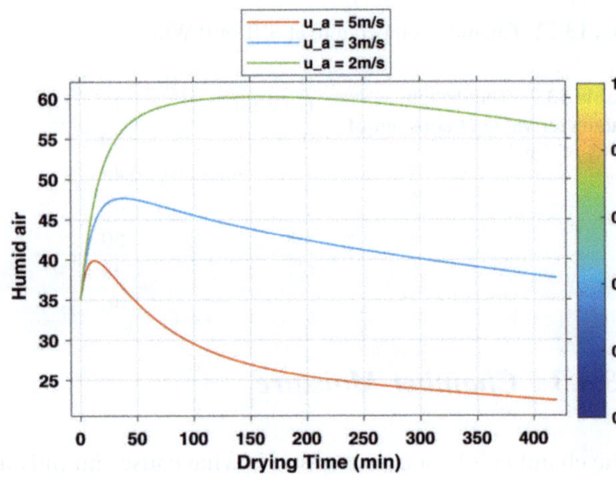

Fig. 13.14 Chamber moisture at the HAS variation

13.4.4.1 Disturbance Signal Source

It is regarded as the sinewave and plus disturbance source signal as depicted in Fig. 13.15, with the parameter value of the disturbance signal being represented in Table 13.8. The unmeasured input signal of environmental disturbances like solar radiation, ambient temperature, and humidity is represented by the disturbance signal.

13.4.4.2 Surface Viewer

As illustrated in Fig. 13.16, the surface viewer presents the three-dimensional curve that represents the mapping from chamber temperature and the difference between desire and chamber temperature to drying or heating temperature. The smooth surface curve is seen as a result of better control rules and no operating range overlap.

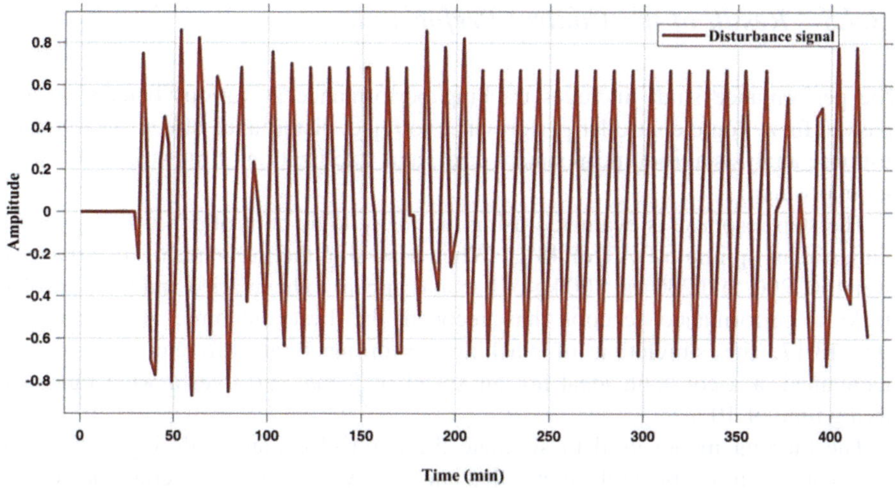

Fig. 13.15 Disturbance signal

Table 13.8 The design parameter value at constant of DAT of 50 °C

Signal	Parameter	Value
Sinewave	Frequency	2π/80(rad)
	Amplitude	25
	Bias	3
Transport time delay	Transport time delay 1	29.5
	Transport time delay 2	29

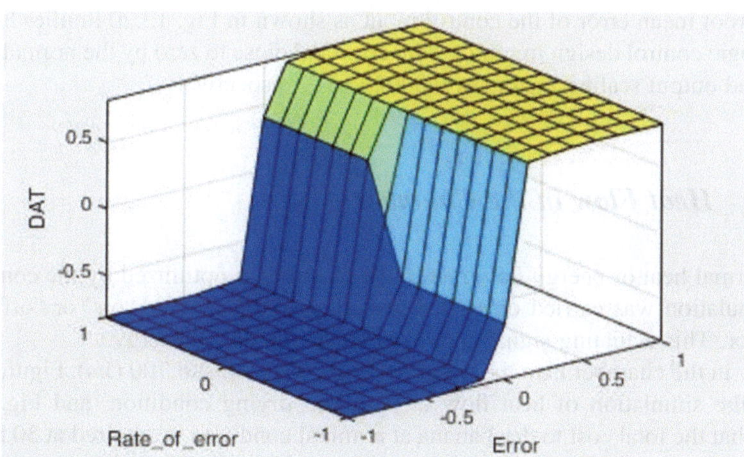

Fig. 13.16 Surface viewer

13.4.5 Result of the Control Performance

The performance study includes tracking performance at various banana drying masses, fuzzy logic controller performance to reject disturbance effect, steady-state response analysis at root mean square error, and stabilization of moisture content at optimal value.

The step response is evaluated using the fuzzy logic controller's tracking capabilities at an optimum DAT of 48°C. When assessing the influence of disturbances over 150, 200, and 250 kg, 100 kg of bananas appear to have dried and FL control's tracking performance reaction has recoded the best outcome. As shown in Fig. 13.17, the simulation of tracking performance at different masses was maintained at a constant, ideal hot air speed of 5 m/s and over a wide range of drying time of 10 h.

The thermal model used to simulate the controlled banana drying curve was remarkably comparable to the controlled banana drying curve at different masses, as shown in Fig. 13.18. This suggests that the FL control operation's robustness has been accomplished over a variety of different degrees of mass drying circumstances to stabilize the ideal DAT for the drying process of bananas. The stability of moisture content changes slightly throughout the initial drying phase, but after settling for (105–125) minutes, the banana's ideal moisture content reached a value of 0.2 wet matter. The thermal model was validated as the banana drying curve got closer to the optimal ultimate moisture content.

The performance of FL controller at Fig. 13.19 is how to reject the disturbance effect of wind, outside humidity, and ambient temperature fluctuation which affect the drying processes.

The root mean error of the control input as shown in Fig. 13.20 implies how the fuzzy logic control design minimizes the error and close to zero by the normalization input and output scaling factors of 0.0002 and 2, respectively.

13.4.6 Heat Flow in the Chamber

The thermal heat or energy flow rate in the chamber is optimized by the controller, and simulation was carried on by the actuating signal used as "on" or "off" relay contactor. This actuating signal is used to set on the desire energy.

Flow in the chamber is in the range between 300 (J/s) and 700 (J/s). Figure 13.21 shows the simulation of heat flow at optimum drying condition, and Fig. 13.22 shows that the total cost to dry banana at nominal condition is required at 30 birr per total drying of 7 h. The optimum drying condition is called nominal operational condition for plant at drying mass of 100 kg and at forced conviction HAS of 5 m/s.

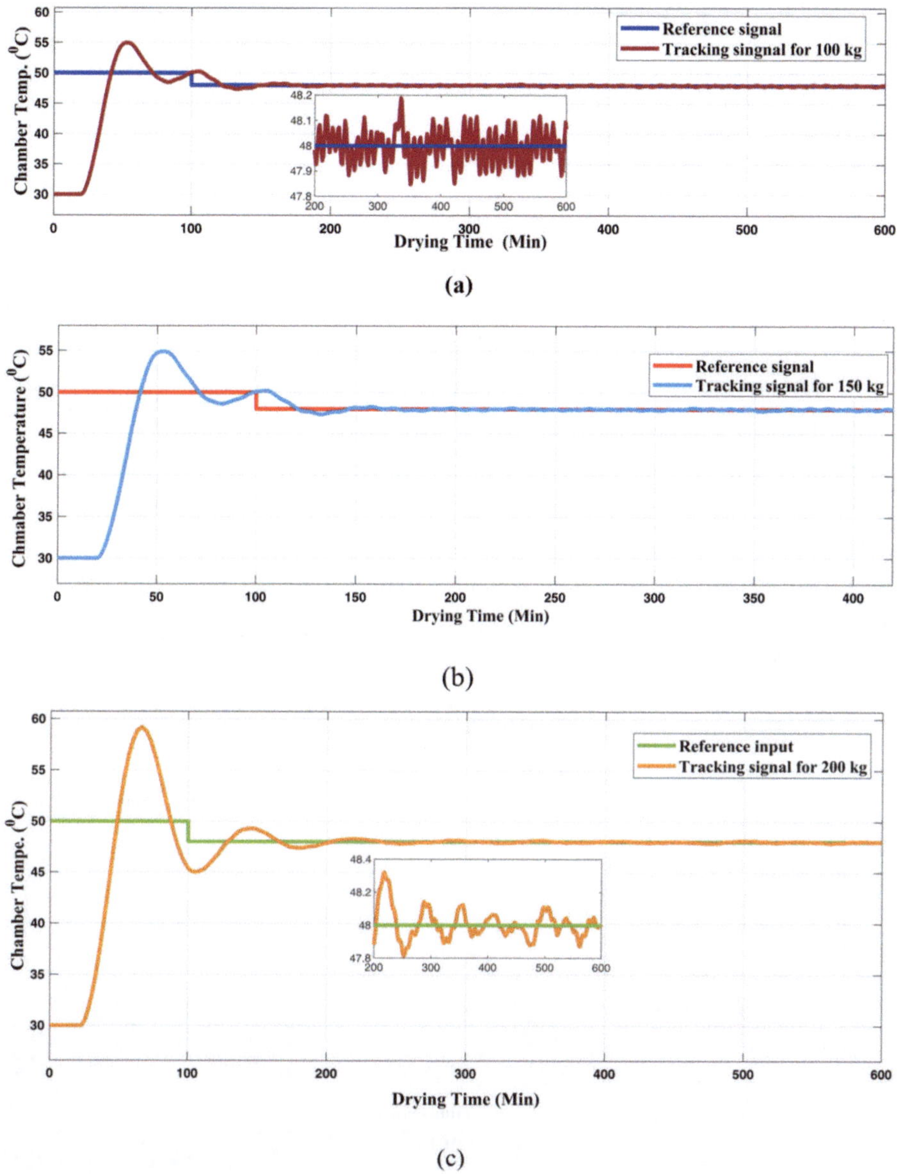

Fig. 13.17 Tracking performance of FLC

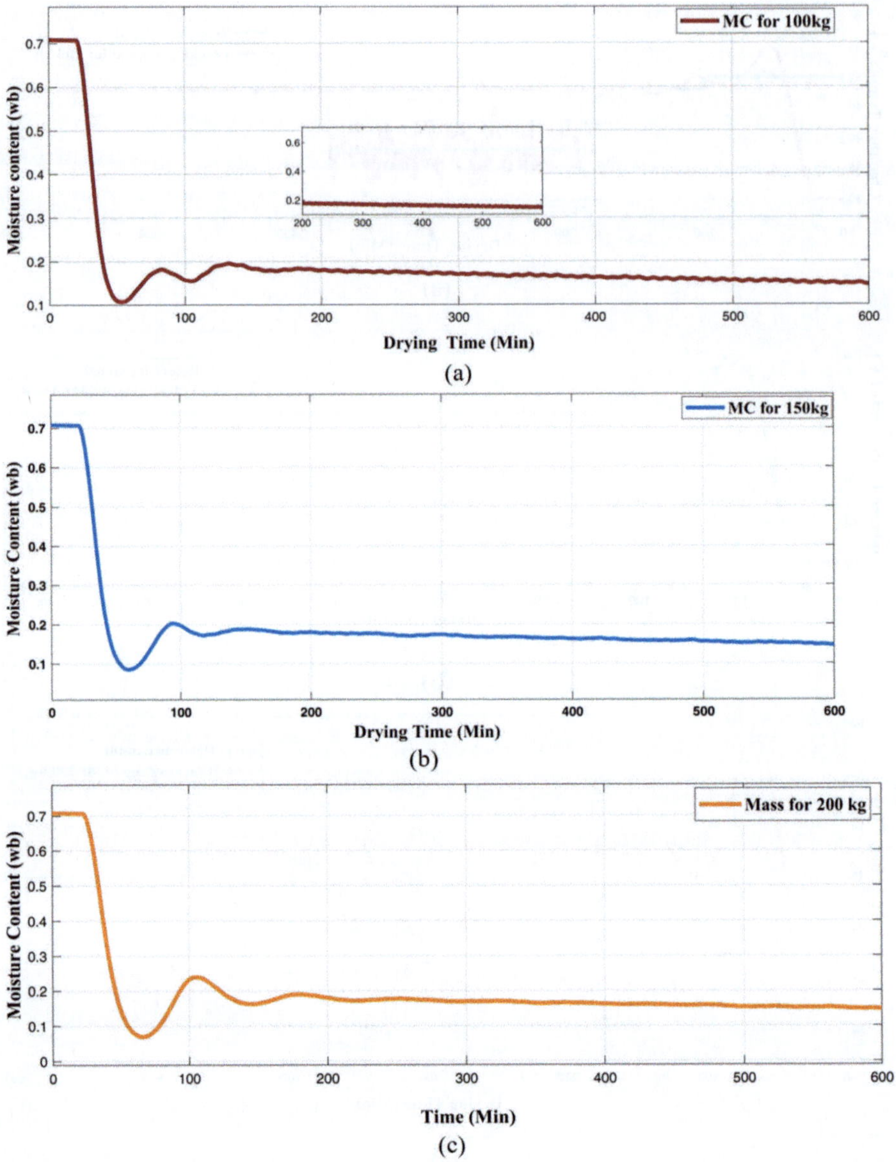

Fig. 13.18 The controlled drying curve

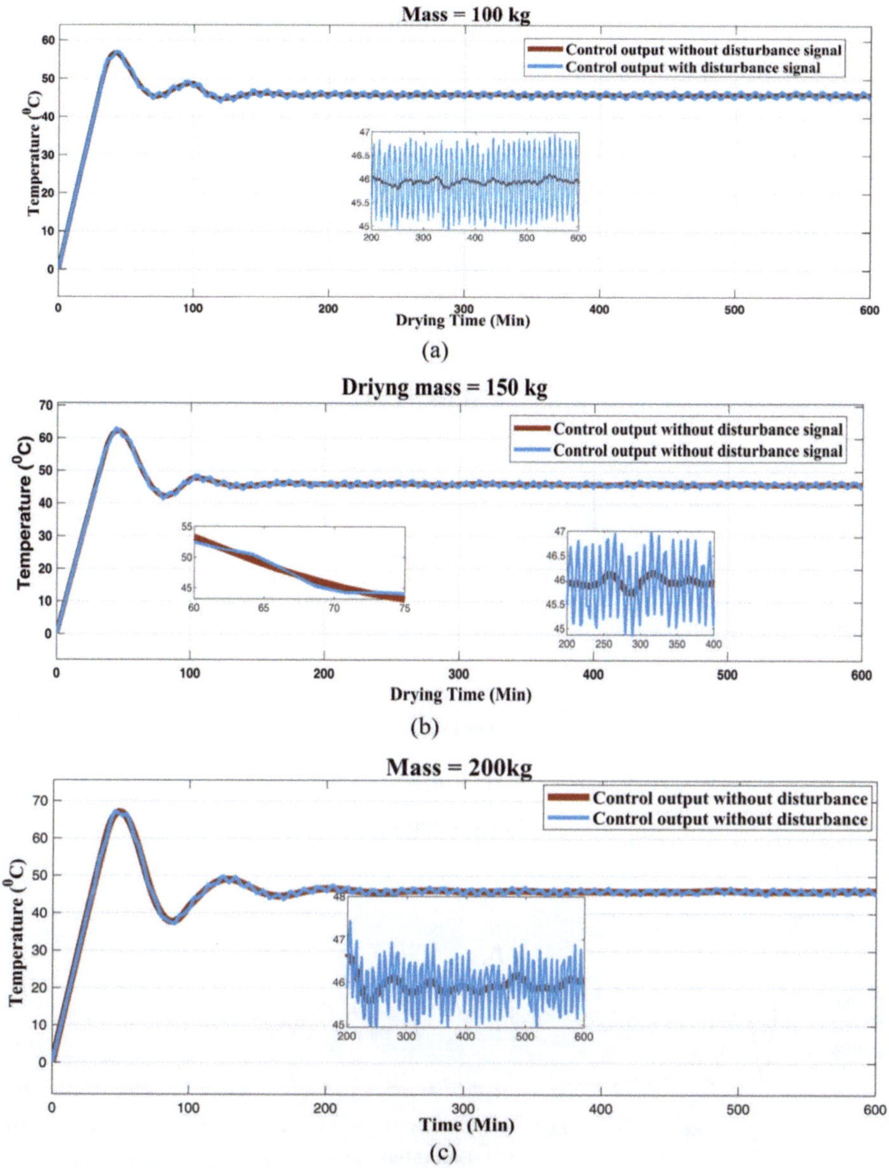

Fig. 13.19 Control output with and without disturbance

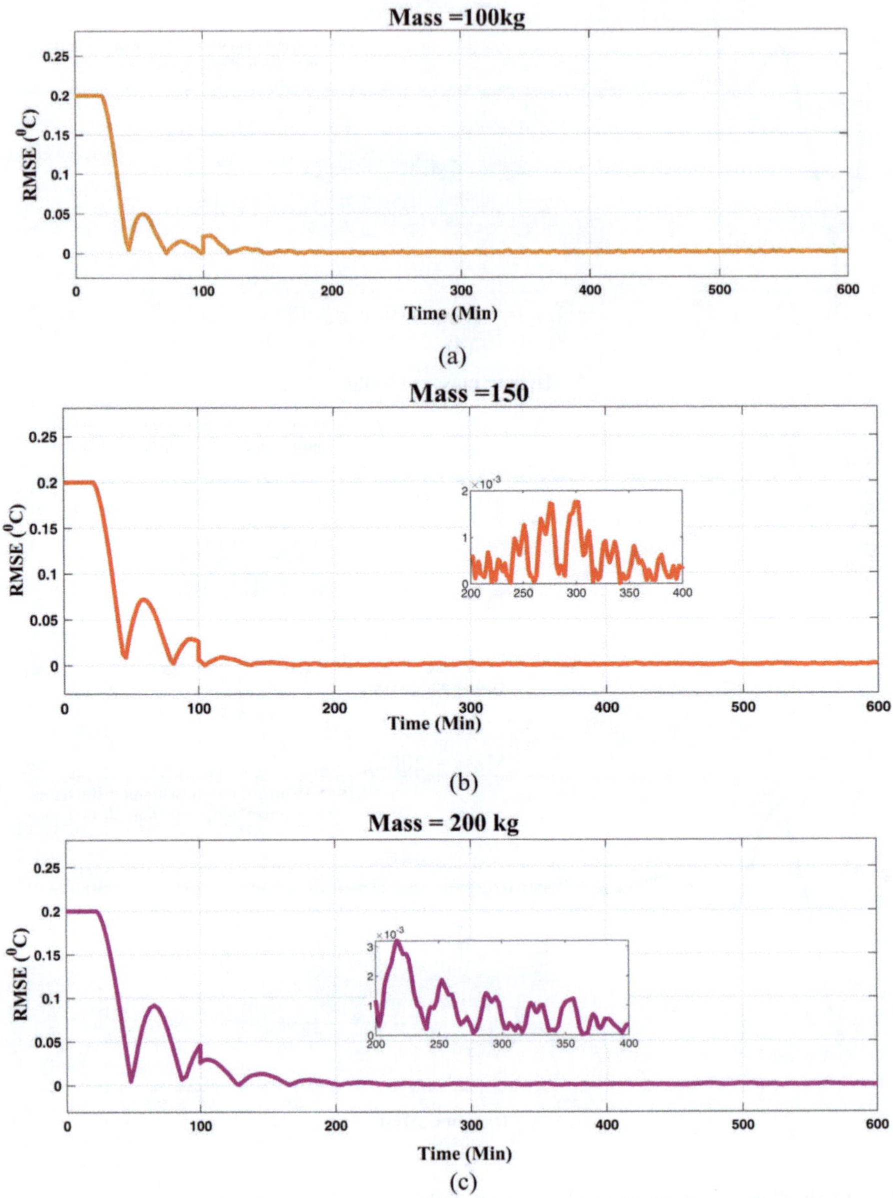

Fig. 13.20 Root mean square error

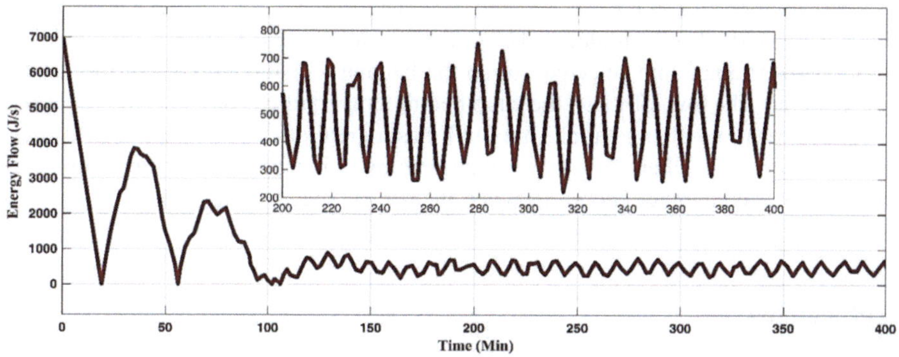

Fig. 13.21 Heat flow in the chamber

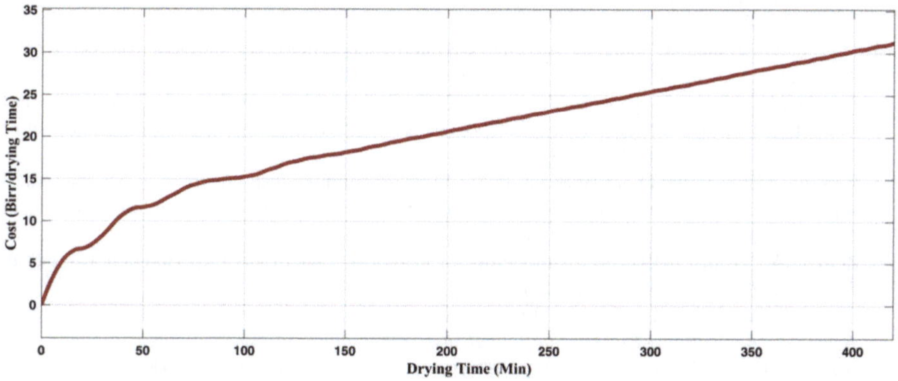

Fig. 13.22 Cost per drying time

13.5 Conclusion

The dynamic behavior of the thermal banana drying processes was described by nonlinear differential equations. The bond graph method was developed and applied as an appropriate tool for modeling techniques. This is due to the heat balance of chamber air and the mass balance of moisture air as state-controlled variables. The chamber and product temperature, as well as heat balance, were studied, and they varies when the DAT and HAS vary.

FLC was applied to control the chamber air temperature at the optimal value of drying air temperature while reducing the coupling signal of chamber moisture. The moisture is controlled at the desired chamber temperature optimum of 46–50 °C and at a higher voltage of 5 volts. The amount of moisture in the chamber is under good control at a range of 36–25%. The controller design tracks and responds to the desired temperature with the performance of disturbance rejection.

The variation of DAT at 46, 48, and 50°C and the variation of HAS at 2, 3, and 5 m/s were studied and simulated. The optimum value of DAT of 48 °C and HAS of 5 m/s is determined and selected due to the banana moisture content of the drying curve. In fact, the low effect on the chamber air temperature is due to HAS variation; it doesn't mean no change at all, but DAT has a great effect. The chamber temperature is increased when the drying temperature is increased, and the amount of moisture in the chamber decreases when the hot air speed is increased.

References

1. Lingayat, A., Chandramohan, V., & Raju, V. (2017). Design, development and performance of indirect type solar dryer for banana drying. *Energy Procedia, 109*, 409–416.
2. Al-Neama, M., & Farkas, I. (2016). Influencing of solar drying performance by chimney effect. *Hungarian Agricultural Engineering, 30*, 11–16.
3. Ma, H., W. Zhang, & Yang, S. X. (2015). Systems modeling and intelligent control of meat drying process. In *2015 10th system of systems engineering conference (SoSE)*. IEEE.
4. Oueslati, H., Mabrouk, S. B., & Mami, A. (2012). System design, mathematical modelling and simulation of process drying in a solar-gas convective tunnel dryer. *International Journal of Scientific & Engineering Research, 3*(5), 1–8.
5. Wakjira, M., Adugna, D., & Berecha, G. (2011). Determining slice thickness of banana (Musa spp.) for enclosed solar drying using solar cabinet dryer under Ethiopian condition. *American Journal of Food Technology, 6*(7), 568–580.
6. Zoukit, A., et al. (2019). Fuzzy modeling of a hybrid solar dryer: Experimental validation. *Journal of Energy Systems, 3*(1), 1–13.
7. Gatea, A. A. (2010). Design, construction and performance evaluation of solar maize dryer. *Journal of Agricultural Biotechnology and sustainable development.*
8. Bolaji, B. O., & Olalusi, A. P. (2008). Performance evaluation of a mixed-mode solar dryer. *AU Journal of Technology, 11*(4), 225–231.
9. Macedo, L. L., et al. (2020). Effect of drying air temperature on drying kinetics and physicochemical characteristics of dried banana. *Journal of Food Process Engineering, 43*(9), e13451.
10. Olawoye, B. T., Kadiri, O., & Babalola, T. R. (2017). Modelling of thin-layer drying characteristic of unripe Cardaba banana (Musa ABB) slices. *Cogent Food & Agriculture, 3*(1), 1290013.
11. Vafai, K., & Hadim, H. (2018). Overview of current computational studies of heat transfer in porous media and their applications-natural and mixed convection. *Advances in Numerical Heat Transfer, 2*, 331–369.
12. Savart, M., et al. (1987). Association of calpains 1 and 2 with protein kinase C activities. *FEBS Letters, 216*(1), 22–26.
13. Li, H.-X., & Gatland, H. (1995). A new methodology for designing a fuzzy logic controller. *IEEE Transactions on Systems, Man, and Cybernetics, 25*(3), 505–512.

Chapter 14
Standalone Solar PV System Design and Analysis Using PVsyst Software: A Case of Bahir Dar, Ethiopia

Hailemariam Mulugeta Wassie, Muluken Zegeye Getie, Bimrew Tamrat Admasu, Mulat Shitye Alem, Molla Asmare Alemu, Aschale Getnet Alemu, Atrsaw Jejaw Bezie, Tewodros Simeneh Ayalewu, and Assefa Beyene Ayalew

Nomenclature

AC	Alternative current
DC	Direct current
SAPV	Standalone PV system
BOS	Balance of system
PV	Photovoltaic
MPPT	Maximum point tracking
IEC	International Electrotechnical Commission

H. M. Wassie (✉)
Faculty of Mechanical and Industrial Engineering, Bahir Dar Institute of Technology, Bahir Dar University, Bahir Dar, Ethiopia

Bahir Dar Energy Center, Bahir Dar institute of Technology, Bahir Dar University, Bahir Dar, Ethiopia

M. Z. Getie · B. T. Admasu · M. S. Alem · A. G. Alemu · A. J. Bezie · T. S. Ayalewu
A. B. Ayalew
Faculty of Mechanical and Industrial Engineering, Bahir Dar Institute of Technology, Bahir Dar University, Bahir Dar, Ethiopia

M. A. Alemu
Bahir Dar Energy Center, Bahir Dar institute of Technology, Bahir Dar University, Bahir Dar, Ethiopia

K. Mequanint et al. (eds.), *Sustainable Development Research in Materials and Energy*, Green Energy and Technology,
https://doi.org/10.1007/978-3-031-69860-6_14

14.1 Introduction

The fundamental issue facing both developed and developing nations today is how to consistently produce enough electricity to fulfill the demand, which is steadily rising [1]. The stress on the electrical infrastructure is also increased by the expanding demand [2]. Due to population growth and rising electricity demand, conventional energy sources, such as fuel, oil, and gas, are exhausting in nature and release hazardous pollutants into the atmosphere that contribute to global warming [3]. Whatever the accessibility of the energy source, different factors determine the choice of energy use. The determinant factors for the choice of household energy have been investigated in Ethiopian urban areas [4]. The research finding disclosed that the education and type of residence are the first two most determinant factors.

Sectors in Ethiopia such as households and business institutions that are fed by the grid electricity are frequently experienced to power fluctuation and interruption. According to World Bank studies, power disruptions affected 80% of the firms studied in the country. These outages happened on average eight times per month and lasted 6 h each outage [5, 6]. Poor availability of electricity can reduce overall industrial production, raise the costs of production, and influencing their technological decisions away from energy-efficient equipment. The use of alternative methods by businesses, which lowers product quality, stops manufacturing, and delays order delivery, further harms a company's ability to compete. The placement of businesses and investment choices are also impacted by the availability of energy [7]. Moreover, the household sector which is based on the most traditional and inefficient sources of energy use the major share of energy in Ethiopia. Cooking consume nearly 89% of the total energy in Ethiopia [8]. It has been pointed out that in the country, injera baking took the larger share from the total cooking energy consumption.

Significant environmental issues have been expressed throughout the long-term operation of coal-based energy plants [9]. Solar PV technology is one of the most dependable, hygienic, and eco-friendly alternative to turn solar energy into power [1, 10]. For the reasons mentioned above, nations are making considerable efforts to utilize renewable energy [11]. Solar energy is the greatest and sole sources of almost all renewable sources, and the sun is the planet's primary energy source [12]. It can be used for many applications, including heating water and producing energy [13]. PV module development is continuing to expand at a 40% annual rate [13, 14]. One of the industries with the quickest growth over the past 10 years has been PV [10, 15].

In Ethiopia various researches have been conducted on solar energy conversion and utilization. An experimental box-type solar cooker has been studied, and the effects of different reflectors have been investigated [16]. The researcher disclosed that a practically worthy performance level has been achieved using such cookers with mirror glass reflectors to cook most Ethiopian cuisine. Moreover, a parabolic dish solar concentrator and rock-bed thermal storage material have been studied experimentally [17].

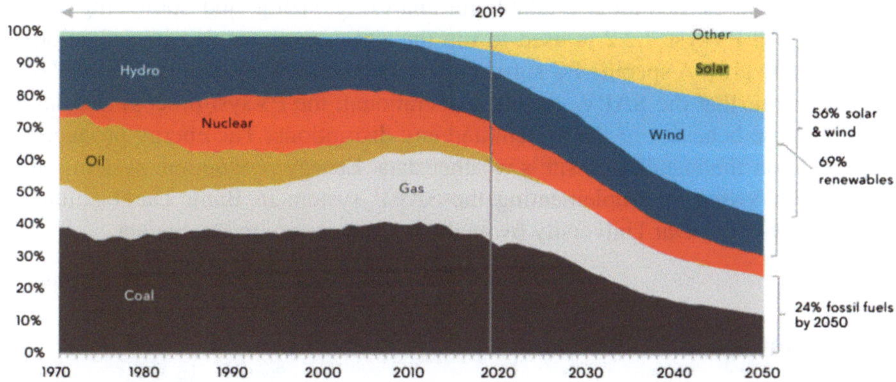

Fig. 14.1 Global electricity generation mix [20]

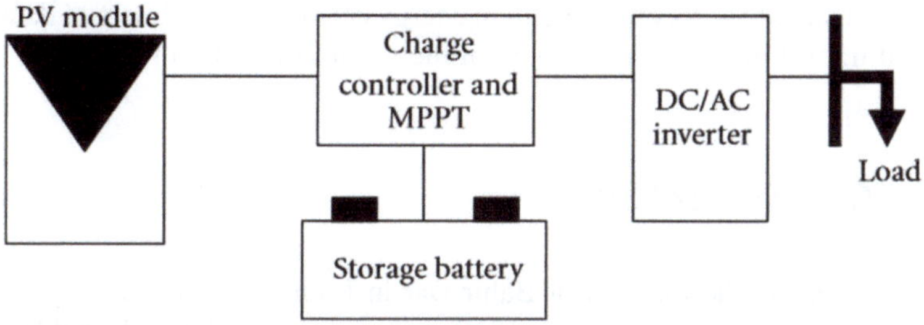

Fig. 14.2 Typical PV system components [23]

The annual change in energy production rate from 1970 to 2019 is shown in the Fig. 14.1. As seen in Fig. 14.1, the amount of energy produced by conventional sources is declining, while the amount produced by renewable sources, particularly solar energy, is quickly overtaking the conventional energy sources [18, 19]. Hydropower, solar, wind, geothermal, wave, tidal, and bioenergy are examples of renewable energy sources. As can be observed, the utilization of renewable energy is generally on the rise [11, 20].

A standalone PV (SAPV) system is an autonomous system that converts solar energy directly into electricity and distribute to the load independent to the grid [21]. Currently, the use of standalone PV power systems is being studied as an alternative energy source for isolated places [10]. Although the load is directly connected to the generated electricity, a storage device is still required for almost all of the cases. PV array and the balance of system (BOS) component are the two key components that make up a fully functional SAPV system [22]. The general PV system components are presented in as shown in Fig. 14.2. As the solar PV industry showing speedy growth, effective sizing of the PV system is crucial for designing and estimating yearly energy output. A trustworthy tool is needed by system designers and installers to design and forecast system performance, including total energy output under actual conditions.

There are numerous tools available for analyzing, sizing, and optimizing SAPV systems [24]. PVsyst 7.2.2 is used to run the simulations for this investigation as illustrated by [10]. A specific PC software program for a PV system is called PVsyst. For each hour that the SAPV system is operational, the PVsyst process simulation calculates the behavior of the system and any disruptions. The energy produced is dependent on the installation site's weather data. Energy production, system setup, and system losses for implementing the SAPV system in Bahir Dar Institute of Technology-Bahir Dar University by analyzing the energy resource data.

14.2 Database Collection

Finding the essential information that affect the system's performance is crucial for the development of a trustworthy SAPV system. The data for this investigation were geographical site of the location and total energy consumed. Both sets of data have a significant impact on the establishment of the economic system [13].

14.2.1 Geographical Site

Figure 14.3 depicts the location of Bahir Dar in Ethiopia's Amhara region, where this study has been based. The site is 1810 meters above sea level, 11.57°N latitude, and 37.36°E longitude. The site offers an excellent solar radiation potential. In addition, the fundamental element in producing a trustworthy PV system is the potential of the ambient temperature and wind resources.

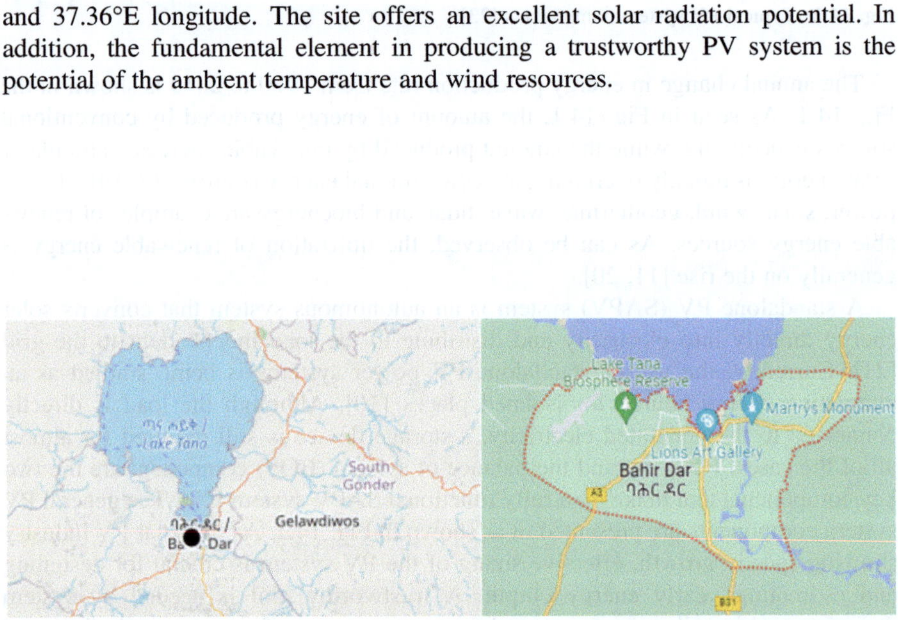

Fig. 14.3 Site selection

Table 14.1 Appliance energy consumption

Appliance used	Quantity	Power (W)	Use (hour/day)	Energy (Wh/day)
Lamps (LED or fluorescence)	950.0	120 W/lamp	17.0	1938000.0
TV/PC/mobile	185.0	150 W/app	24.0	666000.0
Domestic appliances	100.0	350 W/app	14.5	507500.0
Dish and cloth-washer	100.0	320 W/app	8	256000.0
Standby consumers			24.0	24.0
Total daily energy demand				3367524
Annual energy demand				40410288 Wh/year

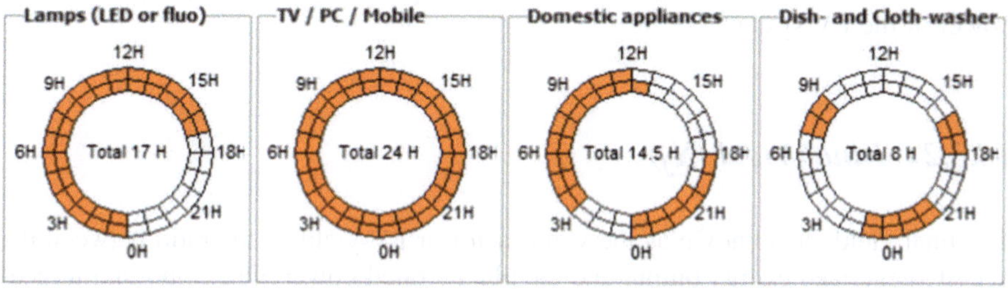

Fig. 14.4 Energy demand consumption

14.2.2 Energy Consumption Analysis

To prevent oversizing of the power system in a SAPV system, which could result in additional costs, the energy demand must be calculated. The energy demand consumption is calculated based on the daily power usage and the duration of operation, as shown in Table 14.1. Moreover, the duration of using different appliances is presented in Fig. 14.4.

14.3 Performance Indicators

The International Electrotechnical Commission (IEC) 61724 standard was released by the IEC to offer a comprehensive set of guidelines and limitations for the performance assessment of any solar PV system. This standard has been adopted worldwide. In 1998, the Bureau of Indian Standards also adopted this standard. Below is a list of some of the crucial specifications provided by the standard:

14.3.1 Array Yield (Y_a)

The energy generated by a PV array on a per-kW basis after it is installed is known as the array yield. It represents the plant's output of energy and is unaffected by the size of the plant. Also, it represents the relationship between a PV plant's rated power and its production of DC-generated energy. Array yield can be formulated in Eq. (14.1).

$$Y_a = \frac{E_{DC}}{P_{PVrated}} \tag{14.1}$$

where Y_a is the array field, E_{DC} is the solar PV output energy, and $P_{PVrated}$ is the rated power of the PV system.

14.3.2 Final Yield (Y_f)

The final yield, also known as the yield factor or PV yield, is the ratio between the net AC-generated energy output and the plant's rated power. One important indicator of plant operating parameters, such as the tilt angle of the PV array and mounting structure, is the yield factor. When the value of Y_f is close to one, adequate technology has been applied. The final yield can be expressed as in Eq. (14.2).

$$Y_f = \frac{E_{AC}}{P_{PVrated}} \tag{14.2}$$

er, energy needed by thwhere E_{AC} is the amount of AC energy that is readily available to the grid.

14.3.3 Reference Yield (Y_r)

Reference yield is calculated as the ratio between reference radiation on the Earth's surface, Ht (kWh/m^2/day), and the total global radiation (Gi.ref. (kWh/m^2)) accessible at the sea's surface. Since incident solar radiation on the Earth's surface differs depending on where you are in the world, it is location-dependent. One way to express reference yield is expressed in Eq. (14.3).

$$Y_r = \frac{H_t}{G_{i,ref}} \tag{14.3}$$

14.3.4 Performance Ratio (PR)

The ratio of the reference yield to the final yield is termed the performance ratio (PR) [18, 25]. The PV plant's overall loss is displayed by the performance ratio. The performance ratio under ideal circumstances is 1 (100%). However, in reality, the PR is less than one. Given that the performance ratio is independent of the plant's rating, the PR is the most important parameter for every plant. A small PV plant, for instance, can operate more effectively and demonstrate a greater performance ratio than a big PV plant, which can operate inefficiently and produce a lower performance ratio. The performance ratio can be formulated using Eq. (14.4).

$$\mathrm{PR} = \frac{Y_\mathrm{f}}{Y_\mathrm{r}} \tag{14.4}$$

14.3.5 Parameter Selection and System Configuration

To fulfill the system setup, every component's selection parameter is crucial. All components were chosen based on the load demand's, energy requirements, and the possibilities for meteorological data at the installation site.

14.4 Solar Horizon and Sun Height Paths

The SAPV solar power plant is designed for Bahir Dar, Ethiopia, at longitude 37.36° E, latitude 11.57°N, and elevation 1810 m. Figure 14.5 depicts the movement and trajectory of the sun's path at the Bahir Dar site throughout the year, subdivided into angles based on tilt and azimuth values. The horizon part also shows how much solar radiation there is each day of the year. In addition, Fig. 14.6 illustrates the amount of irradiance available within each month in the case study area.

14.4.1 Tilt Angle

The system must provide its greatest output; thus, the tilt angle must be just right as illustrated by [18]. The angle between the collector plane and the south/north is known as the azimuth angle. To receive the maximum solar energy, the PV array's tilt angle is maintained at a value equal to the associated location's latitude [26]. A manual seasonal tilt mechanism was developed by many researchers in order to increase solar radiation absorption and power output. However, people might not

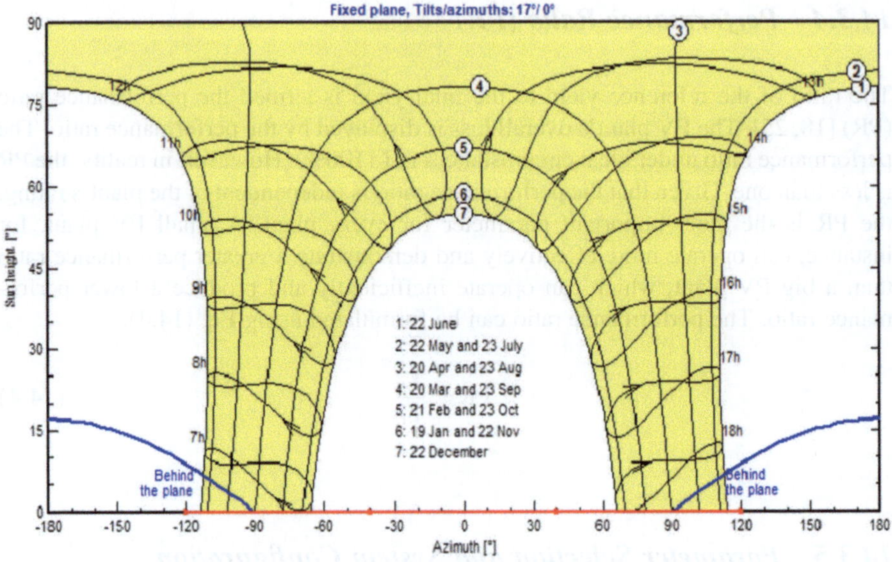

Fig. 14.5 Sun paths height/azimuth plot using PVsyst an Bahir Dar, Ethiopia

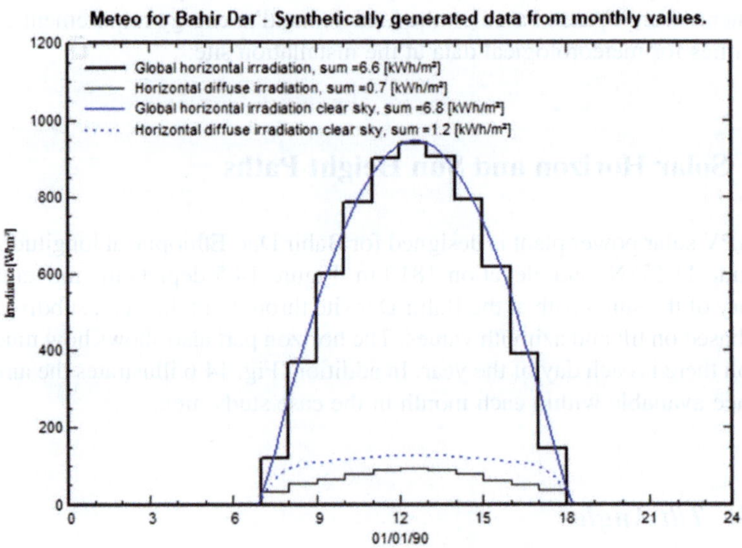

Fig. 14.6 Irradiance data

find manual adjustment practical, so in this study the fixed tilt angle was chosen using PVsyst software using the criteria of 0% loss relative to the ideal and with the highest value of global solar insolation on the collector (see Fig. 14.7).

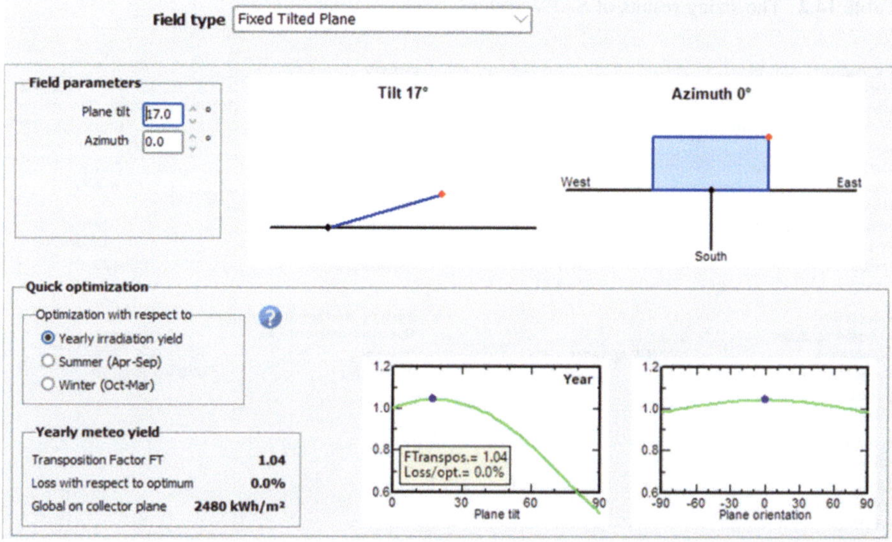

Fig. 14.7 SAPV optimum tilt angle

14.5 Simulation Result

The SAPV system simulation results are examined in this section by showing several simulation outcomes, including system output power, system loss, and performance ratio (PR). Inverter rating, series-parallel PV module configuration, geographical information, etc. were a few of the elements of the actual PV plant that were incorporated into the PVsyst software. To study the overall electrical performance of the installed SAPV plant, some of the most significant parameters, like reference yield, final yield, overall performance ratio, total electricity generation, etc., were investigated using PVsyst software. The standalone PV system, which was simulated using the PVsyst software, comprises 3078 solar panels with a power of 250 Wp, a surface area of 5119 m^2. Table 14.2 displays the simulation settings and characteristics for each system part. Moreover, various consequences drawn from the overall performance evaluation are listed in this section.

14.5.1 Sun Path

In assessing a photovoltaic PV plant's overall performance, shading is an important factor. The SAPV used in this investigation has no shading effect. Figure 14.5 illustrates the plant's solar path graph. The sun deviates more during the summer than it does during the winter. The sun's path changes during the course of the year, as seen in Fig. 14.8. The sun's azimuth fluctuates from roughly −60 to +60°; its

Table 14.2 The sizing results of SAPV system

PV module		**Battery**	
Manufacturer	Generic	Manufacturer	Generic
Model	60P-250	Model	Battery module Li-Ion, 26V 180 Ah
(Original PVsyst database)		Technology	Lithium-ion, LFP
Unit Nom. Power	250 Wp	Nb. of units	382 in parallel x 9 in series
Number of PV modules	3078 units	Discharging min. SOC	10.0 %
Nominal (STC)	770 kWp	Stored energy	13915.6 kWh
Modules	1539 Strings x 2 In series	**Battery Pack Characteristics**	
At operating cond. (50°C)		Voltage	230 V
Pmpp	686 kWp	Nominal Capacity	68760 Ah (C10)
U mpp	54 V	Temperature	Fixed 20 °C
I mpp	12621 A		
Controller		**Battery Management control**	
Universal controller		Threshold commands as	SOC calculation
Technology	MPPT converter	Charging	SOC = 0.96 / 0.80
Temp coeff.	-5.0 mV/°C/Elem.	Discharging	SOC = 0.10 / 0.35
Converter			
Maxi and EURO efficiencies	97.0 / 95.0 %		
Total PV power			
Nominal (STC)	770 kWp		
Total	3078 modules		
Module area	5119 m²		
Cell area	4494 m²		

Thermal Loss factor		**DC wiring losses**		**Serie Diode Loss**	
Module temperature according to irradiance		Global array res.	0.074 mΩ	Voltage drop	0.7 V
Uc (const)	20.0 W/m²K	Loss Fraction	1.5 % at STC	Loss Fraction	1.1 % at STC
Uv (wind)	0.0 W/m²K/m/s				
Module Quality Loss		**Module mismatch losses**		**Strings Mismatch loss**	
Loss Fraction	1.5 %	Loss Fraction	0.6 % at MPP	Loss Fraction	0.1 %
IAM loss factor					
ASHRAE Param.: IAM = 1 - bo (1/cosi -1)					
bo Param.	0.05				

height is shallow (approximately 54°); and its duration is also relatively short on the peak winter days, such as December 22. But as summer approaches, the sun's azimuth is increasing, along with its height and the amount of time it shines. In the middle of the summer, on a day like June 22, the sun's azimuth ranges from −120 to +120 degrees, its height is around 80°, and its length is very long. The blue line in the graph also shows that the plant faces extremely light shading in the morning and evening throughout the summer session. Because of this, we can conclude that the plant does not experience year-round shadowing.

From the PVsyst simulation result, the optimum fixed tilt and azimuth angle for Bahir Dar, Ethiopia, is 17° and 0°, respectively, having the maximum value of global solar radiation on the collector, which is 2480 kw/m^2, and the minimum value of loss with respect to optimum as shown in Fig. 14.8.

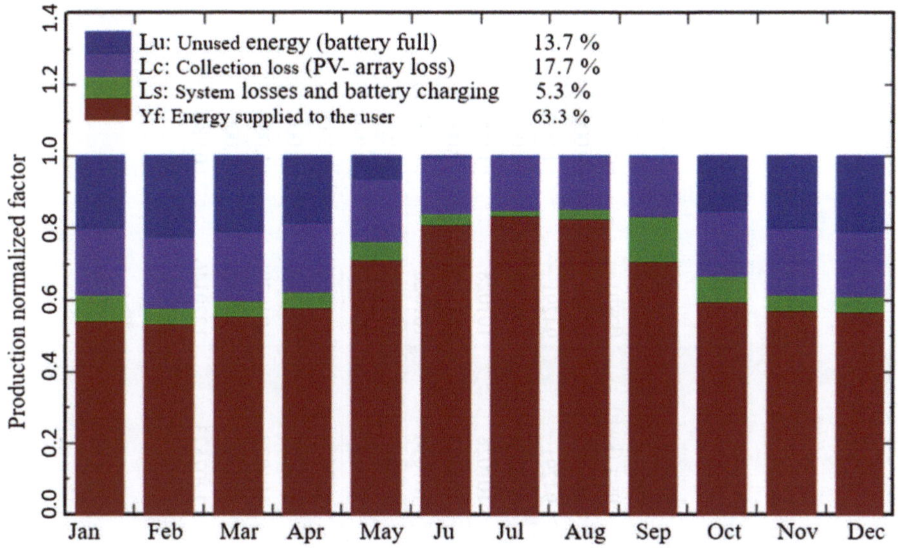

Fig. 14.8 Normalized production and loss factor over the year

14.5.2 Balance Evaluation

Table 14.3 illustrates balances and their primary impacts using data such as ambient average temperature, global horizontal radiation on a horizontal plane, global radiation on a collector plane, and effective global photovoltaic radiation taking losses into account (soiling and shading). For the proposed site location, the global horizontal radiation on the horizontal plane is 2381 kwh/m^2, and the effective global radiation with optical losses is 2417.9 kWh/m^2. The average available solar radiation is 1517116 kwh, whereas the average unused energy is 261980 kwh. Furthermore, energy missed, energy supplied to the user, energy needed by the user (load), and solar fraction are 21574, 1207573, 1229146 kwh, and 0.982, respectively.

Losses that could happen in the system while solar energy is converted from incident to electric energy can be calculated by deducting the total loss from the incident energy on the panel [27]. Users can then compute the amount of energy converted into electricity by the system. These losses are a possible cause of the PV module's performance dropping off [25]. Figure 14.8 displays the normalized loss and production factors. The maximum array losses were recorded from January to March as a result of those months' high temperatures and noticeably weak wind speed. The graph between normalized power production and loss factor, which is yield on an annual basis, is shown in Fig. 14.8. Here, the amount of energy delivered to the user is 63.3%, collection loss (losses on the PV array) is 17.7%, system losses and battery charging are 5.3%, and the amount of unused energy (energy left over when the battery is full) is 13.7%.

Table 14.3 Results of simulation in PVsyst software

Month	Horizontal global irradiation (kwh/m^2)	Effective global irradiation (kwh/m^2)	Available solar energy (kwh)	Unused energy (battery full) (kwh)	Missing energy (kwh)	Energy supplied to the user (kwh)	Energy needs of the user/load (kwh)	Solar fraction (energy used/ energy load)
January	210.8	245.4	152297	38673	0	104393	104393	1.000
February	204.9	224.6	137804	40057	0	94291	94291	1.000
March	238.5	240.2	147654	39408	0	104393	104393	1.000
April	238.0	221.8	137014	32304	0	101028	101028	1.000
May	208.8	185.0	117293	9588	0	104393	104393	1.000
June	180.5	156.8	101031	414	0	101026	101026	1.000
July	166.5	146.9	95061	12	6900	97494	104393	0.934
August	156.9	144.6	938.39	14	9375	95018	104393	0.910
September	175.4	171.0	108151	9	5299	95757	101026	0.948
October	209.6	222.8	139409	26920	0	104393	104393	1.000
November	197.9	225.8	140841	38057	0	101028	101028	1.000
December	197.2	233.6	146722	38524	0	104393	104393	1.000
Year	2381.0	2417.9	1517116	261980	21574	1207573	1229146	0.982

14.5.3 Computation of Losses

Figure 14.9 illustrates the various losses that were calculated from the simulation analysis. These performance indicators, which are outlined in the IEC 61724 standard, are widely used to monitor the efficiency of photovoltaic plants. The array capture loss of the plant, as seen in Fig. 14.9, is 1.2 kwh/kw$_p$/day. The PV plant's system loss is 0.36 kwh/kw$_p$/day, compared to the 4.3 kwh/kw$_p$/day of useful electricity produced and the 0.93 kwh/kw$_p$/day of unused energy.

It can also be observed that the SAPV generates the most electricity in January, February, and March, but losses are also very high. In the rainy season, say June, July, and August, the SAPV generates less electricity, but losses are slightly lower.

14.5.4 Performance Ratio

According to Fig. 14.10, the average performance ratio is 0.633; the highest performance was 82.5% in July, and the lowest performance was 52.5% in February. The graph in Fig. 14.10 shows the annual solar fraction as 0.982, which is calculated by dividing the energy produced by solar PV modules by the energy supplied to the load of ASPV.

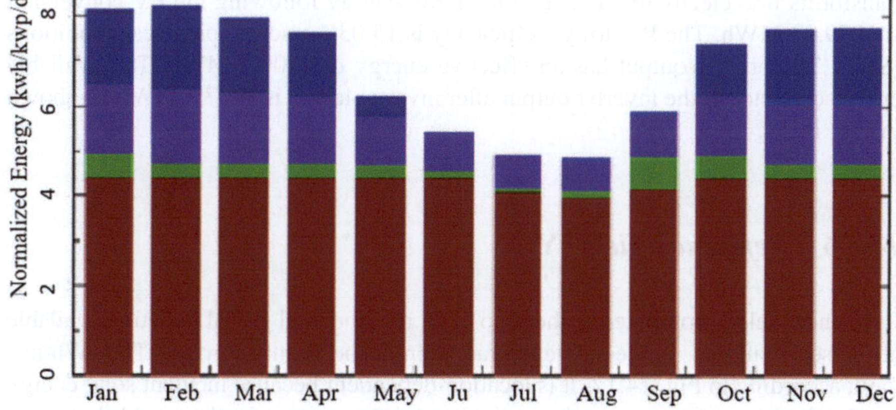

Fig. 14.9 Various losses per installed kWp

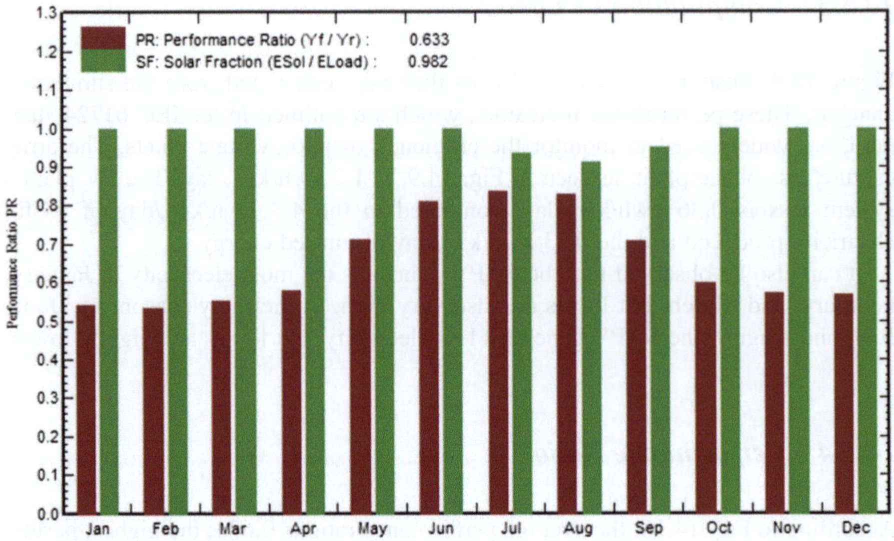

Fig. 14.10 Performance ratio PR and solar fraction

14.5.5 Loss Diagram

The loss diagram illustrates the losses of the solar PV system step by step. In solar PV modules and arrays, a number of losses can occur, including those caused by ohmic wiring, module mismatch, module quality, converter loss during operation, converter loss due to threshold power, battery efficiency loss, charge/discharge current efficiency loss, inverter threshold power loss, etc. as investigated by [28]. The global horizontal irradiance is 2381 kWh/m^2. The collector plane has an effective irradiation of 2418 kWh/m^2. Solar energy strikes the solar panels and transforms into electricity. The nominal array energy following the PV conversion is 1859.92 MWh. The PV array's efficiency is 15.03% under typical test conditions (STC). The array's output has an effective energy of 1309.3 MWh. The available energy collected at the inverter output after inverter losses is 1207.6 MWh, as shown in Fig. 14.11.

14.5.6 Reference Yield (Y_r)

Reference yield is computed as the ratio of Gi.ref., the total global radiation available at the sea's surface, to the reference radiation at the Earth's surface, H_t (kWh/m^2/day), according to Fig. 14.12. It is location-dependent because incident solar energy on the Earth's surface varies depending on where you are in the world. From the result of PVsyst, the average reference incidence energy is 6.795 kwh/m^2/day.

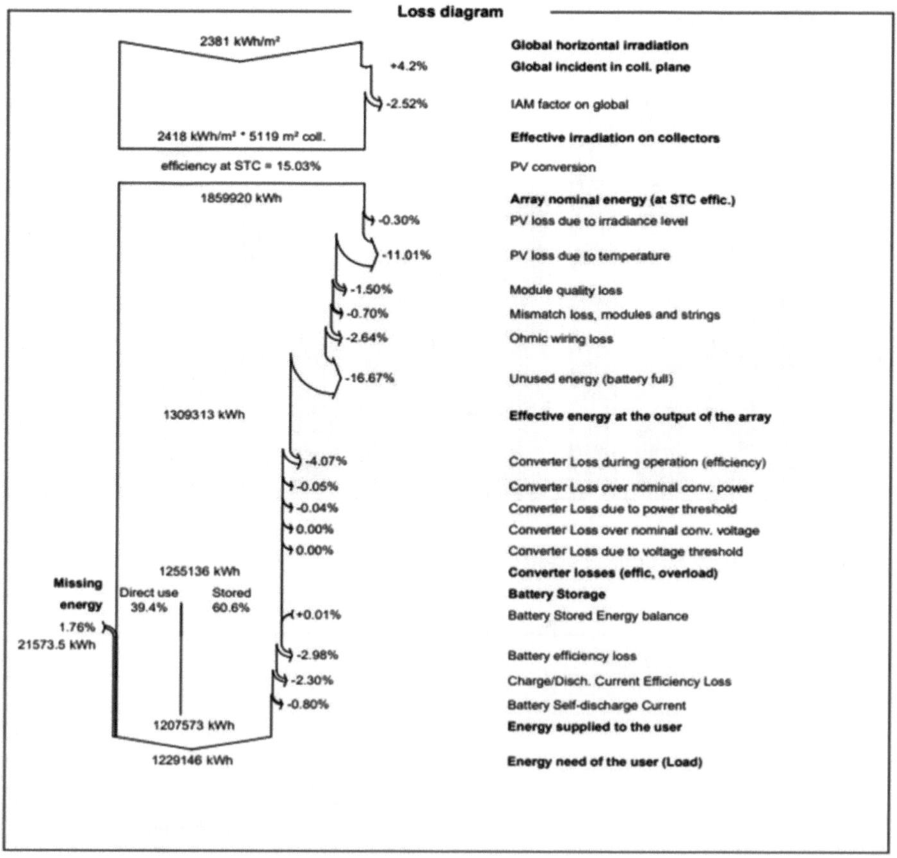

Fig. 14.11 Loss diagram over the entire year

14.5.7 Input/Output Diagram

The effective energy generated by the photoelectric system (kWh/day) as a function of the annual incident radiation (kWh/m^2/day) is shown in the diagram in Figs. 14.13 and 14.14. As more sunlight is incident on the photovoltaic panels' active surface globally, more energy is generated by the PV array. As a result, the system is sized correctly since there is a high correlation between the radiation output and energy output, with little dispersion. The small curve in Fig. 14.13 here represents the temperature effect. In this diagram, some of the spots (days) are off-center at high irradiances, which is a sign of an overload situation.

Figure 14.15 shows the incident irradiation distribution of the global incident collection plane generated annually.

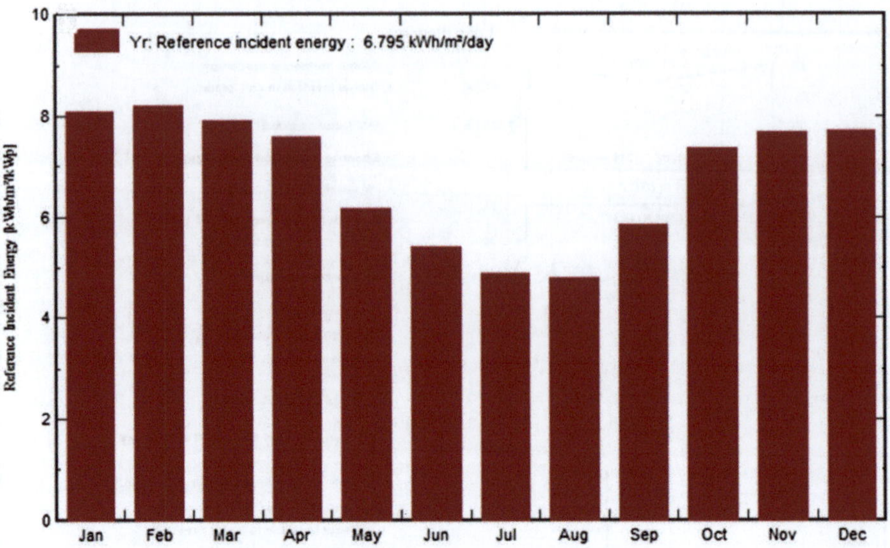

Fig. 14.12 Reference incidence Energy in collector plane

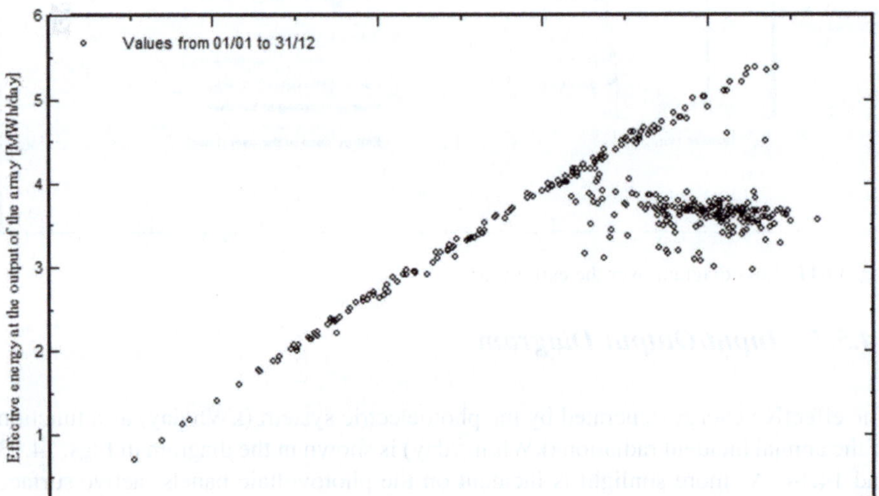

Fig. 14.13 Input/output diagram

The arrays' temperature characteristics against effective irradiance are presented in Fig. 14.15. It can be seen from Fig. 14.16 that the array's temperature ranged from 5 °C during the summer period to as high as about 75 °C during the winter period.

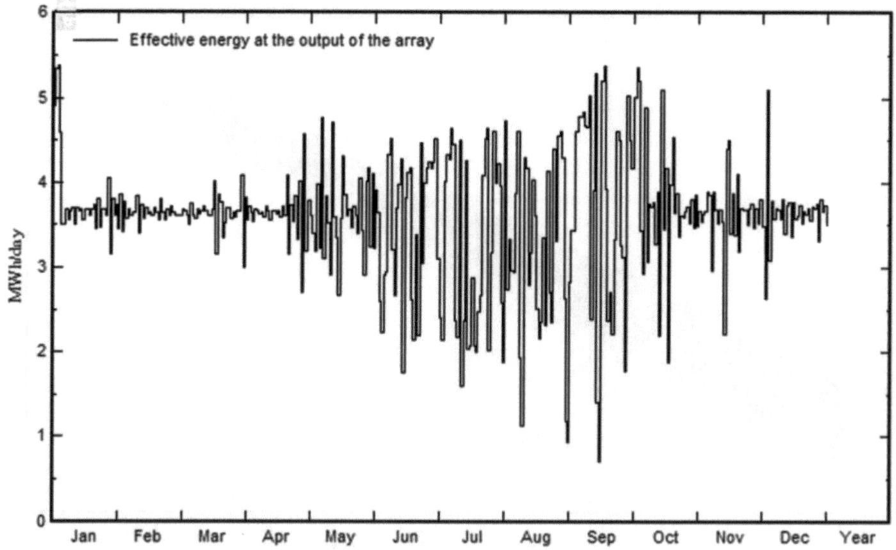

Fig. 14.14 Daily array output energy

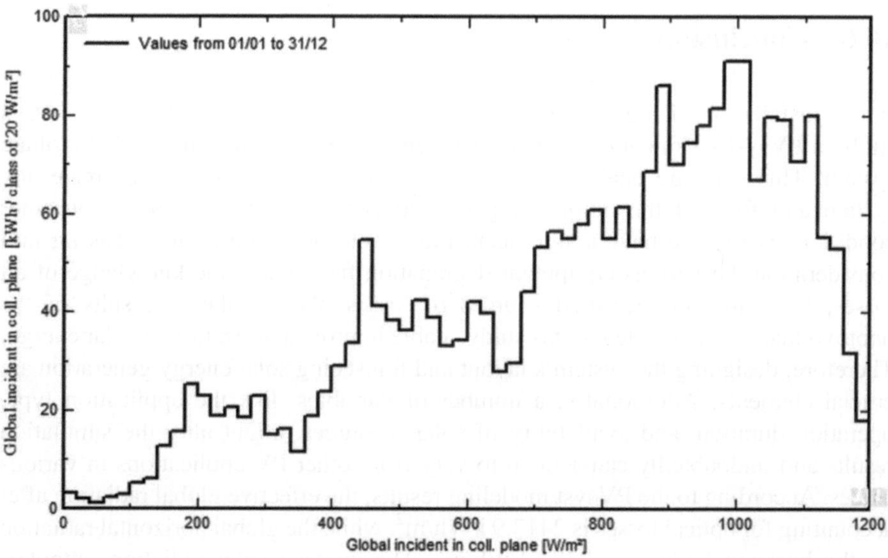

Fig. 14.15 Incidence irradiation distribution

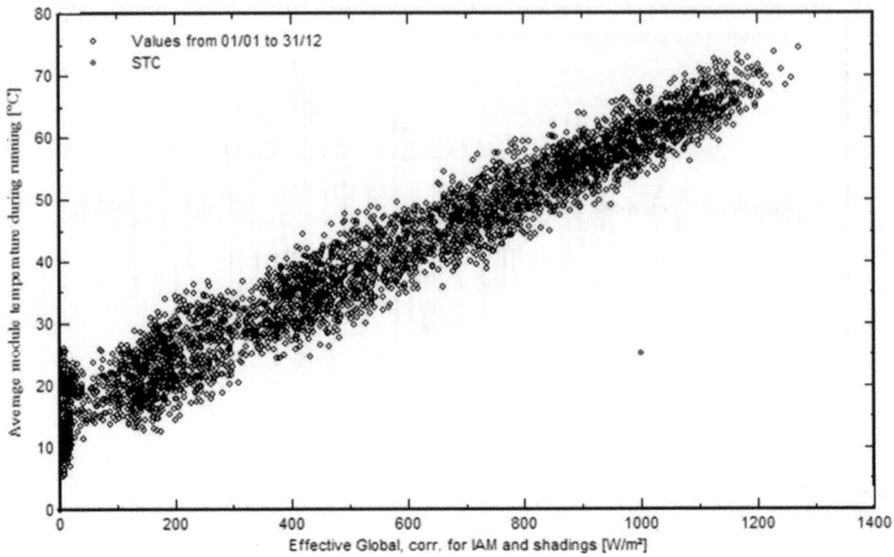

Fig. 14.16 Array temperature VS effective irradiance distribution

14.6 Conclusion

Before installation, sizing and simulation of a solar system are essential steps. In this study, a PVsyst's simulation tools designs and analyzes a standalone photovoltaic system. This software enables solar photovoltaic system designers to create any system and forecast the output of generated energy. The location and weather conditions at the site have a significant impact on the sizing system. Taking into consideration the various equipment degradation factors and the knowledge of all losses, including all identified sources of losses, the simulation results of the photovoltaic system created in this study enable improved utilization of solar energy. Therefore, designing the system's layout and foreseeing solar energy generation are crucial elements. Additionally, a number of variables, like the application type, operation duration, and availability of solar resources, might alter the simulation results and undoubtedly cause them to vary from other PV applications in various places. According to the PVsyst modeling results, the effective global radiation after accounting for optical losses is 2417.9 kWh/m^2, while the global horizontal radiation on the horizontal plane is 2381 kWh/m^2. The average solar radiation output is 1517.116 MWh, whereas the average energy waste is 261.980 MWh. In addition, the following values are given: energy missing, energy supplied to the user, energy required by the user (load), and solar fraction, which are, respectively, 21.574 MWh, 1207.573 MWh, 1229.146 MWh, and 0.982.

Acknowledgments The research has been funded by faculty of Mechanical and Industrial Engineering, Bahir Dar Institute of Technology, Bahir Dar University, Ethiopia.

References

1. Srivastava, R., Amir, M., Ahmad, F., Agrawal, S. K., Dwivedi, A., & Yadav, A. K. (2022). Performance evaluation of grid connected solar powered microgrid: A case study. *Frontiers in Energy Research, 10*, 1–17. https://doi.org/10.3389/fenrg.2022.1044651
2. Rajashekar, J. (2008). Study on design and performance analysis of solar PV rooftop standalone and on grid system using PVSYST. *International Research Journal of Engineering and Technology*, 41.
3. Agyekum, E. B., Mehmood, U., Kamel, S., Shouran, M., Elgamli, E., & Adebayo, T. S. (2022). Technical performance prediction and employment potential of solar PV systems in cold countries – A case study of the Sverdlovsk region of Russia. *Sustainability, 14*(6). https://doi.org/10.3390/su14063546
4. Getie, M. Z., & Degefa, A. M. (2019). Assessment of trend and determinant factors for household energy utilization choice in urban areas of Ethiopia: Case of eastern Amhara. *Journal of Energy Management and Technology, 4*(1), 67–74.
5. Energy for Access. (2021). *Off- and weak-grid solar appliance market*, pp. 1–20.
6. Fakih, A., Ghazalian, P., & Ghazzawi, N. (2020). The effects of power outages on the performance of manufacturing firms in the MENA region. *Review of Middle East Economics and Finance, 16*(3), 1–28. https://doi.org/10.1515/rmeef-2020-0011
7. Abdisa, L. T. (2018). Power outages, economic cost, and firm performance: Evidence from Ethiopia. *Utilities Policy, 53*, 111–120. https://doi.org/10.1016/j.jup.2018.06.009
8. Alem, M. S., Admasu, B. T., Minale, T. A., Getie, M. Z., & Wassie, H. M. (2023). Experimental investigation of solar cooker using parabolic dish collector for indoor cooking application. In B. H. Woldegiorgis, K. Mequanint, M. Z. Getie, E. G. Mulat, & A. Alemayehu Assegie (Eds.), *Advancement of science and technology: Materials and energy* (pp. 225–241). Springer Nature. https://doi.org/10.1007/978-3-031-33610-2_12
9. Satish, M., Santhosh, S., & Yadav, A. (2020). Simulation of a Dubai based 200 KW power plant using PVsyst software. In *2020 7th international conference on signal processing and integrated networks networks, SPIN 2020* (pp. 824–827), https://doi.org/10.1109/SPIN48934.2020.9071135.
10. Shrivastava, A., Sharma, R., Kumar Saxena, M., Shanmugasundaram, V., & Rinawa, M. L. (2021). Solar energy capacity assessment and performance evaluation of a standalone PV system using PVSYST. *Materials Today Proceedings, 12*(13), 2384–2390. https://doi.org/10.1016/j.matpr.2021.07.258
11. Siregar, Y., & Hutahuruk, Y. (2020). Optimization design and simulating solar PV system using PVSyst software. *2020 4th international conference on electrical, telecommunication and computer engineering (ELTICOM 2020)* (pp. 219–223) https://doi.org/10.1109/ELTICOM50775.2020.9230474.
12. Ahsan, S., Javed, K., Rana, A. S., & Zeeshan, M. (2016). Design and cost analysis of 1kW photovoltaic system based on actual performance in Indian scenario. *Perspectives on Science, 8*, 642–644. https://doi.org/10.1016/j.pisc.2016.06.044
13. Spea, S. R., & Khattab, H. A. (2019). Design sizing and performance analysis of stand-Alone PV system using PVSyst software for a location in Egypt. In *2019 21st international middle east power systems conference (MEPCON 2019)* (pp. 927–932) https://doi.org/10.1109/MEPCON47431.2019.9008058.
14. Mehdipour, C., & Mohammadi, F. (2019). Design and analysis of a stand-alone photovoltaic system for footbridge lighting. *Journal of Solar Energy Research, 4*(2), 85–91. https://doi.org/10.22059/jser.2019.283926.1120
15. Tyagi, V. V., Rahim, N. A. A., Rahim, N. A., & Selvaraj, J. A. L. (2013). Progress in solar PV technology: Research and achievement. *Renewable and Sustainable Energy Reviews, 20*, 443–461. https://doi.org/10.1016/j.rser.2012.09.028

16. Wassie, H. M., Getie, M. Z., Alem, M. S., Kotu, T. B., & Salehdress, Z. M. (2022). Experimental investigation of the effect of reflectors on thermal performance of box type solar cooker. *Heliyon, 8*(12), e12324. https://doi.org/10.1016/j.heliyon.2022.e12324
17. Wassie, H. M., Admasu, B. T., Getie, M. Z., & Alem, M. S. (2023). Experimental investigation of parabolic solar dish concentrator-based solar dryer assisted with thermal energy storage system. In B. H. Woldegiorgis, K. Mequanint, M. Z. Getie, E. G. Mulat, & A. Alemayehu Assegie (Eds.), *Advancement of science and technology: Materials and energy* (pp. 263–283). Springer Nature. https://doi.org/10.1007/978-3-031-33610-2_15
18. Belmahdi, B., & El Bouardi, A. (2020). Solar potential assessment using PVsyst software in the northern zone of Morocco. *Procedia Manufacturing, 46*(2019), 738–745. https://doi.org/10.1016/j.promfg.2020.03.104
19. Alnoosani, A., Oreijah, M., Alhazmi, M., Samkari, Y., & Faqeha, H. (2018). Design of 100MW solar PV on-grid connected power plant using (PVsyst) in Umm Al-Qura University. *International Journal of Scientific Research, 8*(11), 356–363. https://doi.org/10.21275/30101901
20. Alzahrani, A., Rajeh, M., & Banjar, S. (2021). Design and simulation of a standalone pv system for a mosque in neom city. In *International conference on industrial engineering and operations management (ICIEOM)* (pp. 634–633).
21. Irwan, Y. M., et al. (2015). Stand-alone photovoltaic (SAPV) system assessment using PVSYST software. *Energy Procedia, 79*. https://doi.org/10.1016/j.egypro.2015.11.539
22. Sami, S. M., & Mahmood, A. L. (2021). Design and simulation of stand-alone photovoltaic system supplying bts in Iraq. *International Journal of Power Electronics and Drive Systems, 12*(1), 463–473. https://doi.org/10.11591/ijpeds.v12.i1.pp463-473
23. Khatib, T., Mohamed, A., Sopian, K., & Mahmoud, M. (2012). A new approach for optimal sizing of standalone photovoltaic systems. *International Journal of Photoenergy, 2012*, 1. https://doi.org/10.1155/2012/391213
24. Kumar, R., Rajoria, C. S., Sharma, A., & Suhag, S. (2020). Design and simulation of standalone solar PV system using PVsyst software: A case study. *Materials Today Proceedings, 46*, 5322–5328. https://doi.org/10.1016/j.matpr.2020.08.785
25. Shiva Kumar, B., & Sudhakar, K. (2015). Performance evaluation of 10 MW grid connected solar photovoltaic power plant in India. *Energy Reports, 1*, 184–192. https://doi.org/10.1016/j.egyr.2015.10.001
26. Labed, S., & Lorenzo, E. (2004). The impact of solar radiation variability and data discrepancies on the design of PV systems. *Renewable Energy, 29*(7), 1007–1022. https://doi.org/10.1016/j.renene.2003.12.009
27. Sudhakar, N. K. L. K. (2015). Theoretical design and simulation analysis of PV based pumping system for domestic applications in Bhopal, M.P, India. *International Journal of Scientific Research, 4*(4), 2293–2297.
28. Rout, K. C., & Kulkarni, P. S. (2020). Design and performance evaluation of proposed 2 kW solar PV rooftop on grid system in Odisha using PVsyst. In *2020 IEEE international students' conference on electrical,electronics and computer science (SCEECS)*. https://doi.org/10.1109/SCEECS48394.2020.124.

Chapter 15
Solar Energy Resource Potential Assessment and the Impacts of Solar PV System Integration on the Performance of Radial Distribution Network

Estifanos Abeje Sharew

15.1 Introduction

An electric power distribution system delivers electric power to individual consumers from the transmission system [1]. Currently, the existing distribution system cannot meet the desired power demand in terms of both quality and reliability. This problem can be mitigated by integrating nonconventional renewable energy resources like solar and wind which are clean and safe for the surroundings using efficient technology [2, 3]. The optimal integration of these renewable energy resources with the distribution network can enhance power loss and voltage profile [2–5]. Photovoltaic (PV) generation is the best and has a strong potential among different renewable generations because the sun is a powerful infinite resource that can easily generate electric energy by the photoelectric effect [6]. The solar PV system integration into the distribution network improves the performance of the network [7–10].

The solar insolation level, ambient temperature, and sunshine hour are the basic factors that determine the performance of solar PV systems [11]. Numerous PV cells are connected in series and in parallel to generate high power output [12]. The modules are also integrated together to produce the desired level of power [13]. The standard test conditions are irradiance of 1000 W/m^2, solar spectrum of AM 1.5, and module temperature at 25 °C [13, 14].

The penetration level of solar PV systems determines the network performance. When the solar PV system supplies power beyond the load power demand, the extra power flows to the grid [15, 16]. However, when produced power from the solar is

E. A. Sharew (✉)
Faculty of Electrical and Computer Engineering, Bahir Dar Institute of Technology,
Bahir Dar University, Bahir Dar, Ethiopia

K. Mequanint et al. (eds.), *Sustainable Development Research in Materials and Energy*, Green Energy and Technology,
https://doi.org/10.1007/978-3-031-69860-6_15

not enough for load demand, the grid provides power to the load. Up to a certain amount of penetration level, the voltage profile improves and the power loss reduces, but increasing the PV production beyond the maximum penetration capacity causes poor system operation and high system power losses [17]. When the PV system is integrated with the distribution network, harmonics are generated due to the conversion of DC to AC by an inverter. The production of harmonic distortion also has drawbacks to the performance of the network leading to the failure of various electrical equipment [12].

The PV penetration level can be determined using Eq. 15.1 [5, 18].

$$\text{PV Penetration Level } (\%) = \frac{\text{Total generated PV power(MW)}}{\text{Total system power (MW)}} \times 100\% \quad (15.1)$$

15.2 Related Works

A few decades ago, different scholars discussed various kinds of literature about the impacts of solar photovoltaic system integration on distribution networks.

Solanki et al. [19] proposed an analysis of high-penetration solar PV on a distribution system at different PV penetration levels (0, 10, 30, and 50%) using OpenDSS. It is found that the effect of PV integration on the distribution system based on profile and losses depends upon the amount of solar PV generation.

Enock Mulenga [20] studied the impacts of solar PV connect with the grid for different PV penetration levels. The study was conducted without PV, and with different percentages of PV penetration levels (0, 30, 60, and 90%) using NEPLAN, Paladin DesignBase, and PVsyst software. The voltage profile increases after solar system integration.

Amoura et al. [17] suggested the performance of photovoltaics in terms of power loss. The impact of various penetration level PV systems on radial feeders in terms of power loss is evaluated by DIgSILENT power factory software. The system power loss is improved, but it depends on the level of PV penetration.

Balamurugan et al. [21] proposed the performance of distributed solar photovoltaic generators connected to a distribution network using power loss as a parameter. It is found that increasing the penetration level of solar PV systems reduces active and reactive power losses up to the maximum possible capacity of PV integration.

Farhoodnea et al. [22] presented the effect of highly penetrated solar PV in distribution systems. It is found that the voltage profile increases over the standard limit due to the injection of high active power of the PV.

Shafiullah et al. [23] presented the impact of solar PV integration with the distribution network in terms of harmonics. Zero percent, 50%, and 100% PV amounts are taken. The harmonic distortion increases as the solar PV injects high power into the distribution grid.

Rachita Sahoo et al. [3] proposed the influence of PV integration on the harmonics distribution networks. The amount of photovoltaics (10–60%) is taken into consideration. The harmonic analysis also indicated that up to 38.2% PV penetration level, it was found that the harmonics were in the accepted range of operation.

The research reviewed in [17–19] suggested the influence of different amounts of solar PV integration on the distribution system and determined that the integration of PV systems has either a positive or negative influence depending on the level of solar PV penetration. The various level of a photovoltaic system is integrated with the selected Bahir Dar distribution feeder called BATA for this research to study it's impact.

15.3 Assessment of the Solar Energy Resource Potential of the Area

It is important to assess whether the solar resource is feasible or not in the specified location under study before proceeding to photovoltaic power installation for both grid-connected and standalone PV systems [24]. Photovoltaic is a technique of producing electrical power by changing radiant light energy into electricity using semiconductor materials. Solar photovoltaic (PV) system generates energy using the sunshine hours. The sunshine hour is the length of the sun shining in a given area at a specified time. It is considered as 4–6 h per day as expected enough value in most of the design issues [25]. It is mostly given as an average of different years. The sunshine hour data of Bahir Dar City for 10 consecutive years are given in Table 15.1. as recorded from the metrology agency.

From Table 15.1, the monthly average sunshine hour of Bahir Dar City meets the minimum sunshine hour required for the designing of the solar power system. It implied that the area has untouched plenty of solar resources. The radiation of the

Table 15.1 Monthly average sunshine hour data

	Region: Gojjam						Station: Bahir Dar					
	Element: monthly sunshine hour											
Month	Jan	Feb	Mar	Apr	May	June	July	Aug	Sept	Oct	Nov	Dec
2010	9.2	10.0	8.7	8.6	8.7	7.1	4.8	5.1	6.2	8.8	10	10.3
2011	9.8	10.0	8.8	8.8	7.7	7.0	4.7	4.1	5.7	8.8	9.6	9.6
2012	9.2	8.6	9.6	8.9	8.2	6.6	4.2	4.5	5.8	8.8	9.3	10.3
2013	9.0	9.8	10.2	10.7	8.0	8.0	5.1	4.2	5.8	8.5	9.7	9.4
2014	8.0	9.0	9.2	9.0	7.2	7.6	4.1	5.1	7.7	8.2	9.0	8.5
2015	8.7	8.3	8.6	8.8	8.4	7.2	4.5	4.8	6.4	9.0	9.5	9.3
2016	9.2	10.5	8.0	9.9	8.2	7.1	5.2	4.3	6.1	9.1	9.3	9.9
2017	10.1	10.1	9.2	9.4	8.2	5.6	4.5	4.0	5.9	9.5	8.6	9.9
2018	9.7	10.1	9.3	9.8	8.6	6.6	4.2	4.3	6.8	8.3	9.7	9.9
2019	9.8	9.8	9.1	7.9	7.7	6.9	4.9	4.7	6.0	8.6	9.7	9.7

area is then calculated using the registered average duration of the sunshine hour [3, 18]. The Angstrom-Prescott estimation model is applied to determine the value of the monthly average daily solar radiation of the site [26].

$$H_o = \frac{24}{\pi} G_{SC}\left(1 + \cos \frac{360nd}{365}\right) * \left(\cos \varphi \cos \delta \sin \omega_s + \frac{\pi \omega_s}{180^0} \sin \varphi \sin \delta\right) \tag{15.2}$$

where

$$\delta = 23.45^\circ \sin\left(360^\circ \frac{284 + n_d}{365}\right)$$

$$\mathrm{Cos}\omega_s = -\tan \varphi \tan \delta$$

$$H = H_{0\left(a + \frac{bn}{N}\right)}$$

where

G_{sc} is the solar constant $= 1.376$ kW/m^2
$\varphi$is the latitude in degree
$\delta$is the solar declination in degree
ω_s is the sunset hour angle in degree
nd is a day of the year from January 1 to December 31 taking January 1 as 1
N is the monthly average of the maximum possible hours of sunshine
n is the monthly average daily hours of sunshine
H_o is the monthly average daily extraterrestrial solar radiation
H is the monthly average daily global solar radiation
a and *b* are empirical coefficients

The calculated monthly average radiation of the site is shown in Table 15.2. As presented in Table 15.2, the solar radiation of the city is determined to be in the acceptable limit (4–6 kWh/m^2/day) which implied that the area is rich in solar resources [18].

Table 15.2 The average solar irradiance of Bahir Dar City

Months	Jan	Feb	Mar	Apr	May	Jun	Jul	Aug	Sep	Oct	Nov	Dec
Solar irradiance (kWh/m^2/day)	4.76	5.2	5.5	5.6	5.26	4.74	4.02	4.04	4.67	4.87	5.01	4.67

15.4 Optimal Site and Size Determination of Solar PV System

15.4.1 Algorithm of Particle Swarm Optimization

Particle swarm optimization (PSO) is applied in this research for determination of optimal site and size of the PV system. PSO is a way of determining optimal values of the required parameters of a system using swarms of insects. This algorithm initializes particles as a proposed solution for specified problem and trains to searching location [27]. PSO can determine the solution within a short execution time and iterations which makes it preferable [9, 18, 28]. The particles update their position as shown in Fig. 15.1. [23, 29, 30].

Xi is the current position of the given particle.
Vi is the current velocity of the given particle.
The distance between the personal best (Pbest) and current position of the given particle.
The distance between the global best (PGbest) and current position of the given particle.

$$V_{ij}(t+1) = w^{*}v_{ij}(t) + r_1c_1\left(p_{ij}(t) - x_{ij}(t)\right) + r_2c_2\left(g(t) - x_{ij}(t)\right) \tag{15.3}$$

$$x_{ij}(t+1) = v_{ij}(t+1) + x_{ij}(t) \tag{15.4}$$

where

w is the inertia weight and r_1 and r_2 are random numbers between 0 and 1 which uniformly distributed.

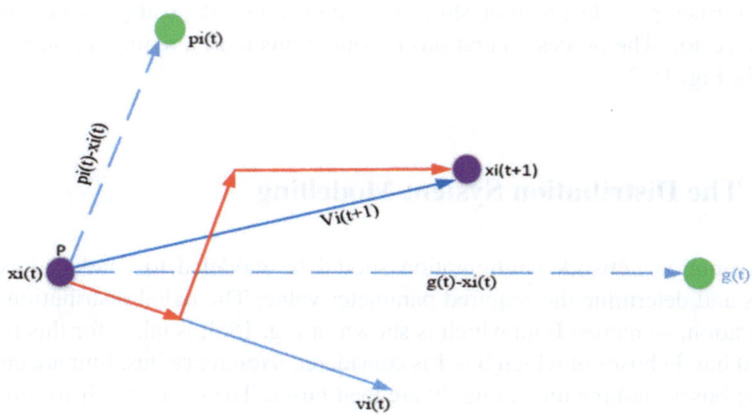

Fig. 15.1 Figurative expression for PSO [18]

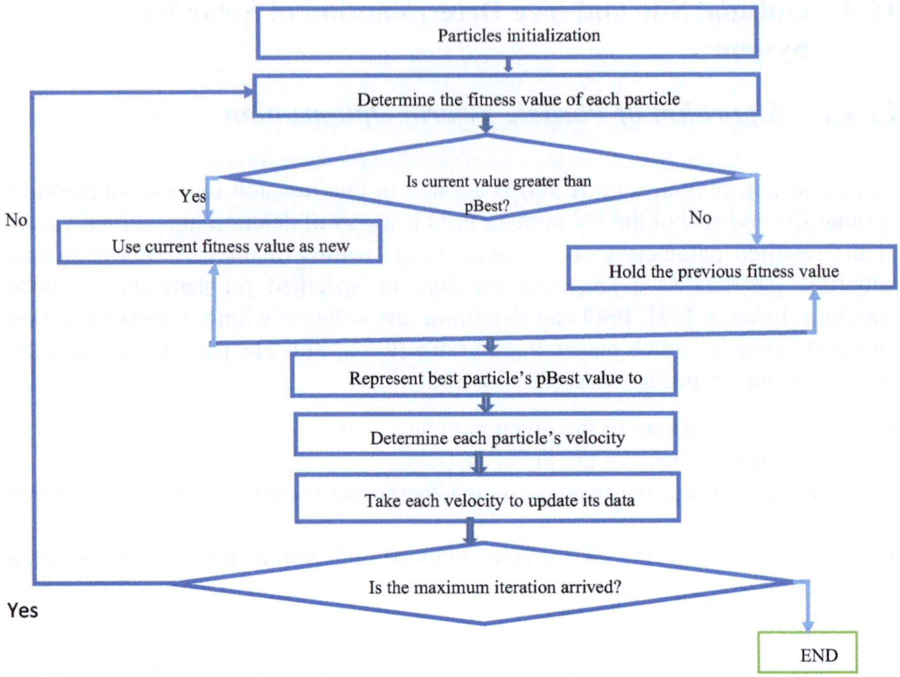

Fig. 15.2 PSO algorithm flowchart

The update velocity has three parts as shown in Eq. (15.3). These are the following:

- The inertia part: $w * v_{ij}(t.)$
- The cognitive part: $r_1 c_1 (p_{ij}(t) - x_{ij}(t))$
- The social part: $r_2 c_2 (g(t) - x_{ij}(t))$

The former particle position shifts to a new updated position using the new velocity vector. The process (iteration) is continuous until a required point is set as shown in Fig. 15.2.

15.5 The Distribution System Modelling

The distribution network configuration should be modeled to conduct load flow analysis and determine the required parameter value. The radial distribution feeder configuration, so named Bata which is shown in Fig. 15.3, is taken for this research article. It has 34 buses of which bus 1 is considered a reference bus, four are common coupling buses, and the remaining 29 are load buses. The overhead distribution line has a total length of 18.77 km. All 34 nodes have 15 kV voltage. 1.86 MW active and 1.25 MVAr reactive powers are supplied into the feeder [18].

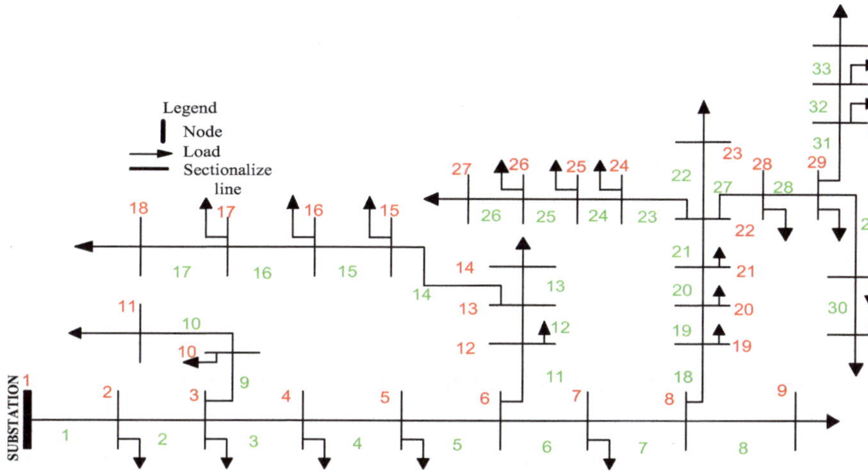

Fig. 15.3 Single-line diagram representation of Bata feeder [18]

Table 15.3 Overhead medium voltage conductor size

Conductor type	Nominal area (mm^2)	Resistance (R_0) (Ω/km)	Reactance (X_0) (Ω/km)
AAC	25	1.1810	0.8345
AAC	50	0.5787	0.4362
AAC	95	0.3085	0.2147

15.5.1 Impedance Calculation of Overhead Line

The overhead transmission line impedance can be determined using the value of the manufacture resistance, perusing length, and the length of the line. The conductors that are used in distribution feeders are stranded conductors. Conductors used in the BATA distribution feeder are stranded conductors and their types are AAC-25, AAC-50, and AAC-95. The impedance for the positive sequence of the line segments is the product of the impedance per kilometer and the length of the segment. The shunt admittance of overhead lines is neglected for the short transmission lines. The line data of the network is calculated using the table data depicted in Table 15.3. The resistance per kilometer of each conductor is collected from the manufacturers' data sheet.

The values of RL (calculated resistance) and XL (calculated reactance) can be determined using R0 (ohm/km) and X0 (ohm/km) for each type of distribution line branch as follows:

$$R_L = (R_0 \times \text{length of the line branch}) \tag{15.5}$$

$$X_L = (X_0 \times \text{length of the line branch} \quad (15.6)$$

15.5.2 *Base Case Load Profile*

The calculated values of R_L, X_L, and load data of the BATA feeder are shown in Table 15.4 [29].

15.6 Power Flow Study of Radial Distribution System

A radial type of distribution network is used in this research, and it has its own characteristics. The basic characteristics of this type of feeder are the following: it has a radial structure, multiphase type of conductors, unbalanced load operation, and high R/X ratio [1]. These characteristics of radial feeders cause a traditional type of power flow methods like conventional Gauss-Seidel, Newton-Raphson, and fast decoupled load flow algorithms to have poor convergence characteristics and cannot meet the distribution system requirements [31]. Hence, an appropriate power flow technique must be developed for the load flow calculations in radial systems. The backward-forward sweep load flow analysis is the suitable and preferred technique. The magnitude of voltage and the source phase angle must be fixed, and the load demand at each bus should be given [1]. The backward sweep calculates the line part currents and voltages from the far end of the feeder back to the source using Kirchhoff's current law (KCL) and Kirchhoff's voltage law (KVL). The forward sweep is executed to refresh the node voltages along the feeder when the calculated voltage diverges from the originally specified voltage beyond the acceptable threshold. The procedure continues oscillating between the forward and backward sweeps until the voltage error at the source falls within the acceptable range [1, 9, 31]. The forward and/or backward sweep load flow technique is applied in this work.

15.6.1 *Forward/Backward Sweep Load Flow Analysis*

This technique is constructed by bus injection to the branch current matrix, branch current to the bus voltage matrix, and the equivalent current injection. The load Si of each bus is represented by [9, 18]:

$$S_{Li} = P_{Li} + jQ_{Li} \; i = 1 \ldots N \quad (15.7)$$

Step 1: Backward sweep load flow algorithm

Table 15.4 The feeder line and load data [18]

Sending node (i)	Receiving node (j)	Conductor type	Length (km)	Resistance RL (Ω)	Reactance XL (Ω)	Receiving end load (kW)	Receiving end load (kVAr)
1	2	AAC95	1.8599	0.57378	0.39932	24.6659	12.4079
2	3	AAC95	0.2911	0.08980	0.06250	9.959	6.689
3	4	AAC95	0.3809	0.11751	0.08178	69.335	62.259
4	5	AAC95	0.6789	0.20944	0.14576	25.769	10.589
5	6	AAC95	0.2899	0.08943	0.06224	0.000	0.000
6	7	AAC95	1.1061	0.34123	0.23748	25.9649	15.85
7	8	AAC95	0.5549	0.17119	0.11914	0.000	0.00
8	9	AAC95	0.4741	0.14626	0.10179	47.99	25.799
3	10	AAC50	0.9031	0.52262	0.39393	81.699	71.109
10	11	AAC50	0.2731	0.15804	0.11913	136.739	63.839
6	12	AAC25	0.4989	0.58920	0.41633	75.448	54.219
12	13	AAC95	0.7111	0.21937	0.15267	0.000	0.00
13	14	AAC50	0.457	0.26447	0.19934	63.899	44.648
13	15	AAC50	1.2759	0.73836	0.55655	80.399	52.797
15	16	AAC25	0.1031	0.12176	0.08604	41.979	21.798
16	17	AAC25	0.7061	0.83390	0.58924	88.279	66.929
17	18	AAC50	0.3471	0.20087	0.15141	5.079	3.059
8	19	AAC50	0.2451	0.14184	0.10691	112.99	46.819
19	20	AAC95	1.1231	0.34648	0.24113	85.069	62.195
20	21	AAC25	0.5181	0.61188	0.43236	55.599	38.989
21	22	AAC50	0.5110	0.29572	0.22291	0.000	0.00
22	23	AAC50	0.4241	0.24543	0.18499	27.219	12.019
22	24	AAC50	0.2810	0.16262	0.12257	28.039	12.589
24	25	AAC50	2.1210	1.22742	0.92518	55.809	37.959

(continued)

Table 15.4 (continued)

Sending node (i)	Receiving node (j)	Conductor type	Length (km)	Resistance RL (Ω)	Reactance XL (Ω)	Receiving end load (kW)	Receiving end load (kVAr)
25	26	AAC25	0.1880	0.22203	0.15689	74.599	55.229
26	27	AAC25	0.2751	0.32489	0.22957	46.519	24.539
22	28	AAC25	0.6310	0.74521	0.52657	48.395	93.79
28	29	AAC50	0.3279	0.18976	0.14303	9.299	5.499
29	30	AAC50	0.3181	0.18409	0.13876	34.449	17.368
30	31	AAC50	0.1971	0.11406	0.08598	25.779	48.739
29	32	AAC50	0.2419	0.13999	0.10552	78.499	47.7968
32	33	AAC50	0.3790	0.21933	0.165320	98.00	66.00
33	34	AAC95	0.0910	0.028074	0.0195377	93.499	86.00

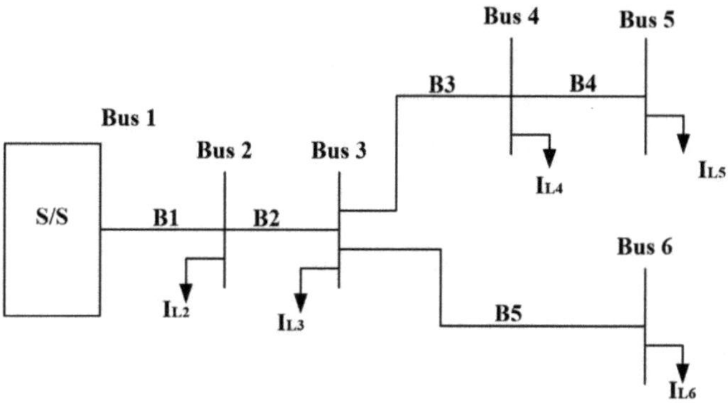

Fig. 15.4 Sample distribution system

The ith bus current injection at its kth iteration is given by

$$I_i^k = I_i^r(V_i^k) + jI_i^i(V_i^k) = \left(\frac{S_i}{V_i^k}\right)^* = \left(\frac{P_i + jQ_i}{V_i^k}\right)^* \quad (15.8)$$

where

V_i^k is the voltage of the *i*th bus at the *k*th iteration,
I_i^r is the real current injection of *i*th node at *k*th iteration, and
I_i^i is the imaginary current injection of *i*th node at *k*th iteration [18].

The relation of the matrix is developed using the sample distribution model shown in Fig. 15.4:

From Eq. (15.8), the currents injected are determined. By utilizing Kirchhoff's current law (KCL) on the distribution network, the branch currents can be determined. Branch currents can subsequently be expressed as functions of the equivalent current injection. The branch currents B_1, B_2, B_3, B_4, and B_5 are formulated as [9]

$$B_1 = I_{L2} + I_{L3} + I_{L4} + I_{L5} + I_{L6} \quad (15.9)$$

$$B_2 = I_{L3} + I_{L4} + I_{L5} + I_{L6} \quad (15.10)$$

$$B_3 = I_{L4} + I_{L5} \quad (15.11)$$

$$B_4 = I_{L5} \quad (15.12)$$

$$B_5 = I_{L6} \quad (15.13)$$

Hence, the association between the bus current injection and the branch currents can be represented as

$$\begin{bmatrix} B_1 \\ B_2 \\ B_3 \\ B_4 \\ B_5 \end{bmatrix} = \begin{bmatrix} 1 & 1 & 1 & 1 & 1 \\ 0 & 1 & 1 & 1 & 1 \\ 0 & 0 & 1 & 1 & 0 \\ 0 & 0 & 0 & 1 & 0 \\ 0 & 0 & 0 & 0 & 1 \end{bmatrix} \begin{bmatrix} I_{L2} \\ I_{L3} \\ I_{L4} \\ I_{L5} \\ I_{L6} \end{bmatrix} \tag{15.14}$$

Equation (15.14) can be expressed as follows

$$[B] = [BIBC][I] \tag{15.15}$$

where

BIBC-BIBC is a matrix that converts bus injection to branch current and it is an upper triangular matrix that contains only zeros and ones.

Step 2: Forward sweep load flow algorithm

The nodal voltage vector V is updated from the source to the load in accordance with Kirchhoff's voltage law (KVL), utilizing the previously determined branch current vector B and the branch current to bus voltage (BCBV) transformation.

The correlation between branch current and bus voltage, as depicted in Fig. 15.3, can be formulated as [9]

$$\begin{aligned} V_2 &= V_1 - B_1 Z_{12} \\ V_3 &= V_2 - B_2 Z_{23} = V_1 - B_1 Z_{12} - B_2 Z_{23} \\ V_4 &= V_3 - B_3 Z_{34} = V_1 - B_1 Z_{12} - B_2 Z_{23} - B_3 Z_{34} \\ V_5 &= V_4 - B_4 Z_{45} = V_1 - B_1 Z_{12} - B_2 Z_{23} - B_3 Z_{34} - B_4 Z_{45} \\ V_6 &= V_3 - B_5 Z_{56} = V_1 - B_1 Z_{12} - B_2 Z_{23} - B_5 Z_{56} \end{aligned} \tag{15.16}$$

where

V_i is bus i's voltage and

Z_{ij} is the line impedance between buses i and bus j.

It is evident from Eq. (15.16) that the substation voltage, branch currents, and line characteristics may all be used to represent the bus voltage. Since additional buses can undergo similar operations, the relationship between branch currents and bus voltages can be stated as follows:

$$\begin{bmatrix} V_2 \\ V_3 \\ V_4 \\ V_5 \\ V_6 \end{bmatrix} = \begin{bmatrix} V_1 \\ V_1 \\ V_1 \\ V_1 \\ V_1 \end{bmatrix} - \begin{bmatrix} Z_{12} & 0 & 0 & 0 & 0 \\ Z_{12} & Z_{23} & 0 & 0 & 0 \\ Z_{12} & Z_{23} & Z_{34} & 0 & 0 \\ Z_{12} & Z_{23} & Z_{34} & Z_{45} & 0 \\ Z_{12} & Z_{23} & 0 & 0 & Z_{56} \end{bmatrix} \begin{bmatrix} B_1 \\ B_2 \\ B_3 \\ B_4 \\ B_5 \end{bmatrix} \tag{15.17}$$

where

BCBV or the branch current to bus voltage is provided by the sample network.

$$[\mathrm{BCBV}] = \begin{bmatrix} Z_{12} & 0 & 0 & 0 & 0 \\ Z_{12} & Z_{23} & 0 & 0 & 0 \\ Z_{12} & Z_{23} & Z_{34} & 0 & 0 \\ Z_{12} & Z_{23} & Z_{34} & Z_{45} & 0 \\ Z_{12} & Z_{23} & 0 & 0 & Z_{56} \end{bmatrix} \tag{15.18}$$

The bus voltage at the $(k + 1)$th iteration can be stated in general form as

$$\left[V^{k+1}\right] = [V_1] - [\mathrm{BCBV}][B] \tag{15.19}$$

Generally, the node and iteration number are indicated by the letters i and k, respectively:

$$\begin{aligned} I_{i-1,i}^{k} &= I_i^k + I_{i,i+1}^k \\ V_i^k &= V_{i-1}^k - Z_{i-1,i}{}^{*} I_{i-1,i}^{k-1} \end{aligned} \tag{15.20}$$

The following formulas can be used to express the distribution system's total active and reactive power losses:

$$P_{\mathrm{loss}} = \sum_{i=1}^{\mathrm{nb}} |I(i)|^2 R(i) \tag{15.21}$$

$$Q_{\mathrm{loss}} = \sum_{i=1}^{\mathrm{nb}} |I(i)|^2 X(i) \tag{15.22}$$

where

nb is the total number of branches in the distribution system.

15.7 Formulation of the Problem

An equation is created between the buses in the network depicted in Fig. 15.5 using a simple line diagram.

The following formulas can be used to determine the active and reactive power between buses i and $i + 1$ prior to PV system integration.

$$\begin{aligned} P_{i+1} &= P_i - P_{\text{loss},i} - P_{\text{loss},i+1} \\ &= P_i - \frac{R_i}{V_i^2}\left(P_i^2 + Q_i^2\right) - P_{\text{loss},i+1} \end{aligned} \tag{15.23}$$

$$\begin{aligned} Q_{i+1} &= Q_i - Q_{\text{loss},i} - Q_{\text{loss},i+1} \\ &= Q_i - \frac{X_i}{V_i^2}\left(P_i^2 + Q_i^2\right) - Q_{\text{loss},i+1} \end{aligned} \tag{15.24}$$

After that, the total power losses of the feeder, Ploss, and Qloss can be calculated by adding the losses of each feeder line section, which is provided as

$$P_{T,\text{loss}} = \sum_{i=1}^{N} \frac{R_i}{V_i^2}\left(P_i^2 + Q_i^2\right) \tag{15.25}$$

$$Q_{T,\text{loss}} = \sum_{i=1}^{N} \frac{X_i}{V_i^2}\left(P_i^2 + Q_i^2\right) \tag{15.26}$$

After PV integration, let the apparent power flow be $P_{pv,\ \text{Loss}(i,\, i\,+\,1)}$ and $Q_{pv,\ \text{Loss}(i,\, i\,+\,1)}$. Using the same methods as previously demonstrated, the power losses between these buses are calculated [18].

$$P_{pv,\ \text{Loss}(i,i+1)} = \frac{\left(P_{i-P_{pv}}\right)^2 + Q_i^2}{V_i^2} R_i \tag{15.27}$$

$$Q_{pv,\ \text{Loss}(i,i+1)} = \frac{\left(P_{i-P_{pv}}\right)^2 + Q_i^2}{V_i^2} X_i \tag{15.28}$$

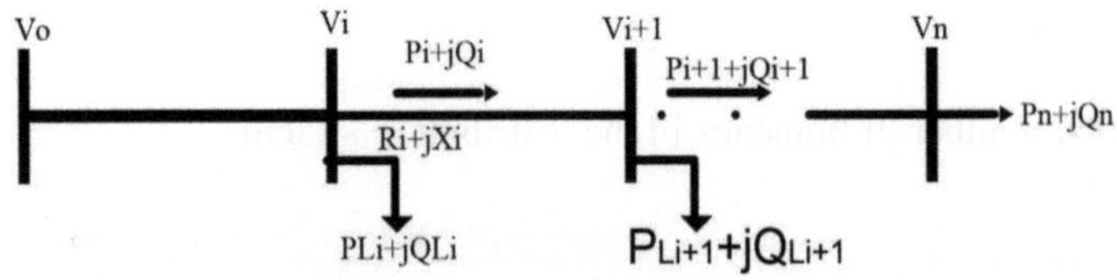

Fig. 15.5 A main feeder's single-line schematic [18]

15.7.1 Working Specifications for Integration of Solar PV

When solar PV systems are connected to the distribution system, the network must meet the following operating range (limiting values) [18].

Voltage constraint: The IEEE agreement states that the bus voltage values (pu) should not exceed the range of 0.95 pu to 1.05 pu.

Harmonic constraint: According to the IEEE's approved operating range, the overall and individual voltage harmonic distortion levels are limited.

$$\mathrm{THD}_v \leq \mathrm{THD}_v^{\max}$$

$$\mathrm{IHD}_v \leq \mathrm{IHD}_v^{\max}$$

The objective function is manipulated as presented in Eq. 15.29 [18]:
Maximize

$$F = W_1{}^* \Delta\ P_{\mathrm{loss}}^{PV} + W_2{}^* \Delta\ Q_{\mathrm{loss}}^{PV} + W_3{}^* \sum\nolimits_{i=1}^{n} (V_i - 1)^2 \tag{15.29}$$

Subjected to:

$$0.95 \leq V_i \leq 1.05$$

$$P_V \leq P_V{}^{\max}$$

$$Q_V \leq Q_V{}^{\max}$$

$$W_1 + W_2 + W_3 = 1$$

where

W_1, W_2, and W_3 are the inertia weights.

With ETAP software, one can model the distribution network with a connected solar PV system, as shown in Fig. 15.6.

15.8 Evaluation of the Radial Distribution Network's Harmonic Level

Through the use of ETAP software, the effects of voltage harmonic distortion on the radial distribution network integrated with varying levels of solar PV system penetration are ascertained in this study.

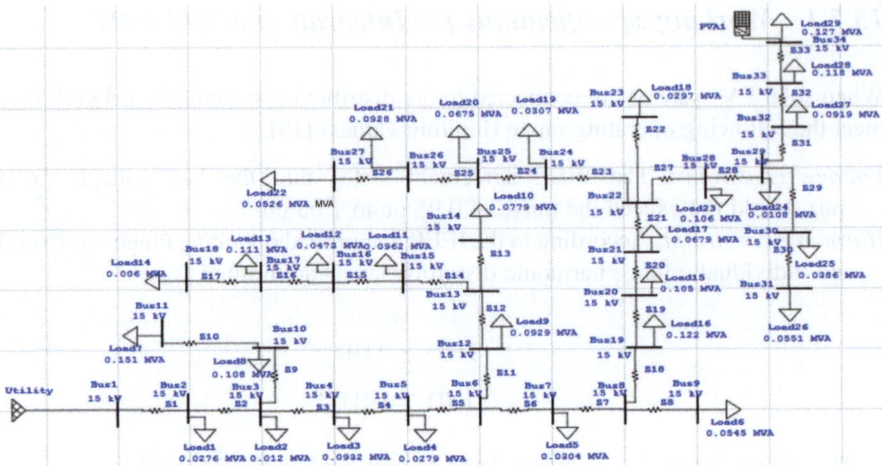

Fig. 15.6 PV-integrated BATA feeder single-line diagram for harmonic analysis [18]

15.8.1 Voltage Harmonic Distortion

This paper examines the integration of solar photovoltaic systems with the distribution system, leading to the generation of harmonics that impact the network's performance and power quality. The harmonics produced in the network are an integral product of the fundamental frequency. Extra high harmonic generation in the network can damage static and rotating electrical devices [11, 18, 32].

The formula to determine the harmonic number is defined by Eq. 15.30 [18, 33].

$$h = (n * p) \pm 1 \tag{15.30}$$

where

h denotes for the harmonic number,
n can be any number, and,
p denotes the circuit's pulse count.

The total RMS value of the signal's harmonic components divided by the signal's fundamental value is known as total harmonic distortion. Equations 15.31 and 15.32, respectively, can be used to define the individual and total voltage harmonic distortions at each bus, as the following illustration illustrates [2, 18].

$$\mathrm{IHD}_v(\%) = \frac{V_{h,\mathrm{rms}}}{V1,_{\mathrm{rms}}} * 100 \tag{15.31}$$

$$\mathrm{THD}_V(\%) = \sqrt{\textstyle\sum_{\mathrm{n}=2}^{\infty} (Vn,_{\mathrm{rms}})^2} / (V1,_{\mathrm{rms}}) * 100 \tag{15.32}$$

Equation 15.32 gives us the overall harmonic distortion of voltage:

Table 15.5 Suggested distortion limits for voltage harmonics [18]

Bus voltage V at PCC	Individual harmonic distortion IHD (%)	Total harmonic distortion THD (%)
$V \leq 1.0$ kV	5.0	8.0
1 kV $< V \leq$ 69 kV	3.0	5.0
69 kV $< V \leq$ 161 kV	1.5	2.5
161 kV $< V$	1.0	1.5

$$V_{\text{rms}} = V_{dc}$$

$$V1,_{\text{rms}} = \frac{4V_{dc}}{\sqrt{2}\Pi}$$

where

Vh,rms is the voltage at h order's rms value
Vn,rms is the voltage's rms value, and
V1,rms is the voltage's root mean square (rms) at fundamental.

In Table 15.5. are the approved harmonic distortion ranges at the point of common coupling (PCC) [34].

15.9 Results and Discussion of the Simulation

The MATLAB R2013a software is utilized to perform the backward-forward sweep load flow analysis in order to ascertain the voltage profile and base case power losses. Based on power loss, voltage profile, and harmonic distortion metrics, the impact of integrating a solar PV system into the distribution network is examined at various PV system penetration levels. Base MVA = 100 and base KV = 15 are used as system input data to convert into per-unit values. The simulation is conducted in five separate cases (scenarios) (base case, 20% PV integration, 40% PV integration (at the optimal level), 60% PV integration, and 80% PV integration).

15.9.1 Base Case Simulation Results

In order to determine the voltage profile at each bus and the real and reactive power losses in each line, a load flow analysis employing the backward-forward sweep algorithm has been put into place for the base case.

The majority of buses have voltage profiles that are below the 0.95 pu minimal threshold. The implication was that the chosen distribution network has an undervoltage issue. Bus 34 exhibits a minimum voltage profile of 0.8393 pu, with a voltage deviation that represents 16.07% of the nominal voltage. The base case results in active and reactive power losses for the system of 179.17 kW and 111.80 kVAr, respectively.

The particle swarm optimization (PSO) algorithm detects the best spot for the solar PV at bus 34.

15.9.2 Results of the Simulation for Each Scenario

The network's voltage profiles for each scenario are shown in Fig. 15.7.

Scenario I and scenario II have minimum voltage profiles that are below the minimum threshold value (0.95 pu), as can be seen in Fig. 15.7. Because some bus voltages are higher than the maximum threshold value of 1.05 pu, scenarios IV and V breached the acceptable working limit. The voltage profiles are found in the IEEE acceptable standard limit only in the third scenario (at the optimal level).

The network's power losses for each scenario are shown in Fig. 15.8.

Figure 15.8 illustrates how the degree of solar PV system integration affects the power loss reductions. Scenario III, which represents the optimal level of PV

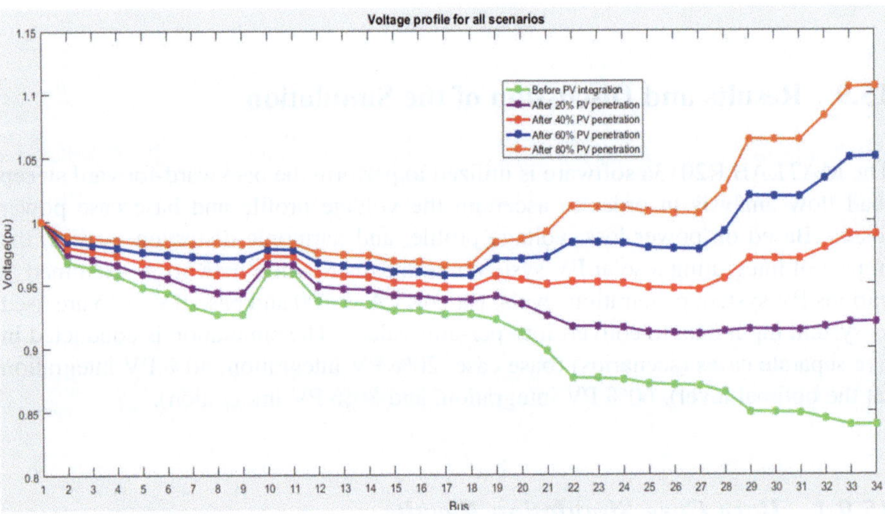

Fig. 15.7 Comparison of network voltage profiles for all scenarios

Fig. 15.8 Power loss results for all scenarios

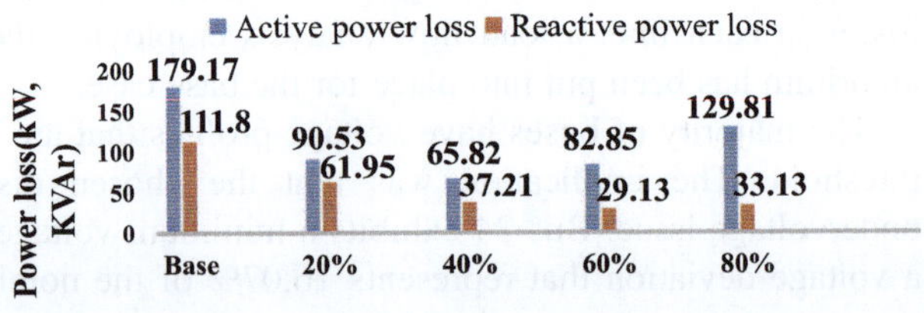

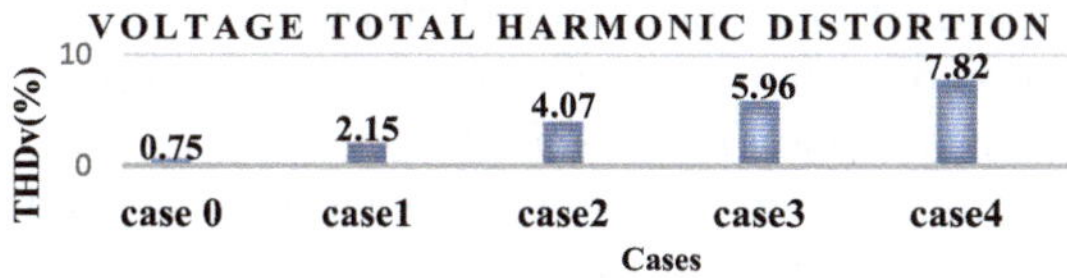

Fig. 15.9 THDv in various PCC scenarios

integration at 40%, yields the lowest power losses. In scenarios IV and V (60 and 80% respectively), the losses are higher than the optimal level.

Using PSO, it is determined that bus 34 is the best location for PV integration. Thus, this bus is regarded as a common coupling point. The degree of PV integration with the distribution network ought to be such that the level of harmonic distortion falls within the permitted range. The voltage harmonic distortion with various penetration level cases is depicted in Fig. 15.9.

Table 15.5 displays the suggested total voltage distortion range at the point of common coupling. This case study uses a bus voltage of 15 kV. Therefore, 5% is the highest allowable total bus voltage harmonic distortion in this scenario at PCC.

The value of 0.75% is established as the base case total voltage harmonic distortion. Figure 15.9 illustrates that cases 1 and 2 fall within the permitted ranges. But the remaining cases are beyond the standard ranges. Similar to how PV power introduced into the system increases, voltage harmonic distortion also rises. The overall voltage harmonic distortion level decreased to 4.07%, which is within IEEE standards, when the 40% optimal size PV system (744 kW) was integrated.

15.10 Conclusion

The performance level of integrating solar PV systems into the distribution system in terms of power losses, voltage profile, and harmonic distortion is studied in this paper using one of the 34-bus Bahir Dar radial distribution feeders, namely, BATA. MATLAB software is utilized to calculate the bus voltage profile and branch power losses through the application of the backward-forward sweep algorithm. Particle swarm optimization is used to determine that bus 34 is the best location for solar system integration. The harmonic distortion level is analyzed using ETAP software. To investigate the impact of solar PV on the distribution system with respect to power losses, voltage profile, and harmonic distortion, the various penetration levels of solar PV system integration into the feeder are taken into consideration. According to the evaluation of solar resources, solar energy is abundant and practical in the case study region. Taking power losses and voltage profile into account, the optimal PV system size is found to be 744 kW (or 40%). Overvoltage issues and significant power loss resulted from the PV system's integration beyond the optimal PV size in the chosen feeder. The integration of the optimal PV system results in an improvement of the minimum voltage profile from 0.8393 pu to 0.95 pu and reductions of the active and reactive power loss of 63.26% and 66.72%, respectively. Furthermore, the overall harmonic distortion is discovered to be 4.07%, falling within the IEEE operating limit. High power losses, overvoltage, undervoltage, and the creation of high harmonic distortion were the results of the distribution network's integration of nonoptimal solar PV levels. It was suggested that adding solar PV systems on their own wouldn't be able to improve network performance on its own and might even make things worse. In general, the distribution system performs better when an optimal-sized PV system is integrated at its optimal location.

References

1. Rupa, J. M., & Ganesh, S. (2014). Power flow analysis for radial distribution system using backward/forward sweep method. *International Journal of Electrical, Computer, Electronics and Communication Engineering, 8*(10), 1540–1544.
2. Kadir, A. A., Mohamed, A., & Shareef, H. (2010). Harmonic impact of grid connected photovoltaic inverters on 13.8 kv distribution system. In *Regional engineering postgraduate conference EPC* (pp. 367–373).
3. Rachita, S., & Prakash, K. (2019). Impact of photovoltaic penetration on distribution systems. *International Journal of Computer Applications, 14*(11), 2596–2601.
4. Singh, R., Tripathi, P., & Yatendra, K. (2019). Impact of solar photovoltaic penetration in distribution network. In *2019 3rd international conference on recent developments in control, automation & power engineering (RDCAPE)* (pp. 551–556). IEEE.
5. Eftekharnejad, S., Vittal, V., Heydt, G. T., Keel, B., & Loehr, J. (2013). Impact of increased penetration of photovoltaic generation on power systems. *IEEE Transactions on Power Apparatus and Systems, 28*(2), 893–901.
6. Chang, H., Bo, K., Yoon, S., & Gilsoo, J. (2017). Application of PV inverter on distribution system with high penetration of PV. *Journal of International Council on Electrical Engineering, 7*(1), 263–268.
7. Dilini, A., Sathsara, A., & Janaka, E. (2020). Analysis of rooftop solar impacts on distribution networks. *Ceylon Journal of Science, 48*(2), 103–112.
8. ElNozahy, M. S., & Salama, M. M. A. (2013). Technical impacts of grid-connected photovoltaic systems on electrical networks – A review. *Journal of Renewable and Sustainable Energy, 5*(3).
9. Devi, S., & Geethanjali, M. (2014). Optimal location and sizing determination of distributed generation and DSTATCOM using particle swarm optimization algorithm. *International Journal of Electrical Power & Energy Systems, 62*, 562–570.
10. Abdel-Salam, M., El-Mohandes, M. T., & Shaker, E. (2016). PSO-based performance improvement of distribution systems using DG sources. In *2016 eighteenth international Middle East power systems conference (MEPCON)* (pp. 866–870). IEEE.
11. Pawar, S., & History, M. (2019). Harmonic analysis of high penetration PV system on grid. *International Journal of Applied Engineering Research, 6*(06), 401–40(8.
12. Lue, X., Mutasim, N., & Eyad, R. (2020). Harmonic analysis of photovoltaic generation in distribution network and design of adaptive filter. *International Journal of Computing and Digital Systems, 9*(1), 77–85.
13. Naveen, N., & Gowri, M. (2016). Study on performance characteristics of PV Array under non-uniform irradiation conditions. *International Journal of Advanced Research in Science, Engineering and Technology, 3*(11), 187–193.
14. Chandrasiri, S. (2017). *Temperature effect on solar photovoltaic power generation*. Master's thesis. University of Sri Jayewardenepura.
15. Kamel, A., & Salman, M. (2020). Impact of rooftop photovoltaics on the distribution system. *Journal of Renewable Energy, 2020*, 4831434.

16. Ibraheem, W. E., Gan, C. K., & Ghani, M. R. A. (2014). Impact of photovoltaic (PV) systems on distribution networks. *International Review on Modelling and Simulations, 2*(4604), 70–74.
17. Al Momani, T., Harb, A., & Amoura, F. (2017). Impact of photovoltaic systems on voltage profile and power losses of distribution networks in Jordan. In *2017 8th international renewable energy congress (IREC)* (pp. 1–6). IEEE.
18. Estifanos, A., Habtemariam, A., & Yalew, G. (2021). Power quality and performance analysis of grid-connected solar PV system based on recent grid integration requirements. *International Journal of Photoenergy, 2021.*
19. Solanki, S. K., Ramachandran, V., & Solanki, J. (2012). Steady state analysis of high penetration PV on utility distribution feeder. In *Pes T&D 2012* (pp. 1–6). IEEE.
20. Mulenga, E. (2015). Impacts of integrating solar PV power to an existing grid.
21. Balamurugan, K., Srinivasan, D., & Reindl, T. (2012). Impact of distributed generation on power distribution systems. *Energy Procedia, 25*, 93–100.
22. Masoud, F., Azah, M., Hussain, S., & Hadi, Z. (2012). Power quality impact of grid-connected photovoltaic generation system in distribution networks. In *2012 IEEE student conference on research and development (SCOReD).*
23. Shafiullah, G. M., & Oo, A. M. (2015). Analysis of harmonics with renewable energy integration into the distribution network. In *2015 IEEE innovative smart grid technologies-Asia (ISGT ASIA)* (pp. 1–6). IEEE.
24. Raimon, B., & Numan, C. (2020). Comprehensive overview of optimizing PV-DG allocation in power system and solar energy resource potential assessments. *Energy Reports, 6*, 173–208.
25. Ibrahim, N. F. B., Bakar, Z. B. A., & Ibrahim, W. S. B. W. (2016). The feasibility study of solar PV lighting. In *2016 IEEE industrial electronics and applications conference (IEACon)* (pp. 92–96). IEEE.
26. Podder, S., Khan, R. S., Mohon, S. M. A. A., Hussain, M. J., & Basher, E. (2014). Solar radiation approximation using sunshine hour at Patenga, Bangladesh. In *8th international conference on electrical and computer engineering* (pp. 321–324). IEEE.
27. Mahmood, A., Hisham, S., Thani, M. A., & Al-Alawi, A. (2017). Optimal allocation of PV systems to minimize losses in distribution networks using GA and PSO: Masirah Island case study. *Journal of Electrical Systems, 13*, 678.
28. AlRashidi, M. R., AlHajri, M. F., Al-Othman, A. K., & El-Naggar, K. M. (2010). Particle swarm optimization and its applications in power systems. In *Computational intelligence in power engineering* (pp. 295–324). Springer.
29. Habtemiam, A., Elias, M., & Kassaye, G. (2021). Optimal design of grid-connected solar photovoltaic system using selective particle swarm optimization. *International Journal of Photoenergy, 2021.*
30. Tahir, M. J., Latiff, I. A., Alam, M., & Mazliham, M. S. (2018). Network reconfiguration using modified particle swarm algorithm. In *2018 2nd international conference on smart sensors and application (ICSSA)* (pp. 1–5). IEEE.
31. Michline, R., & Santhi, G. (2014). Power flow analysis for radial distribution system using backward/forward sweep method. *International Journal of Electrical, Computer, Electronics and Communication Engineering, 8*(10), 1628–1632.
32. Ciprés, E. (2011). *Analysis of the Implementation of A Photovoltaic Plant in Distribution System.* M.S. thesis. Politechnika Łódzka, Catalonia.
33. Duong, M. Q., Pham, T. D., Nguyen, T. T., Doan, A. T., & Tran, H. V. (2019). Determination of optimal location and sizing of solar photovoltaic distribution generation units in radial distribution systems. *Energies, 12*(1), 174.
34. Langella, R., Testa, A., & Alii, E. (2014). IEEE recommended practice and requirements for harmonic control in electric power systems. In *IEEE recommended practice*. IEEE.

Chapter 16
Theoretical Analysis of Working Fluid Inside Solar Thermal Trackers

Solomon Tesfamariam Teferi, Yonatan Tadesse, and Kamil Dino Adem

16.1 Introduction

The performance of a solar receptor (solar panel or concentrator) primarily depends on its orientation. There are several types of active tracking systems used to orient solar receptors toward the sun. Such systems mainly comprise two coplanar and perpendicular linear actuators coupled with a single-linkage arm and pivots [1, 2]. Integrating such a tracking system showed a 30% performance enhancement in comparison with the non-tracked ones. Due to their complexity, active tracking systems are costly in construction and maintenance [3].

Other tracking systems that do not use active power are termed as passive tracking systems. They are relatively cheap and easy to construct. However, their tracking accuracy is lower as compared to the active solar tracking systems. Basically, thermo-passive tracking systems employ weight balance mechanisms, and they are of two types driven by the internally developed thermo-physical forces. The first one uses bimetallic strips controlled by viscous damping. The developed computer model on such systems showed the possibility of up to 23% efficiency improvement [4]. However, the findings revealed that the time required for realigning the tracker changes exponentially with the change in solar radiation. For solar isolation below 200 W/m^2, the system stopped to respond and to track.

S. T. Teferi (✉) · Y. Tadesse
Center for Renewable Energy, Addis Ababa Institute of Technology, Addis Ababa University, Addis Ababa, Ethiopia
e-mail: Solomon.tesfamariam@aau.edu.et

K. D. Adem
School of Mechanical and Industrial Engineering, Addis Ababa Institute of Technology-Addis Ababa University, Addis Ababa, Ethiopia

K. Mequanint et al. (eds.), *Sustainable Development Research in Materials and Energy*, Green Energy and Technology,
https://doi.org/10.1007/978-3-031-69860-6_16

The second passive tracking system operates by expanding low boiling point fluids inside a cylinder piston assembly which generates mechanical power to drive the solar receptor. An experimental and simulation study conducted in India using high pressure expandable working fluids showed the possibility of driving one axis tracking system [5].

The system comprised two partially shaded thanks attached on both sides of the receptor. The tanks are connected with a reciprocating type of cylinder in opposite directions. Partially shading the tanks required to have different amount of solar energy incident on the surface of the tanks. At a particular time, the tank which had been exposed to the sun gained more heat and caused the fluid inside to expand and able to push the piston forward. A similar study on working fluid-based solar tracking system with another design approaches had been also developed [6]. This system used a double-acting cylinder to actuate the tracking mechanism that provides a smooth and controlled motion. It also used copper tubes instead of plastics to connect the double-acting cylinder with the working fluid holders to enhance the expansion process as it passes. Research reports have shown that the effect of stroke to bore ratio on the performance of reciprocating machines [7, 8].

In general, passive solar tracking systems are not widely used. Experimental and theoretical analyses on such system are not as such deeply investigated. Taking this into consideration, the general objective of this paper is to theoretically analyze the effects of change of properties of the working fluid, particularly ammonia, on the piston displacement. The method can be used as a reference while developing a solar thermal tracker and studying it further.

16.2 Methodology

16.2.1 Description of the System

As depicted in Fig. 16.2, the main components of the proposed solar thermal tracker are two: partially shaded heat-absorbing tanks (Tank A, 2 and Tank B, 3) and one double-acting—single-ended cylinder, 6. The top parts of the tanks are covered by selected surface to absorb solar energy, and the remaining parts are thermal insulated to minimize heat losses. Shades 4 and 5 are expected to be mounted above the tanks to regulate sunlight that strikes the top surfaces of the tanks. The reciprocating motion induced during the working fluid's expansion is used to propel the tracker of the solar receptor which can be either a solar PV module or a solar concentrator. In this study, the focus is on the process inside the cylinder pistons assembly 6 (Fig. 16.1).

When ammonia inside the cylinder gets heated, it can easily sense the heat addition and force the piston to move linearly forward. The piston will move to the right if Tank A is heated to a higher temperature than Tank B and to the left if Tank B is heated to a higher temperature. Hence, the target is to move the piston

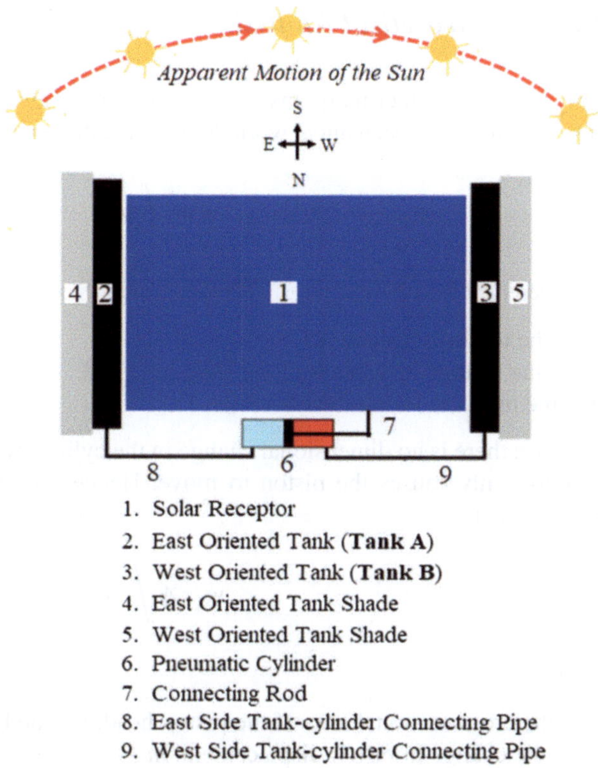

Fig. 16.1 Schematic diagram of the solar thermal tracking system

from the center either in the right or left side depending up on the direction of the sun (the heat source).

16.2.2 Design Parameters

During the evaluation of the solar thermal tracker, the following conditions have been considered.

- Each tank is 1 m long and its inner diameter is 0.5 m.
- Both tanks are directly connected with the two opposite sides of a 0.5-m-long cylinder-piston assembly (double-acting single-ended cylinder). The bore diameter of the cylinder is 0.2 m.
- Ammonia is the working fluid.
- The heating temperature ranges between 20 to 60 °C.
- Heating does not cause thermal expansion on the tanks and cylinder.

16.2.3 Theoretical Analysis

The cylinder is thermally insulated in order to avoid heat exchange with the surrounding. The mechanical work done by the piston can be determined as follows:

$$w = \int pdv \tag{16.1}$$

where

W is the total work done, kJ
dV is the change in volume of the fluid, m^3; and
p is the pressure inside the container, Pa.

Since there is no dimensional change in the cylinder due to the heat addition, heat addition only causes the piston to move. Hence, Eq. 16.1 can be represented by Eq. 16.2 [9].

$$w = A \int pds \tag{16.2}$$

where

A is the cross-sectional area of the piston head, m^2, and
ds is the change in piston displacement, m.

PTC Mathcad Prime 4.0 software was used to simulate the process undergone within the system, and the results are plotted using the Grapher software. The saturated and superheated properties of ammonia during the different temperature intervals of the expansion process are taken from reference books [10].

The fluid at different states are to be discussed. Its properties (temperature, pressure, specific volume) play a great role to do the mechanical work [9, 10].

16.3 Results and Discussion

During the expansion process, once the piston moves and finally reaches its top center or bottom center limit, further heating occurs only at constant volume. As a result, it causes significant pressure to rise in the system. To avoid unnecessary pressurization of the system, the constant volume (isochoric) heat addition processes need to be first properly analyzed.

16.3.1 Isochoric Heating

Figure 16.2 demonstrates all the potential processes that could happen during isochoric heating of ammonia at any phase between 20 and 60 °C. The study revealed that heating wet ammonia may end in a subcooled, wet, or superheated region depending upon its initial conditions and of course the final state.

When wet ammonia at 20 °C and quality which is quality x ranging between 0 and 0.00132 (specific volume ranging between 1.6386 and 1.834 cm^3/g) [10] undergoes isochoric heating, it enters the subcooled (liquid) region before reaching 60 °C. Once it enters the subcooled region, the pressure rises will be significant. On the other hand, if wet ammonia at the same temperature 20 °C and quality above 0.32 (specific volume above 48.8 cm^3/g) undergo isochoric heating, it gets superheated before reaching 60 °C. In this case, once the fluid leaves the wet regions, the rise in pressure is not as significant (may not sufficiently pushing the piston). However, if the entire isochoric heating is accomplished within the wet region having quality x ranging between 0.0132 and 0.32, the pressure increases gently as the fluid temperature rises up to 60 °C. This is the case when the system's specific volume falls within the range of 1.834 and 48.8 cm^3/g (the system is charged within the range of this specific volume). Hence, it can be analyzed that a preliminary assessment to determine the fluid's initial specific volume is required for a safe and smooth expansion process within the cylinder. The result can then be considered as one of the primary criteria for designing or choosing the cylinder piston assembly. The next subsection discusses how the piston is responding when liquid, wet, and vapor ammonia is heated inside the tank.

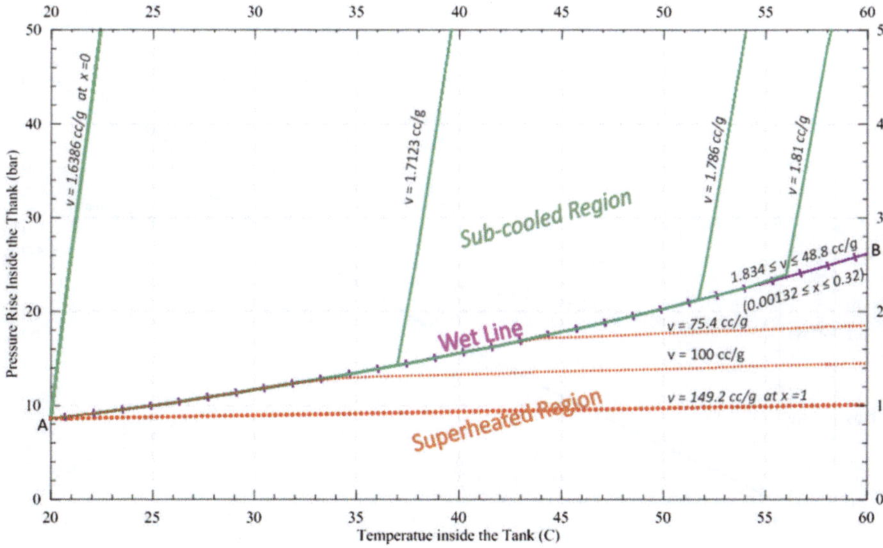

Fig. 16.2 Heating ammonia at constant volume

16.3.2 Piston Movement Achieved by Heating Liquid Ammonia

When subcooled ammonia in one of the tanks gets heated, it does work by pushing the piston forward. Figure 16.3 shows the case when liquid ammonia is heated from 20 to 60 °C at constant pressure. During the process, the density of the fluid drops from 613 to 546 kg/m^3, and the specific enthalpy rises from 275 to 473 kJ/kg. This leads the piston to move from center to 1.543 m forward after doing 0.424 kJ of work by the fluid.

16.3.3 Piston Movement Achieved by Heating Wet Ammonia

When wet ammonia inside one of the tanks gets heated, first the piston will move forward keeping its pressure constant. The process is illustrated in Fig. 16.4. During the process, the density of the fluid drops from 613 to 546 kg/m^3, and the specific enthalpy rises from 275 to 473 kJ/kg. This leads the piston to move from the center to 1.543 m forward. Here, it should be clear that there is no rise in temperature and pressure. The heat addition in the system causes the liquid particles to vaporize.

When the quality changes from its initial value of 0.0016 to 1.0, the piston has the chance to move 273.8 m by doing 7345 kJ of work. However, it should be very clear that high-capacity heat source is required to raise the fluids' specific enthalpy from 276 to 1460 kJ/kg.

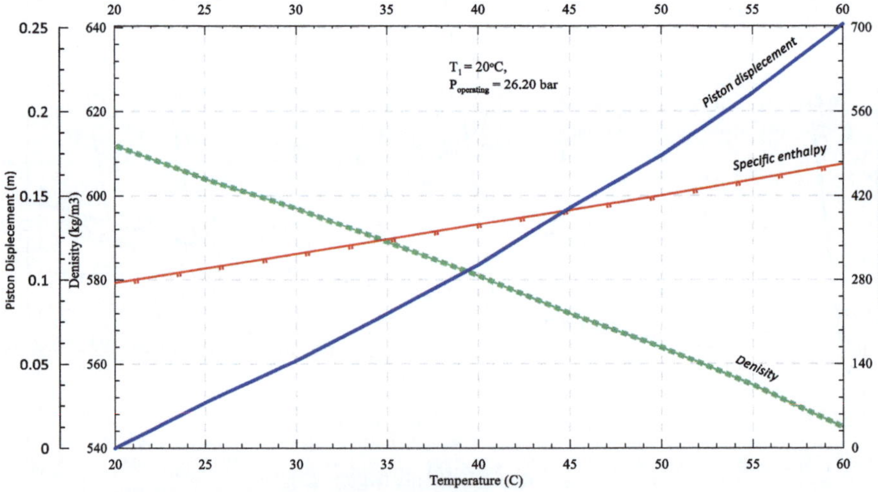

Fig. 16.3 Isobaric heating of ammonia in the subcooled region

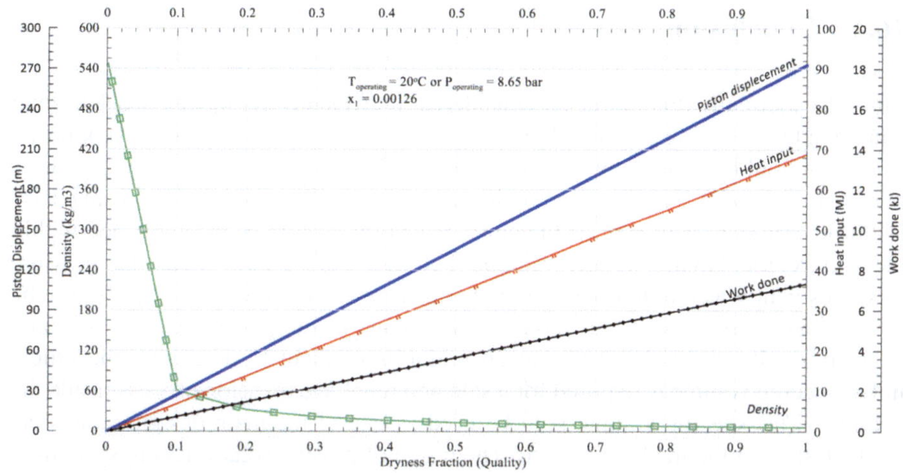

Fig. 16.4 Isobaric heating of ammonia in the wet region

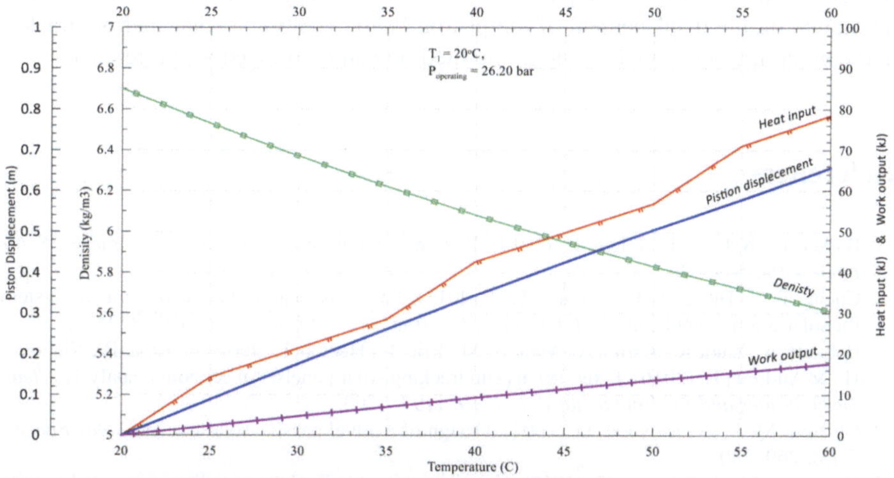

Fig. 16.5 Isobaric heating of ammonia in the superheated region

16.3.4 Piston Movement Achieved by Heating Vapor Ammonia

When superheated ammonia in one of the tanks gets heated, it does a work by pushing the piston forward. Figure 16.5 shows the case when saturated vapor ammonia is heated from 20 to 60 °C at constant pressure. During the process, the density of the fluid drops from 6.7 to 5.61 kg/m^3, and the specific enthalpy rises from 1460 to 1570 kJ/kg. This leads the piston to move from the center to 0.656 m forward by doing 17.58 kJ of work.

16.4 Conclusion

The solar thermally driven tracker considered in this study comprises two 1-m-long ammonia tanks having an inner diameter of 0.5 m each attached to the two opposite sides of a double-acting single-ended cylinder. The cylinder is 0.5 m long with its bore diameter of 0.2 m. The simulation results have been determined using PTC Mathcad Prime 4.0 software. If the entire isochoric heating is accomplished within the wet region having quality ranging between 0.0132 and 0.32, the pressure increases gently as the fluid temperature rises from 20 up to 60 °C. As a result, the system has to be charged to have specific volume ranging between 1.834 and 48.8 cm^3/g. It can be deduced that an initial assessment to determine the fluid's initial specific volume is required for a safe and reliable expansion process inside the cylinder. This specific volume is to be used as one of the main design parameters while designing the tracking system. The usage of passive solar tracking systems is quite unusual. On such systems, experimental and theoretical studies have not been looked at as thoroughly. Taking this into account, the overall goal of this study was to theoretically analyze the impacts of change of working fluid's properties, notably ammonia, on the displacement of the piston. While creating a solar thermal tracker and conducting additional research, the method can be utilized as a reference.

References

1. Barker, L., Neber, M., & Lee, H. (2013). Design of a low-profile two-axis solar tracker. *Solar Energy, 97*, 569–576.
2. Cheng, T. C., Hung, W. C., & Fang, T. H. (2013). Two-axis solar heat collection tracker system for solar thermal applications. *International Journal of Photoenergy, 2013*(1), 803457.
3. Bahram, A., Zaid, A., Semih, A., Marcus, M., Eric, L., Heejoo, L., Jesus, F., Kyle, R., Nicholas, H., & Andrew, B. (2016). Fixed versus sun tracking solar panels: An economic analysis. *Clean Technologies and Environmental Policy, 19*, 1197–1203.
4. Clifford, M. J., & Eastwood, D. (2004). Design of a novel passive solar tracker. *Solar Energy, 77*(3), 269–280.
5. Natarajan, M., & Srinivas, T. (2017). Experimental and simulation studies on a novel gravity based passive tracking system for a linear solar concentrating collector. *Renewable Energy, 105*, 312–323.
6. Kumar, S. S., & Vijayan, S. N. (2016). Solar tracking system using a refrigerant as working medium for solar. *Energy Conversion, 10*(8), 1610.
7. Rane, M. V., & Uphade, D. B. (2017). Novel long stroke reciprocating compressor for energy efficient Jaggery making. *IOP Conference Series: Materials Science and Engineering, 232*(1), 012004.
8. Bloch, H. P., & Godse, A. (2006). *Compressors and modern process applications*. Wiley.
9. Moran, M. J., et al. (2006). *Fundamentals of engineering thermodynamics* (5th ed.). Wiley.
10. Goodenough, W., & Earl Mosher, G. (1913). *The properties of saturated and superheated properties of ammonia vapour*, 66, Illinois, University of Illinois at Urbana-Champaign Library Large-scale Digitization Project.

Chapter 17
Pre-feasibility Study of Wind Power and Site Suitability Analysis in Dire Dawa, Ethiopia

Aleazar Abraham Wollebo, Getachew Shunki Tibba, and Endeshaw Alemu Bekele

17.1 Introduction

These days, sustainability is a milestone for climate conditions as well as the growth of a country. The utilization of renewable energy is one of the most promising strategies for sustainability. Wind energy is the most well-known and boundless of the renewable energy sources that have been employed as alternatives to nuclear and fossil fuels over the past 20 years [1]. Additionally, it will act as the cornerstone and driving force behind the quick development of a world energy system driven by renewable energy sources [2].

Ethiopia has experienced a significant rise in the need for energy generation as a result of both industrialization and population growth. Therefore, diversification of the power-producing system is necessary to provide supply security [3]. To address short-term demand and environmental effects, a fast-track execution capacity expansion is needed, with wind energy being the key to improving the lives of those in Ethiopia without immediate access to electricity. With a potential for 1350 GW of wind energy [4], the nation has enormous renewable energy resources that are

A. A. Wollebo (✉)
College of Mechanical & Industrial Engineering, Institute of Technology, Dire Dawa University, Dire Dawa, Ethiopia
e-mail: aleazar.abraham@ddu.edu.et

G. S. Tibba
Menschen fuer Menschen Foundation, Agro Technical and Technology College, Harar, Ethiopia

E. A. Bekele
Department of Thermal Technology, Silesian University of Technology, Gliwice, Poland

K. Mequanint et al. (eds.), *Sustainable Development Research in Materials and Energy*, Green Energy and Technology,
https://doi.org/10.1007/978-3-031-69860-6_17

dispersed throughout all regions and make it a great place for investors and power development.

Due to economic and financial constraints, Ethiopia's central government is unable to provide renewable energy to nearly half a million people in the country's eastern region, such as Dire Dawa. The region's wind potential is reasonable, but research and development are slipping behind expectations [5]. The Dire Dawa location offers a potential of 50 MW in 40 square kilometers, making it a promising option for wind farm plants [6]. However, there is a lack of site suitability and feasibility studies, making it crucial to consider them before designing a wind farm.

The majority of these issues have been thoroughly investigated as part of the overall micro-siting problem, including wind behavior research, wind turbine interaction analysis, auxiliary facility design (foundations, electrical infrastructure, access roads), economic concerns, environmental assessment, and reliability [7].

Many studies have been conducted on pre-feasibility studies and wind farm site selection. For instance, Cicek [8] used a GIS/spatial statistic-based approach to study wind power project site selection in Turkey, emphasizing the significance of tailored solutions in pre-feasibility studies.

Similarly, Arcidiacono [9] employed GIS for a site potential study for small-scale wind power plants, while Al-Shabeeb et al. [10] used GIS for a preliminary site selection and suitability map of wind turbines in Jordan's northwest. Researchers Issa and Saleous [11] described the studies and evaluations made possible in the GIS environment with precise and trustworthy data, which are then converted into information and guarantee effective outcomes. However, Siamak Moradi [12] prescribed that the site selection criteria and multi-criteria decision-making (MCDM) approach for wind resource assessment using ArcGIS should be created and specified for the study area. Díaz-Cuevas et al. [13] provided a methodology able to evaluate wind farm installation at the regional level. Likewise, Bertsiou et al. [14] presented a technique for determining whether locations are suitable for installing wind turbines based on various combinations of parameters.

The researcher, Bastidas-Salamanca and Bayona, worked on a pre-feasibility study for locating new offshore wind projects in Colombia's Caribbean. This study reviewed the factors to determine when they could be analyzed within three major stages (pre-feasibility, feasibility, and final decision) [15]. In various research works, three software packages are used during the pre-feasibility and feasibility study stages: RET Screen, HOMER, and SAM. The pre-feasibility study is conducted early in the process of analyzing a potential wind farm scenario. The goals of this study is to evaluate the wind energy potential, identify a suitable site, and simulate a hypothetical wind farm near Dire Dawa utilizing data from the National Meteorological Agency, NASA's MERRA reanalysis at different heights, and vortex data. Wind energy potential and wind turbine site selection in the study area were determined using a 3D animated wind turbine, a GIS (geographical information system), and a RETScreen expert. Through a wind energy map that includes a compressive pre-feasibility study and site suitability analysis in the region, it can draw in wind energy investors from donor countries and abroad. It also provides enough data for the government or agencies to set up a framework and policy for applying and utilizing the potential on a large scale.

17.2 Methods and Materials

17.2.1 Description of Study Area

The study was carried out in Dire Dawa, one of Ethiopia's two chartered cities (the other being Addis Ababa). It is located at latitude and longitude 9°36′ N and 41° 52′ E, and its elevation is 1276 m above sea level. A map of the study area is shown in Fig. 17.1.

17.2.2 Wind Resource Assessment

17.2.2.1 Wind Data Analysis

The mean wind speed, mean wind power density, and wind frequency rise were examined in this section. To calculate the potential wind energy of a place, the wind data gathered from the area should be thoroughly examined and analyzed. The wind resource has been categorized into seven wind classes based on the energy produced by a specific wind speed as depicted in Table 17.1.

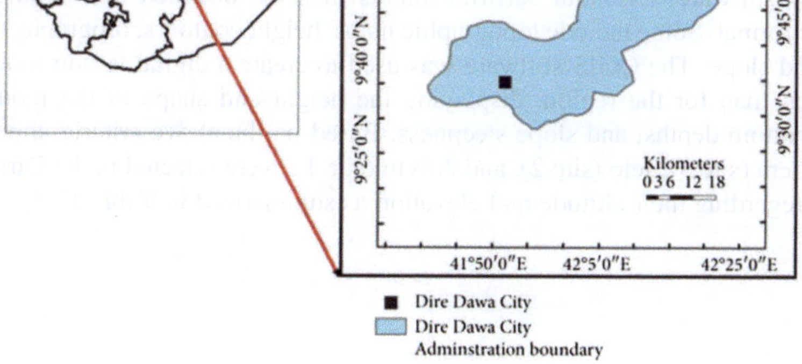

Fig. 17.1 Map of the study area, Dire Dawa

Table 17.1 Standard classes of wind power [16]

Various heights		At 10 m heights		At 50 m heights		At 100 m heights	
S. no	Resource class	m/s	W/m ^ 2	m/s	W/m ^ 2	m/s	W/m ^ 2
1	Poor	0–0.44	0–100	0–5.1	0–160	0–5.4	0–200
2	Marginal	4.4–51	100–150	5.1–5.9	160–240	5.4–6.2	200–300
3	Moderate	5.1–5.6	150–200	5.9–6.5	240–320	6.2–6.9	300–400
4	Good	5.6–6.0	200–250	6.5–7	320–400	6.9–7.4	400–500
5	Excellent	6.0–6.4	250–300	7–7.4	400–480	7.4–7.8	500–600
6	Excellent	6.4–7.0	300–400	7.4–8.2	480–640	7.8–8.6	600–800
7	Excellent	>7.0	>400	8.2–11	640–1600	>8.6	>800

Table 17.2 Selection criteria justification [10]

Criteria	Justifications
Wind speed (W)	Wind turbines depend on having enough wind speed to rotate the wind turbine blades to generate electricity
Rainfall (R)	Areas with more rainfall will generate more runoff, which will affect the wind turbine sites
Altitude (A)	Areas with higher altitudes have the potential of getting more winds than their surrounding
Slope (S)	Areas with a high slope will generate more runoff. Also, high slopes cause technical difficulties when installing wind turbines
Land use (L)	Land uses are essential for environmental and socioeconomic reasons. It is imperative not to degrade the economic value of the land that has a fundamental use

17.2.2.2 Site Suitability Analysis

The GIS (geographic information system) was utilized for analyzing wind farm locations based on site screening criteria listed in Table 17.2. GIS is used for spatial decision support in determining the most suitable sites for new wind farm development.

Once a site is carefully chosen, some data must be found before introducing them into any software. Essential software inputs must be obtained and handled in a specific format. Some include topographic maps, height contours, orography, roughness, and slope. The QGIS software was used to create a digital terrain model and elevation map for the region, displaying the height and shape of the mountains, ocean bottom depths, and slope steepness. Based on the above criteria, three sites, K'ench'era (site 3), Jelo (site 2), and Afretu (site 1), were selected in the Dire Dawa region regarding their altitude and elevation as summarized in Table 17.3.

Table 17.3 Selected sites in Dire Dawa

Sites	Location	Latitude	Longitude	Elevation (ft)
K'ench'era (site 1)	Dire Dawa	9°53′08.06″N	41°83′3.15″N	4549
Jelo (site 2)	Dire Dawa	9°49′08.71″N	41°08′0.43″N	5100
Afretu (site 3)	Dire Dawa	9°54′03.79″N	41°99′0.82″N	5511

17.3 Result and Discussion

To ascertain the proportion of time the wind blew from each direction, the gathered wind data was then categorized by direction of wind. Twelve identical arc segments, each measuring 30°, are often used to sort the wind direction data. The maximum wind speed for the region blown from each segment during the logging period is shown (Fig. 17.2), allowing for a better understanding of wind patterns.

Figure 17.2 shows that the hardest-hitting wind frequency comes from the south (S) and southeast south (SES), as determined by the global wind atlas, vortex, and National Meteorological Agency data. Hence, the turbine should be installed in the south and southeast south direction to get maximum wind frequency.

As altitude increases, the pressure gradient between warm and cold air increases, resulting in a stronger wind due to a larger pressure differential at greater slant. The elevation of Dire Dawa ranges from 996 to 3360 meters above sea level, expressing the region's high altitude. In Fig. 17.3, blue indicates higher elevation, whereas red indicates lower elevation.

A wind index is the most often used data source to calculate the long-term energy production of wind turbines. As Fig. 17.4 illustrates, the wind speed index in Dire Dawa increased toward more notable and stable values between 2015 and 2018. The confirmed data indicates that the wind speed index was lower in 2010 and 2014, while it was moderate in the other years. The indices allow owners of wind turbines to determine whether changes in energy production are caused by wind below projected levels or by defects in wind turbine operation. In comparison to the long-term baseline, the monthly energy yields are relative values [17].

The monthly wind data collected concerning a single reference height is also used to determine wind index areas. Based on various data inputs, the monthly wind data index from May to August and the hourly wind speed index from 6 to 18 h logged at Dire Dawa got lower, as labeled in Fig. 17.5a, b. On the other hand, the remaining monthly and hourly wind speeds in indexes constantly change.

Wind speed significantly impacts the electricity generated by a turbine, as higher winds allow faster blade rotation, resulting in greater power. Most mechanical and electrical power from the generator comes from faster rotation [18]. The turbine's maximum power is produced at this point. The average wind speed in the 10% windiest area in the selected region (Dire Dawa region) is 8.24 m/s, as shown by the percentage of the windiest area in Fig. 17.6b. The measurements were taken at a height of 150 meters using reliable sources like the global wind atlas, vortex, and NASA's MERRA-2 data for predicting wind speed.

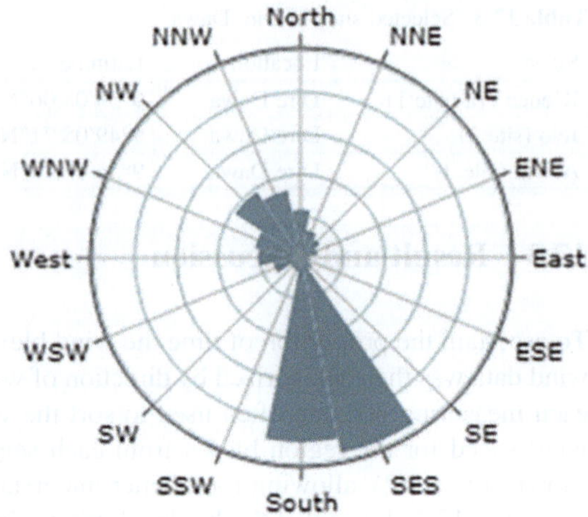

Fig. 17.2 Wind rose of Dire Dawa

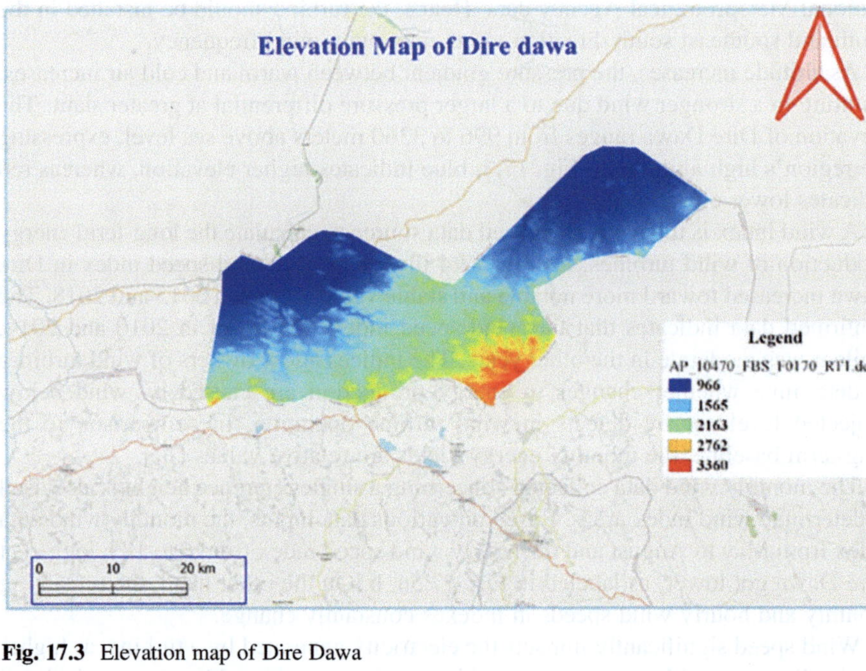

Fig. 17.3 Elevation map of Dire Dawa

The amount of energy that can be converted by a wind turbine at a certain location is indicated by the mean wind speed power density. A greater power density is correlated with an increase in wind speed as height (elevation) rises [19]. The mean wind speed for the 10% windiest area in the selected region (Dire Dawa wind farm) reaches 682 W/m ^ 2 and 804 W/m ^ 2 at 100 m and 150 m, respectively, as illustrated in Fig. 17.7a, b.

Fig. 17.4 Inter-annual data

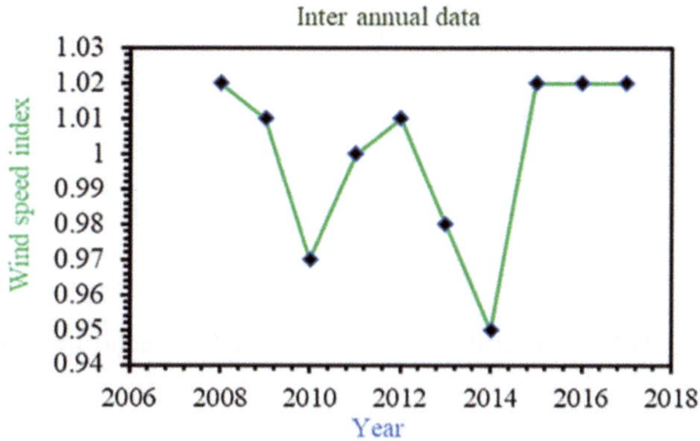

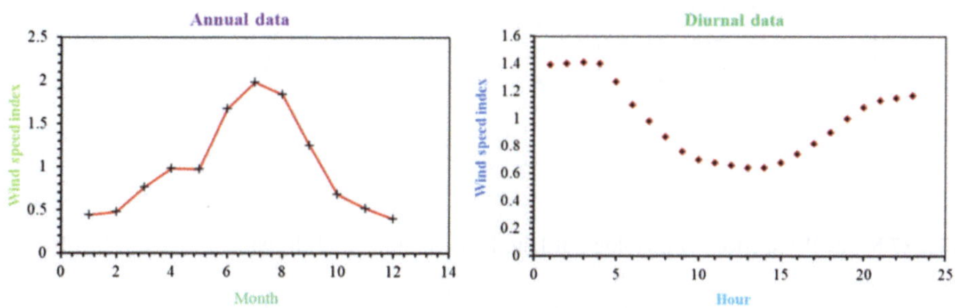

Fig. 17.5 (**a**) and (**b**) Annual and diurnal wind data of Dire Dawa

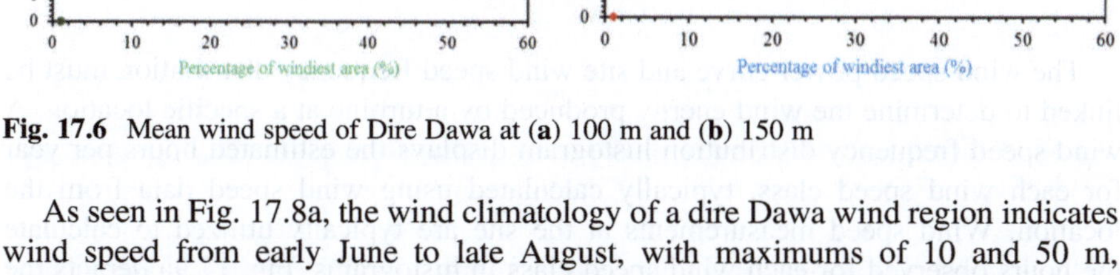

Fig. 17.6 Mean wind speed of Dire Dawa at (**a**) 100 m and (**b**) 150 m

As seen in Fig. 17.8a, the wind climatology of a dire Dawa wind region indicates wind speed from early June to late August, with maximums of 10 and 50 m, respectively. In the same manner, Fig. 17.8b depicts the wind speed distribution in the Dire Dawa area, which is common at 4 m/s and tolerable at 5 and 6 m/s. For a certain place and time, the performance of wind energy systems may be precisely predicted using the wind speed distribution.

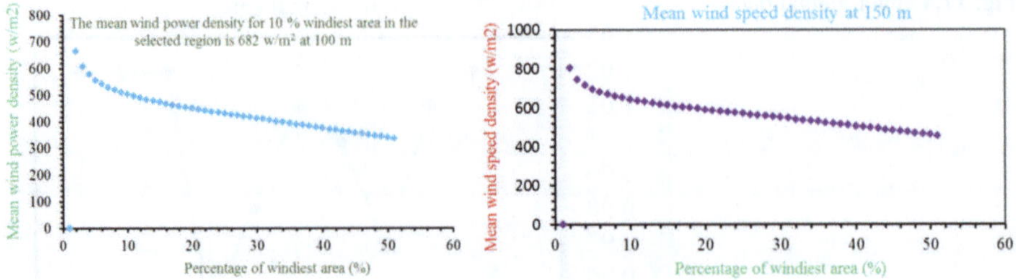

Fig. 17.7 Mean wind power density of Dire Dawa at (**a**) 100 m and (**b**) 150 m

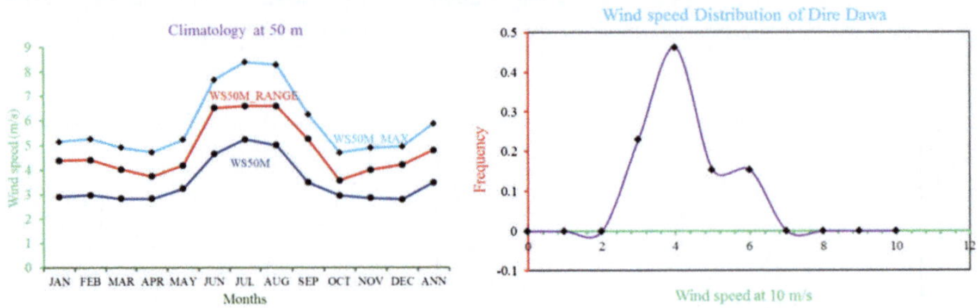

Fig. 17.8 (**a**) Wind speed climatology of Dire Dawa at 50 m and (**b**) wind frequency distribution

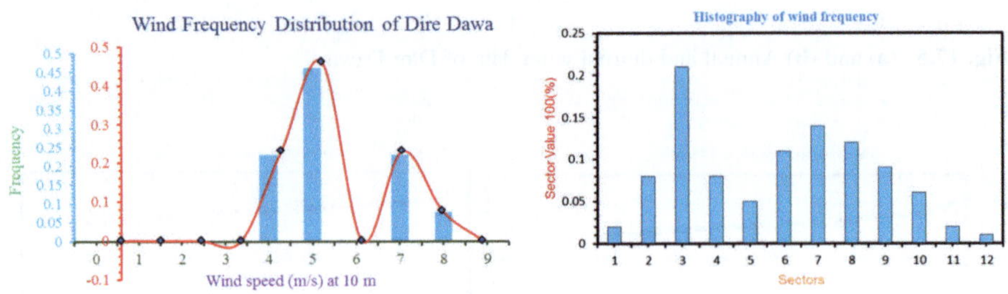

Fig. 17.9 (**a**) Wind frequency distribution and (**b**) histograph of wind frequency

The wind speed power curve and site wind speed frequency distribution must be linked to determine the wind energy produced by a turbine at a specific location. A wind speed frequency distribution histogram displays the estimated hours per year for each wind speed class, typically calculated using wind speed data from the location. Wind speed measurements at the site are typically utilized to calculate the hours observed for each wind speed class in histograms. Fig. 17.9a *d*epicts the distribution of wind frequencies in the Dire Dawa, which is frequent at 4 m/s and moderate at 5 and 6 m/s. Likewise, Fig. 17.9b displays that most months have wind speeds of 2.91–5.25 m/s with a higher frequency. The higher value of wind speed, 5.24–8.41 m/s, with some noticeable frequencies, is found in May and September.

Fig. 17.10 K'ench'era site 3 map

The turbines on hilly terrain encounter larger wind loads than those on flat land; therefore, selecting an appropriate site for the turbine installation is crucial. As a consequence, wind travels from turbine to turbine across hilly terrain, resulting in a faster recovery of its power potential. Hence, turbine rows on hillsides can be more closely spaced. Based on different standards, three sites have been selected in Dire Dawa (Table 17.2) for this paper (Fig. 17.10).

According to the findings, K′ench'era (site 3), which has a mean wind speed of 6.4 m/s, a mean wind power density of 82 W/m ^ 2 , and is located 11.16 Km from the major access road, has a much lower wind energy potential than the other two locations. However, it may be suggested that there is typically capacity for a large, independent wind farm (Fig. 17.11).

In addition, Jelo (site 2), which is located further from the main access road than K′ench'era (site 3) and has a mean wind speed of 7.0 m/s and a mean wind power density of 780 W/m ^ 2, could be exceptional, according to the National Renewable Energy Laboratory (NREL) (Fig. 17.12).

Above and beyond, Afretu (site 1) showed a mean wind speed of 7.4 m/s and a mean wind power density of 804 W/m ^ 2 noticeably higher than the two locations, which could be exceptional based on the National Renewable Energy Laboratory. Additionally, the site is located far from Dire Dawa, at a distance of 28.83 Km from the main access road. The results suggest that Jelo (site 2) and Afretu (site 1) are better candidates than K′ench'era (site 3).

Fig. 17.11 Jelo site turbine placement map

Fig. 17.12 Afretu site turbine placement map

17.4 Conclusion

This study aims to evaluate wind energy potential, identify suitable sites, and simulate a hypothetical wind farm near Dire Dawa. Three sites, K′ench'era, Jelo, and Afretu, are chosen based on wind speed, rainfall, altitude, slope, and land use. The results show that the wind energy potential of K′ench'era (site 1) is significantly lower than the other two sites, with a mean wind speed of 6.4 m/s and a wind power density of 682 W/m ^ 2. However, it has been concluded that there is potential for a large, independent wind farm. Jelo (site 2) has a mean wind speed of 7.0 m/s and a wind power density of 780 W/m ^ 2 and is 11.36 Km away from the main access road. Afretu (site 3) has a mean wind speed of 7.4 m/sand a wind power density of 804 W/m ^ 2 and is 28.83 Km away from the main access road, making it more accessible to Dire Dawa. The results suggest that Jelo and Afretu are better candidates than K'ench'era.

The research aims to assist urban and energy planners in selecting the best alternative energy source based on site characteristics and constraints. It provides a versatile framework for gathering information. Future studies could utilize computational fluid dynamics (CFD) simulation software to model the wind farm and its feasibility. Advanced simulations like Qblade, FAST, and TurbSim could be used to analyze turbine blade performance and fatigue load impacts from wind conditions on complex terrains.

References

1. Gökçek, M., & Genç, M. S. (2009). Evaluation of electricity generation and energy cost of wind energy conversion systems (WECSs) in Central Turkey. *Applied Energy., 86*(12), 2731–2739.
2. Jami, H. (2014). *World wind resource assessment report*. World Wind Energy Association: Bonn, Germany, vol. 7, pp. 868–78.
3. Yebi, A. (2011). *Techno-economic assessment of wind energy to supply the demand of electricity for a residential community in Ethiopia* [Internet]. Dissertation. Available from: https://urn.kb.se/resolve?urn=urn:nbn:se:kth:diva-40914
4. Gielen, D., Gorini, R., Leme, R., Prakash, G., Wagner, N., Janeiro, L., Collins, S., Kadir, M., Asmelash, E., Ferroukhi, R., et al. (2021). *World energy transitions outlook: 1.5°C pathway*; International Renewable Energy Agency: Abu Dhabi, United Arab Emirates, 2021; Available online: https://www.irena.org/publications/2021/Jun/World-Energy-Transitions-Outlook
5. Ministry Of Foreign Affairs of Denmark. (2020). Concept Note for The Program Committee Meeting on Enhanced Danish Ethiopian Energy Partnership.
6. DSIF Investment Proposal Danida Program Committee. (2020). *Assela 2 wind farm project*. Danida Sustainable Infrastructure Finance, Ethiopia.
7. E. I. Administration. (2008). *Wind energy-energy from moving air*. Renewable Energy Annual 2006.
8. Cicek Oz, C. (2015). *Examining the site selection of wind power plants in Turkey in time: A GIS/spatial statistics-based approach*. Master of Science Thesis. Dokuz Eylul University, Graduate School of Natural and Applied Sciences.

9. Arcidiacono, S. A. (2012). *GIS-based site potential analysis for small-scale wind power plants*. Bachelor Thesis. Carinthia University of Applied Sciences School of Geoinformation and Environmental Technologies.
10. Al-Shabeeb, A. R., Al-Adamat, R., & Mashagbah, A. (2016). AHP with GIS for a preliminary site selection of wind turbines in the north west of Jordan. *International Journal of Geosciences, 7*(10), 1208–1221.
11. Issa, S., & Saleous, N. (2018). Modeling the environment with remote sensing and GIS: Applied case studies from diverse locations of The United Arab Emirates (UAE). *Geographic Information Systems and Science., 14*, 81.
12. Moradi, S., Yousefi, H., Noorollahi, Y., & Rosso, D. (2020). Multi-criteria decision support system for wind farm site selection and sensitivity analysis: Case study of Alborz Province, Iran. *Energy Strategy Reviews, 1*(29), 100478.
13. Díaz-Cuevas, P., Biberacher, M., Domínguez-Bravo, J., & Schardinger, I. (2018). Developing a wind energy potential map on a regional scale using GIS and multi-criteria decision methods: The case of Cadiz (south of Spain). *Clean Technologies and Environmental Policy, 20*(6), 1167–1183.
14. Bertsiou, M. M., Theochari, A. P., & Baltas, E. (2021). Multi-criteria analysis and geographic information systems methods for wind turbine siting in a North Aegean Island. *Energy Science & Engineering, 9*(1), 4–18.
15. Bastidas-Salamanca, M., & Bayona, J. G. (2021). Pre-feasibility assessment for identifying locations of new offshore wind projects in the Colombian Caribbean. *International Journal of Sustainable Energy Planning and Management, 15*(32), 139–154.
16. Baloch, M. H., Abro, S. A., Sarwar Kaloi, G., Mirjat, N. H., Tahir, S., Nadeem, M. H., Gul, M., Memon, Z. A., & Kumar, M. (2017). A research on electricity generation from wind corridors of Pakistan (two provinces): A technical proposal for remote zones. *Sustainability, 9*(9), 1611.
17. Kalmikov, A. (2017). Wind power fundamentals. In *Wind energy engineering* (pp. 17–24). Academic.
18. Ayodele, T. R., Ogunjuyigbe, A. S., Odigie, O., & Jimoh, A. A. (2018). On the most suitable sites for wind farm development in Nigeria. *Data in Brief, 1*(19), 29–41.
19. Badurek, C. A.. (2023). *Wind turbine*. Encyclopedia Britannica, https://www.britannica.com/technology/wind-turbine. Accessed 6 Nov 2023.

Chapter 18
Thermodynamics Modeling and Optimization of Stirling Heat Pump: A Review

Temesgen Assefa Minale, Francois Lanzetta, Sylvie Begot, and Muluken Z. Getie

18.1 Introduction

The rapid increase in global temperature that has been observed in the recent years is mostly the result of increased emissions from fossil fuels and the working fluids in the refrigeration and air-conditioning systems [1].

Heating and cooling are essential for our day-to-day activities, and most of them uses the environmentally hazardous working fluids (CFC and HCFC). Therefore, Montreal and Kyoto Protocol were signed in 1987 and in 1997, respectively, for restricting the usage of CFC, HCFCs, Halon, PFCs, HFCs, and SF6 [2]. For that reason, it will be crucial to design a heat pump system in the future without the need of those working fluids.

Regenerative thermal machines are alternative energy conversion machines for heating and cooling at near ambient temperature. The target temperature (170–400 k)

T. A. Minale (✉)
Université de Franche-Comté, CNRS, Institut FEMTO-ST, Belfort, France

Faculty of Mechanical and Industrial Engineering, Bahir Dar Institute of Technology, Bahir Dar University, Bahir Dar, Ethiopia
e-mail: temesgen.minale@femto-st.fr

F. Lanzetta · S. Begot
Université de Franche-Comté, CNRS, Institut FEMTO-ST, Belfort, France
e-mail: francois.lanzetta@univ-fcomte.fr; sylvie.begot@univ-fcomte.fr

M. Z. Getie
Faculty of Mechanical and Industrial Engineering, Bahir Dar Institute of Technology, Bahir Dar University, Bahir Dar, Ethiopia

K. Mequanint et al. (eds.), *Sustainable Development Research in Materials and Energy*, Green Energy and Technology,
https://doi.org/10.1007/978-3-031-69860-6_18

and many applications between this temperature common in air-conditioning, household and commercial refrigeration, vending machines, space- and water-heating systems, and high-temperature heat pump system for industrial applications [3].

Among these regenerative thermal machines, Stirling heat pump is one alternative that can utilize existing energy more efficiently. A Stirling heat pump is operating on Stirling cycle, which consists of four thermodynamic process: isothermal compression, constant volume heat rejection, isothermal expansion, and constant volume heat addition [4]. During the isothermal compression process, the refrigerant is compressed and releases heat to the high-temperature region. The constant heat rejection process follows, during which refrigerant expands further, cools down, and heat is rejected to the regenerator. During the isothermal expansion process, the refrigerant in the cycle absorbs heat from the low-temperature source and expands. Finally, in the constant volume heat addition process, heat is added to the refrigerant through the regenerator, and the refrigerant is heated up.

The researchers are currently looking for alternatives to improve the performance of Stirling heat pumps and the performance predicting techniques of Stirling heat pump by thermodynamics modeling.

18.2 Objective of Review

As compared to the Stirling engine and cry coolers, there is a little study done on Stirling heat pumps. Nowadays, the demand of developing thermal model for Stirling cycle devices increases in order to precisely forecast their performances. The Stirling heat pump design parameters are also being optimized in the meantime in order to enhance its performance for efficient and practical operation. Therefore, the objective of this review is to study the various thermodynamic modeling techniques of the Stirling heat pump and compare the results of each model with the experimental results.

18.3 Methodology

In order to perform this review, articles on the modeling and optimization of the Stirling cycle machines from peer-reviewed journals and conference proceedings were gathered and carefully examined. Zeroth-fourth-order analyses were used to categorize modeling of Stirling hat pump.

18.4 Thermodynamic Modeling

The model reviewed was grouped based on the basic assumptions in the cyclic process. Martin proposed the following thermodynamic modeling for Stirling heat engine, and the modeling of the Stirling heat pump is adapted from this engine model [5] (Fig. 18.1).

18.4.1 Zero-Order Modeling

The primary goal of this model is to create a mathematical correlation and simple computations to evaluate the power output and performance of Stirling cycle machines. The mathematical equation to determine the engine's power output was initially set out by William Beal:

$$w = kV_{SC}fP_{mean} \tag{18.1}$$

where V_{SC} is the compression space is swept volume, f is frequency, and P_{mean} is the mean pressure of the engine.

West [6] verified the validity of Beal's expression and suggested a modified version as follows:

$$w = 0.25V_{SC}fP_{mean} \tag{18.2}$$

This modeling approach provides a rapid overview of the performance, but cannot be utilized to develop new machines or calculate their power output and work input.

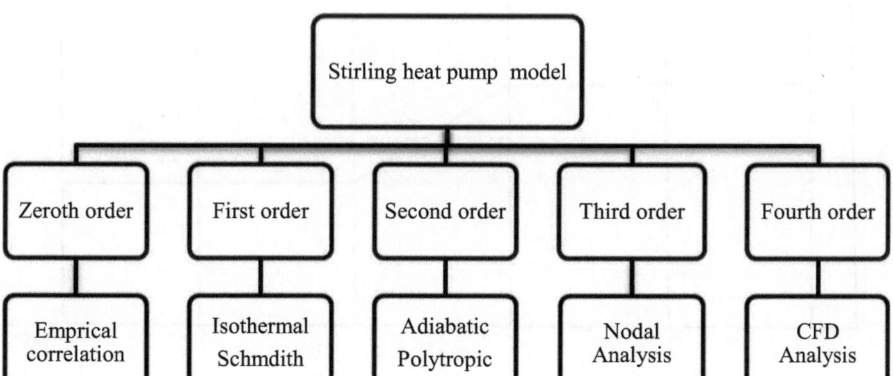

Fig. 18.1 Thermodynamics modeling of Stirling cycle heat pump machines

18.4.2 First-Order Modeling

The first-order model is based on the isothermal analysis, and Schmidt theory was proposed by Gustav Schmidt in 1871 [4]. The main assumption in this study is that the gas in the compression area and heater is at a constant sink temperature, while the gas in the expansion space and cooler is at a constant source temperature. In isothermal analysis, the following assumptions are assumed:

- The mass of working fluid is constant (no mass leakage).
- The equation of state for perfect gas is applied.
- The speed of the machine is constant and steady state is considered.
- The potential and kinetic energy is neglected.
- There is no change in pressure in the system.
- The three heat exchangers (cold heat, cold, and regenerator) are perfect.

The heat pump is configured as five components, namely, compression space (c), heater (h), regenerator (r), cooler (c), and expansion space (e), respectively; see Fig. 18.2. The terms w, T, V, and m p represent the work, temperature of the working gas, volume of the space, mass of working gas, and pressure of the working gas, respectively (Table 18.1).

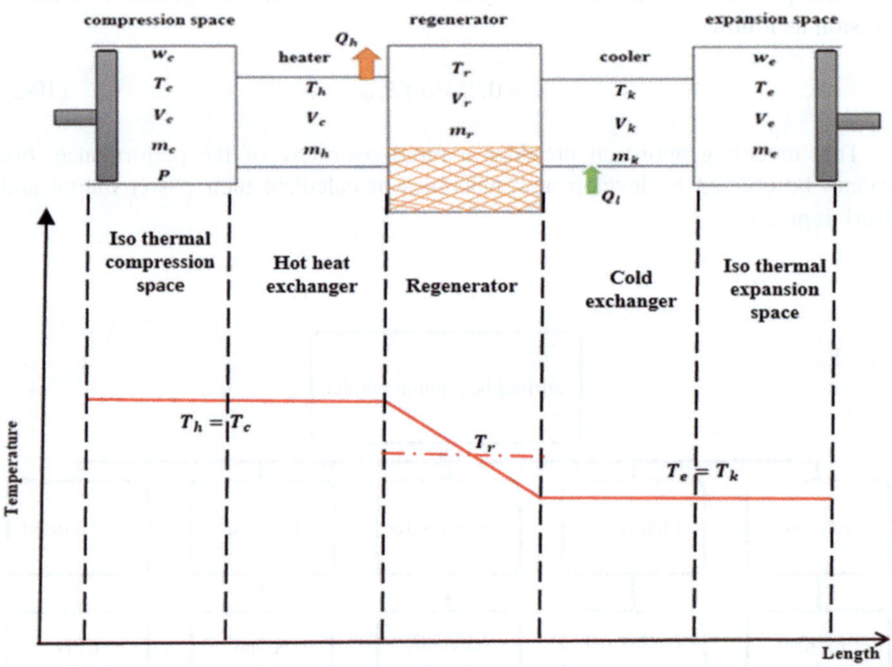

Fig. 18.2 Isothermal analysis of Stirling heat pump

Table 18.1 Ideal isothermal set of equation

$P = MR\left(\frac{V_c}{T_k} + \frac{V_c}{T_c} + \frac{V_r \ln\left(T_c/T_k\right)}{T_c - T_k} + \frac{V_k}{T_k} + \frac{V_e}{T_e}\right)^{-1}$	Pressures
$Q_C = W_C = \oint P \frac{dV_C}{d\theta} d\theta$	
$Q_e = W_e = \oint P \frac{dV_e}{d\theta} d\theta$	Heat transfer
$W = W_e + W_C$	Work input
$COP_{hp} = \frac{Q_h}{W}$	Coefficient of performance

This analysis provides a quick way to estimate the relationship between the overall size and its power output and work input, but it is not very useful as a detailed design tool for Stirling cycle machines. The calculation of power input starts with an ideal loss free analysis (no thermal and power loss), and the simple correlation factor is used to find the break power input. Hence, due to the analysis conducted under ideal conditions, the efficiency values estimated by this model varies extremely from the experimental value.

18.4.3 Second-Order Model

This model is a modification of the first-order analysis, with the compression and expansion space being either adiabatic or polytropic rather than isothermal and including the thermal and power losses depending on the type of second-order model. This model can be classified as follows (Fig. 18.3):

18.4.3.1 Ideal adiabatic Model

An ideal adiabatic analysis is a modification of an isothermal analysis in which the compression space and expansion spaces are adiabatic rather than isothermal; there is no power or thermal loss, and also assumed a perfect heat exchangers [7]. The heat pump is configured by five components serially connected, and the working gas in the heater/hot heat exchanger and cooler/cold heat exchanger is kept under isothermal conditions; see Fig. 18.4. Enthalpy flows across the cells through four interfaces as a result of mass flow being, respectively, compression space/heater, heater/regenerator, regenerator/cooler, and cooler/expansion space.

The assumptions for the ideal adiabatic model analysis are as follows:

- The temperature in the hot and heat exchanger are constant.
- The working gas is constant.
- The compression and expansion space are adiabatic.
- No gas leakage and pressure drop.
- The regenerator is perfect.
- All processes are performed in a steady-state condition.

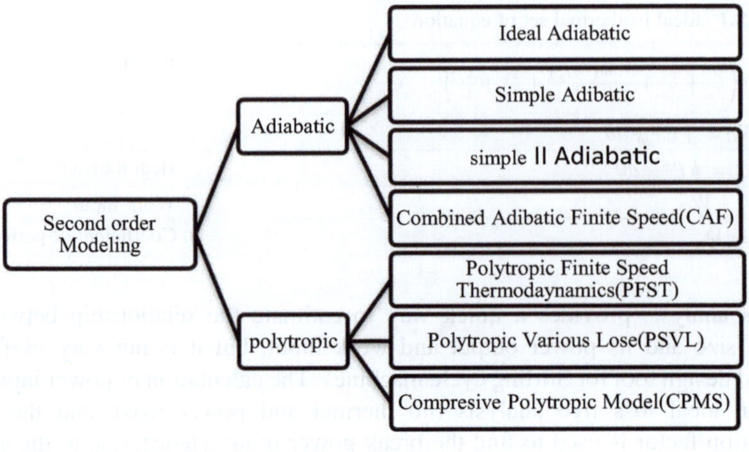

Fig. 18.3 Classification of the second-order analysis

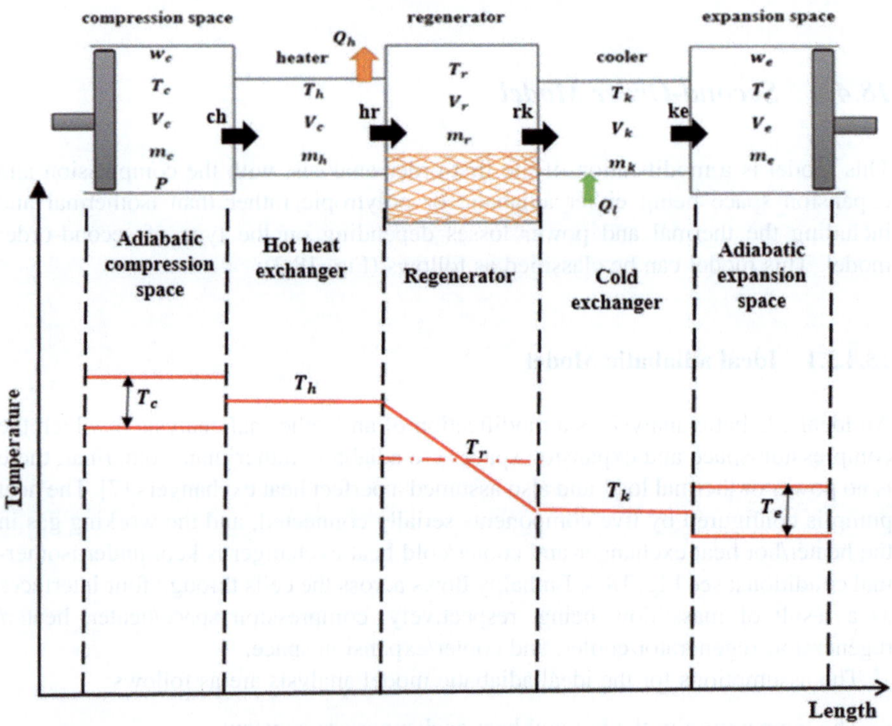

Fig. 18.4 Ideal adiabatic model of Stirling heat pump

- The machine is working at constant angular velocity.
- The temperature of the working fluid in the regenerator is linearly changed.
- The kinetic and potential energies of gas streams are negligible.
- Heat is transferred to the working fluid only in the hot heat exchanger and chiller.
- Heat transfer to the environment is negligible.

18.4.3.2 Simple adiabatic Model (Quasi-Steady Flow Model)

This model is a modification of an ideal adiabatic analysis, in which nonideal heat exchangers are applied for the heat pump analysis (see Fig. 18.5). The heat exchanger wall temperature and gas temperature are considered to be the same in the ideal adiabatic analysis. Thus, the interface temperature of the regenerator is calculated as follows:

$$T_{rh} = \frac{3T_{r1} - T_{r2}}{2} \tag{18.3}$$

$$T_{rr} = \frac{T_{r1} + T_{r2}}{2} \tag{18.4}$$

$$T_{rk} = \frac{3T_{r2} - T_{r1}}{2} \tag{18.5}$$

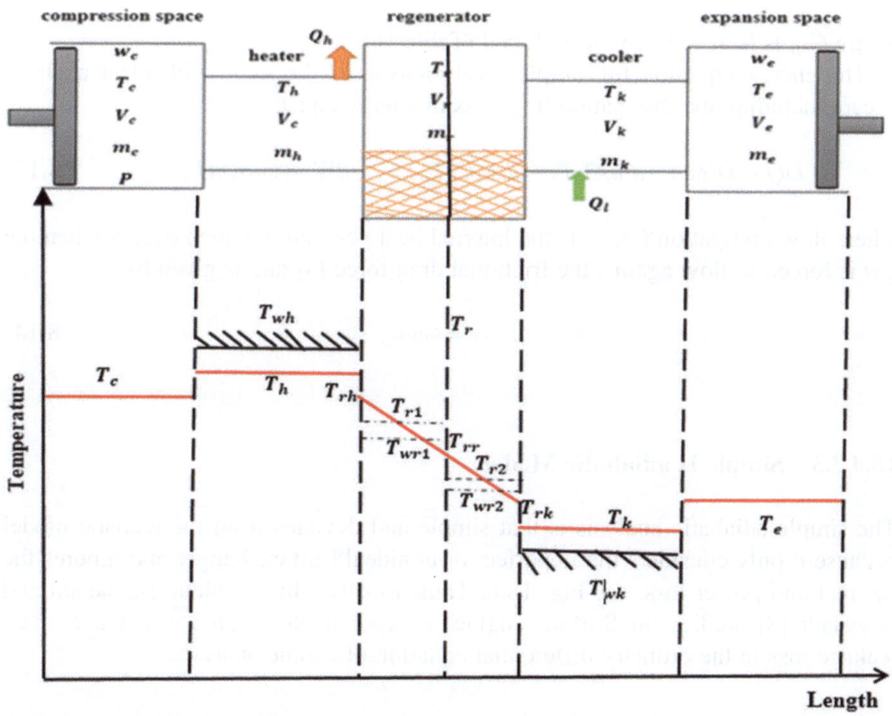

Fig. 18.5 Simple adiabatic model of Stirling heat pump

The heat transfers to the refrigerant or working gas through forced convection, and this convective heat transfer is given by the basic equation:

$$DQ = hA_{wg}(T_w - T) \tag{18.6}$$

where h is convective heat transfer coefficient, T_w is the wall temperature, and T is the gas temperature.

Applying Eq. (18.6) to the four heat exchanger cells, the heat transfer equations for the four heat exchanger cells are as follows:

$$DQ_h = h_h A_{wgh}(T_{wh} - T_h) \tag{18.7}$$

$$DQ_{r1} = h_{r1} A_{wgr1}(T_{wr1} - T_{r1}) \tag{18.8}$$

$$DQ_{r2} = h_{r2} A_{wgr2}(T_{wr2} - T_{r2}) \tag{18.9}$$

$$DQ_k = h_k A_{wgk}(T_{wk} - T_k) \tag{18.10}$$

The heater and cooler wall temperature is maintained isothermally with the sink and source temperature, respectively. The regenerator matrix temperature is given by

$$DT_{wr1} = \frac{DQ_{r1}}{C_{mr}} \tag{18.11}$$

$$DT_{wr2} = \frac{DQ_{r2}}{C_{mr}} \tag{18.12}$$

where C_{mr} is heat capacity of each cell (Table 18.2).

The energy equation for simple ideal analysis is the same with ideal analysis except including the dissipation flow loss in energy equation.

$$DQ - Diss + \left(dm_i C_p T_i - dm_o C_p T_o\right) = -dW + C_v d(mT) \tag{18.13}$$

where flow dissipation (Diss) is the internal heat generation which occurs when the gas is forced to flow against the frictional drag force [4] and is given by

$$Diss = {}^{\Delta\rho g A}/_{\rho} \tag{18.14}$$

18.4.3.3 Simple II adiabatic Model

The simple adiabatic analysis is that simple and deviates from the realistic model because it only considers the imperfect or nonideal heat exchanger and ignores the thermal and power loss; see Fig. 18.6. Thus, to solve this problem, Babaelahi and Sayyaadi [8] studies on Stirling engine incorporate the shuttle effect and mass leakage loss in the ordinary differential equation of simple analysis.

Table 18.2 Ideal adiabatic set of Eq. (18.2)

Parameters	Equations
Pressure Pressure variation	$P = \frac{mR}{\frac{V_c}{T_c}+\frac{V_h}{T_h}+\frac{V_r}{T_r}+\frac{V_k}{T_k}+\frac{V_e}{T_e}}$ $dP = \frac{-P\gamma\left(\frac{DV_C}{T_{Ck}}+\frac{DV_e}{T_{he}}\right)}{\frac{V_c}{T_{ck}}+\gamma\left(\frac{V_h}{T_h}+\frac{V_r}{T_r}+\frac{V_k}{T_k}\right)+\frac{V_e}{T_{he}}}$
Mass	$m_i = \frac{PV_i}{RT_i}$ where i is $c, h, r, k,$ and e
Change of mass	$Dm_i = \frac{m_i DP}{P} = \frac{DP_i}{R}\frac{V_i}{T_i}$ where i is h, r, and k $Dm_c = \frac{PDV_c + \frac{V_c DP}{\gamma}}{RT_{ch}}$ $Dm_e = \frac{PDV_e + \frac{V_e DP}{\gamma}}{RT_{ke}}$
Mass flow	$\dot{m}_{ch} = \frac{dm_c}{dt}$ $\dot{m}_{ke} = \frac{dm_e}{dt}$ $\dot{m}_{hr} = \dot{m}_{ch} - \frac{dm_h}{dt}$ $\dot{m}_{rk} = \dot{m}_{ke} - \frac{dm_k}{dt}$
Conditional temperature	If $\dot{m}_{ch} > 0$ then $T_{ch} = T_c$ else $T_{ch} = T_h$ If $\dot{m}_{ke} > 0$ then $T_{ke} = T_k$ else $T_{ke} = T_e$
Temperature variation	$dT_e = T_e\left(\frac{dP}{P}+\frac{dV_e}{V_e}-\frac{dm_e}{m_e}\right)$ $dT_c = T_c\left(\frac{dP}{P}+\frac{dV_c}{V_c}-\frac{dm_c}{m_c}\right)$
Energy and power	$DQ_{h,ideal} = \frac{V_h DPC_v}{R} - C_p(T_{ch}\dot{m}_{ch} - T_{hr}\dot{m}_{hr})$ $DQ_{r,ideal} = \frac{V_r DPC_v}{R} - C_p(T_{hr}\dot{m}_{hr} - T_{rk}\dot{m}_{rk})$ $DQ_{k,ideal} = \frac{V_h DPC_v}{R} - C_p(T_{rk}\dot{m}_{rk} - T_{ke}\dot{m}_{ke})$ $DW_{in,\ ideal} = -(DW_C + DW_e)$ Where DW_C is PDV_C and DW_e is PDV_e

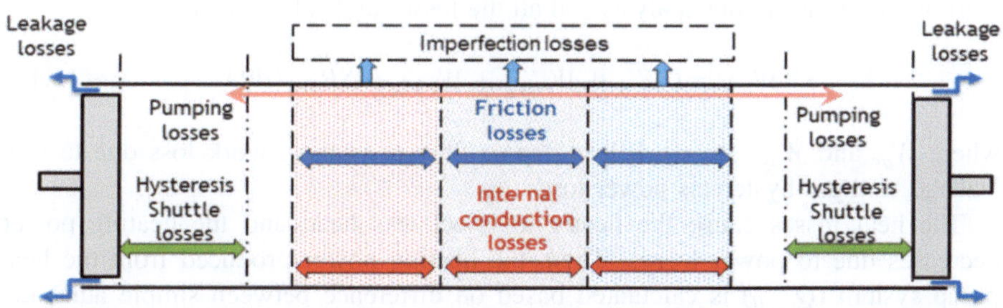

Fig. 18.6 various energy losses in the system [9]

Neglecting the change in potential and kinetic energy, the energy equation for simple II [8] is as follows:

$$dQ + (m_i C_p T_i - m_o C_p T_o) - DQ_{shuttle} = dW + C_v d(mT) \quad (18.15)$$

where $DQ_{shuttle}$ is the heat loss due shuttle effect of the displacer, which is obtained as.

$$DQ_{shuttle} = \frac{\pi S^2 k_g D_d}{8JL_d}(T_C - T_e) \tag{18.16}$$

where k_g, S, L_d, J, and D_d are the thermal conductivity of gas, displacer stroke, length of displacer, cylinder wall, and displacer gap and displacer diameter, respectively.

The working fluid's net mass is assumed to be constant in the adiabatic model and is calculated by [4]

$$M = m_c + m_h + m_r + m_k + m_e - m_{leak} \tag{18.17}$$

where m_{leak} is the leakage of working gas to the crankcase is given by [4]

$$m_{leak} = \pi D \frac{p + p_{buffer}}{4RT_g}\left(u_p J - \frac{J^3}{6\mu}\frac{P - p_{buffer}}{L}\right) \tag{18.18}$$

where Tg, u, J, μ, L, p, p_{buffer}, and D are working fluid temperature, linear velocity of piston, piston-cylinder circular gap, working fluid viscosity, piston length, working space pressure, buffer pressure, and diameter, respectively.

The shuttle and gas leakage loss added to the basic differential equations in the simple adiabatic analysis and obtains new basic differential equations. The set equation considers the effects of pressure loss in heat exchangers, pressure loss due to the piston's finite speed, power loss from mechanical friction, conduction heat loss from the regenerator wall, and nonideal heat transfer in the heat exchangers. The input power requirement ($W_{a,\ II}$) for the heat pump is evaluated as the sum of actual work input in the simple analysis and all the losses and calculated as

$$W_{a,II} = DW_{in,} + W_{mec,f} + W_{FST} + \Delta P_{trott}(dV) \tag{18.19}$$

where W_{pr} and W_{hys} are work loss due to flow resistance, work loss due to seal leakage, and gas hysteresis power loss.

The heat losses cause the heater to reject less heat, and the heating power decreases due to power losses. Then, the heating power produced from the heat pump system ($Q_{a,\ II}$) is calculated based on difference between simple adiabatic analysis heating power and the heat power losses from the system [8] (Table 18.3):

$$Q_{a,II} = DQ_{in,} - (DQ_{shutt} + Q_{rl} + Q_{cond}) \tag{18.20}$$

where Q_{rl}, Q_{cond}, and Q_{shutt} are heat power loss due to regenerator imperfection, conduction heat power loss, and shuttle heat loss, respectively.

$$COP = \frac{Q_{a,II}}{W_{a,II\,p}} \tag{18.21}$$

Table 18.3 Simple II adiabatic set of Eq. (18.8)

Parameters	Equations
Pressure Pressure variation	$P = R\frac{M - m_{leak}}{\frac{V_c}{T_c}+\frac{V_h}{T_h}+\frac{V_r}{T_r}+\frac{V_k}{T_k}+\frac{V_e}{T_e}}$ $dP = \frac{-P\gamma\left(\frac{DV_c}{T_{Ch}}+\frac{DV_e}{T_{ke}}\right)+\gamma R\frac{DQ_{shut}}{C_p}\left(\frac{T_{Ch}-T_{ke}}{T_{Ch}T_{ke}}\right)-2\gamma RDm_{leak}}{\frac{V_c}{T_{ch}}+\gamma\left(\frac{V_h}{T_h}+\frac{V_r}{T_r}+\frac{V_k}{T_k}\right)+\frac{V_e}{T_{ke}}}$
Mass	$m_i = \frac{PV_i}{RT_i}$ where i is c, h, r, k, and e
Change of mass	$Dm_i = \frac{m_i DP}{P} = \frac{DP_i}{R}\frac{V_i}{T_i}$ where $is = h$, r, and k $Dm_c = \frac{PDV_c+\frac{V_c DP}{\gamma}}{RT_{ch}} + \frac{DQ_{shut}}{C_p T_{ch}} + Dm_{leak}$ $Dm_e = \frac{PDV_e+\frac{V_e DP}{\gamma}}{RT_{he}} - \frac{DQ_{shut}}{C_p T_{he}}$
Mass flow	$\dot{m}_{ch} = -Dm_c - \frac{DQ_{shut}}{C_p T_{ch}} - Dm_{leak}$ $\dot{m}_{hr} = m_{hr} - Dm_h, m_{rk} = m_{ke} + Dm_K$ $\dot{m}_{ke} = D\dot{m}_e - \frac{DQ_{shut}}{C_p T_{kc}}$
Conditional temperature	If $\dot{m}_{ch} > 0$ *then* $T_{ch} = T_c$ *else* $T_{ch} = T_h$ If $\dot{m}_{ke} > 0$ *then* $T_{ke} = T_k$ *else* $T_{ke} = T_e$
Temperature variation	$dT_e = T_e\left(\frac{dP}{P} + \frac{dV_e}{V_e} - \frac{dm_e}{m_e}\right)$ $dT_c = T_c\left(\frac{dP}{P} + \frac{dV_c}{V_c} - \frac{dm_c}{m_c}\right)$
Energy and power	$DQ_h = \frac{V_h DPC_v}{R} - C_p(T_{ch})$ $DQ_{h,SII} = \frac{V_h DPC_v}{R} - C_p(T_{ch}\dot{m}_{ch} - T_h\dot{m}_{hr})$ $DQ_{r,SII} = \frac{V_r DPC_v}{R} - C_p(T_h\dot{m}_{hr} - T_k\dot{m}_{rk})$ $DQ_{k,SII} = \frac{V_k DPC_v}{R} - C_p(T_k\dot{m}_{rk} - T_{ke}\dot{m}_{ke})$ $W_{in,\ SII} = -(W_C + W_e)$ $DW_{in,\ SII} = -(DW_C + DW_e)$ Where DW_C *is* PDV_C and DW_e *is* PDV_e

18.4.3.4 Combined Adiabatic Finite Speed (CAF) Model

This approach relies on the combination of finite speed thermodynamics and adiabatic analysis [10]. This model incorporates pressure drop in the three heat exchangers (heater, regenerator, and cooler) into the simple adiabatic analysis of Urieli, the impact of the piston's finite speed mechanical rubbing between the cylinder and piston. The influence of finite speed of the piston and mechanical rubbing is incorporated into the equation of simple adiabatic analysis Eq. (18.10).

$$W_{a,CAF} = DW_{in,} + W_{mec,f} + W_{FST} + \Delta P_{trott}(dV) \tag{18.22}$$

$$\Delta P_{trott} = \Delta P_h + \Delta P_r + \Delta P_k \tag{18.23}$$

The performance of the heat pump is calculated based on Eq. (18.20) by including the mechanical rubbing between the cylinder and piston and finite speed loss in the actual work input, and the working flow diagram of this model is shown in Fig. 18.7.

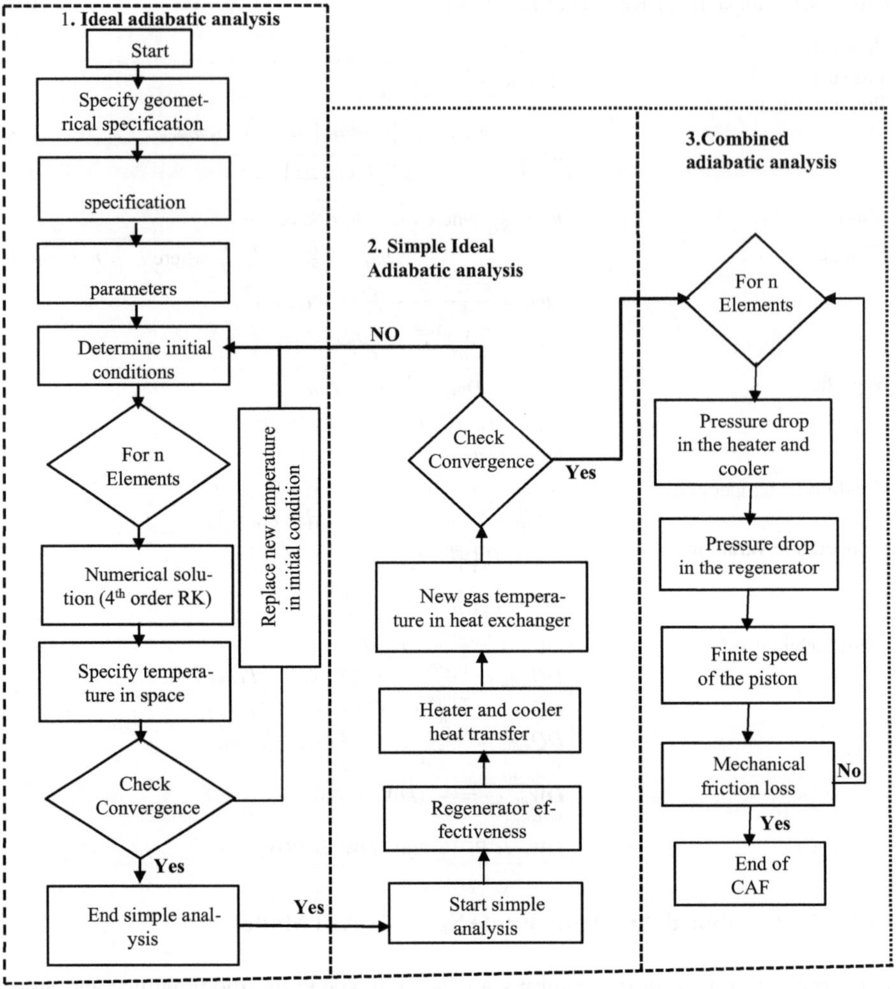

Fig. 18.7 Flow diagram of CAFS model [10]

18.4.4 Polytropic Analysis

On the basis of the modification of the adiabatic analysis model, the polytropic analysis was proposed, and initially Babaelahi and Sayyaadi [11] recommended the polytropic analysis of Stirling engine with various loss mechanisms (PSVL). In the adiabatic analysis, the compression and expansion process is assumed to be adiabatic, but in the polytropic analysis, the compression and expansion space is replaced by polytropic process. In this analysis, the basic differential equation of adiabatic analysis adjusted to include the polytropic compression and expansion process in order to forecast the performance of Stirling heat pump.

The entire engine's volume is divided into five cells similar to the adiabatic analysis, that is, compression space (c), heater (h), regenerator (r), cooler (k), and expansion space (e), and the four interfaces, that is, compression space/heater (ch), heater/regenerator (hr), regenerator/cooler (rk), and cooler/expansion space (ke); see in Fig. 18.4. The basic differential equation changed to include the effect of shuttle loss and polytropic heat transfer given by [12].

$$Q - \left(D_{Polytropic} + DQ_{Shuttle}\right) + \left(dm_i C_p T_i - dm_o C_p T_o\right) = dW + C_v d(mT) \quad (18.24)$$

where $D_{Polytropic}$ is the polytropic heat loss in the compression and expansion space and is given by [13]

$$D_{Polytropic} = C_n(T_a - T) \quad (18.25)$$

where T_a and T are the section temperature and ambient temperature, correspondingly.

C_n is the specific heat capacity of the polytropic and is given by [14]

$$C_n = C_v \frac{n-k}{n-1} \quad (18.27)$$

where n is the polytropic index and is obtained by [11].

$$n = -\frac{VdP}{PdV} \quad (18.28)$$

18.4.4.1 Polytropic with Various Loss (PSVL)

This modeling is based on the polytropic expansion and compression space of the working space. The following losses are considered in this thermodynamics model to enhance its accuracy of the analysis:

- Impact of mass leakage from the working space to buffer space.
- Heat loss (shuttle heat transfer effect).
- Nonideal regenerator.
- Heat exchangers pressure drop.
- Mechanical friction.
- Heat exchangers longitudinal heat conduction loss.

The governing differential equations of this model is the same as Eq. (18.24). The resulting model was applied to the GPU-3 Stirling engine prototype, and the results were compared to those of other developed models and experimental findings. A 24.44% thermal efficiency and helium gas as a refrigerant at heater temperature of 977 K and cooler temperatures of 288 K were found [11].

18.4.4.2 Polytropic Finite Speed Thermodynamics (PFST)

This model has been developed through an integration of finite speed thermodynamics and polytropic analysis. The work input for this model includes the following:

- The pressure drop in the heat exchangers.
- Mechanical friction.
- Finite speed of the piston.

The governing differential equation of this model is a modification of Eq. (18.24) which do not include the shuttle heat loss. The resulting model was applied to the GPU-3 Stirling engine prototype, and the results were compared to those of other developed models and experimental findings. A 23.3% thermal efficiency and helium gas as a refrigerant at heater temperature of 977 K and cooler temperatures of 288 K were found.

18.4.4.3 Compressive Polytropic Model (CPMS)

This modeling technique is a modification of polytropic with various loss (PSVL) analysis. The heater and cooler is considered as non-isothermal instead of isothermal process; see Fig. 18.8.

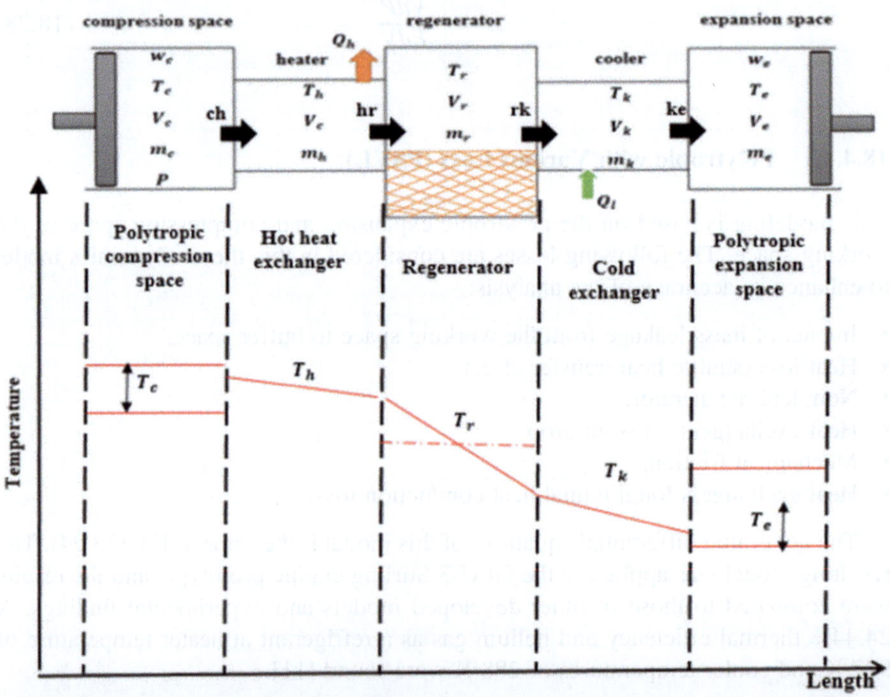

Fig. 18.8 CPMS model of Stirling heat pump

18.4.5 New Nonideal Second-Order Thermal Model

An enhanced nonideal thermodynamic model of the Stirling heat pump with a variety of losses has been created in order to improve the previous second-order models used for the thermodynamic analysis of Stirling devices [15]. In this model, the basic differential equations for the heater and parts of the devices have been coupled with the shuttle heat loss [4]. In addition, the mass leakage into the crankcase and the cooler part were added into the conservation of mass equations, by taking into account the mass leakages across the system's boundaries. Generally, the following losses are considered in this nonideal second-order model [16]:

$$\begin{aligned} &DQ_{ideal} - DQ_{shut} - DQ_{cond} - DQ_{disp} - DQ_{non-ideal} - DQ_{leak} \\ &\quad + C_p(m_i T_i + m_o T_o) = DW_{ideal} + DW_{fs} + DW_{mf} + DW_{pd} + DW_{hys} \\ &\quad + C_v d(mT) \end{aligned} \tag{18.28}$$

The resulting model was applied to the GPU-3 Stirling engine prototype, and the results were compared to those of other developed models and experimental findings. A 20.4% thermal efficiency and helium gas as a refrigerant at heater temperatures of 977 K and cooler temperatures of 288 K were found [15].

18.4.6 Third- and Fourth-Order Model

Based on the mass, momentum, and energy balances of the various control volumes, third-order modeling involves developing partial differential equations, which govern the functioning of the devices. Feurer [17] had previously developed a Stirling engine using the third-order concept. To create an effective regenerator for the Stirling engine, Andersen [18] extensively examined the impact of regenerator matrix temperature oscillations using the third-order model.

Toghyani [19] developed a third-order analysis to evaluate the GPU-3 Stirling's suitable heat source temperature, frequency, engine stroke, and effective mean pressure in to obtain the best output power and performance. However, this model did not produce results that were superior compared to those of the adiabatic models already in use due to simplification of the process in order to increase the computational speed.

The fourth-order analysis uses three-dimensional computational fluid dynamics to solving complex flow equations at each node of the created mesh of the modeling [20]. This model is very useful for analyzing the flow field and heat transfer of the working fluid inside the heat pump. Ahmed et al. [21] conducted an extensive investigation to assess the energy effectiveness of the different types of Stirling engines .

Marek and Jan [22] study shows that in a three-dimensional computational fluid dynamics modeling, a dynamic mesh was used to map the Stirling engine's different volumes. The researchers compared the adiabatic model findings with the results produced using a fourth-order modeling strategy. According to their findings, second-order analysis is better for design and optimization due to the shorter computational time.

18.5 Model Validation

The thermal methods have been studied with a standard Stirling engine that has been implemented as a subject of investigation for several studies. A General Power Unit-3 (GPU-3) Stirling engine is used in order to validate each thermodynamic analysis of a Stirling engine, heat pump, and refrigerator. Comparisons between the thermal efficiency of Stirling engines with different thermal approaches are illustrated through Fig. 18.9.

The efficiency of various second-order thermodynamic modeling techniques of the Stirling engine is shown in Fig. 18.9. Therefore, it can be concluded that the nonideal second-order model is more accurate than all current second-order thermal models to forecast the performance of the GPU-3 engine at the design average effective pressure of 4.14 MPa, heater temperature of 977 K, and cooler temperature of 288 K.

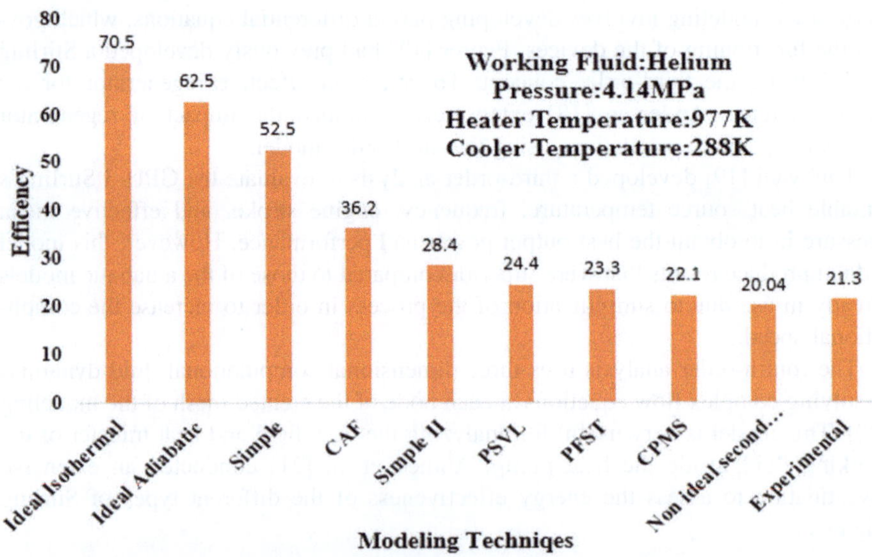

Fig. 18.9 Comparison of various second-order engine models

18.6 Conclusion

In this paper, the thermodynamic analysis of the Stirling heat pump, which is adapted from the Stirling engine model, is extensively reviewed and compared with the experimental value of GPU-3 Stirling engine. Those models start from the simple zero-order model up to the fourth-order model. The zero-order models are empirical approaches that use experimental parameters for predicting the performance of the engine, whereas the first-order models concentrate on closed-form mathematical models. On the other hand, the second-, third-, and fourth-order models are numerical models with increasing degrees of accuracy.

The fourth-order model compared to the other numerical models had enhanced the distribution of losses and the flow fields in the engine; however, the outcomes of this method were not considerably superior to those of second-order models. Furthermore, compared to second-order numerical analyses, fourth-order studies need a lot of computing time.

As a result, the review highlights considerable value in the second-order model, which is better than the other model for thermal analysis of Stirling heat pumps in terms of predicting precise and consistent findings in less computational time. A comparison between the thermal efficiency of each second-order model and the experimental approaches is illustrated. The result of this comparison showed that the nonideal second-order with additional loss effects is better than the other methods.

Acknowledgments This work has been supported by the EIPHI Graduate School (contract ANR-17-EURE-0002) and the Region Bourgogne-Franche-Comté, the Embassy of France to Ethiopia, the African Union, Bahir Dar University, and the Ministry of Education of Ethiopia.

References

1. Karl, T. R., Melillo, J. M., Peterson, T. C., & Hassol, S. J. (2009). *Global climate change impacts in the United States*. Cambridge University Press.
2. Getie, M. Z. (2021). Numerical modeling and optimization of a regenerative Stirling refrigerating machine for moderate cooling applications.
3. Kagawa, N. (2000). *Regenerative thermal machines (Stirling and Vuilleumier cycle machines) for heating and cooling*. International Institute of Refrigeration.
4. Urieli, I., & Berchowitz, D. M. (1984). Stirling cycle engine analysis.
5. Martini, W. R. (1983). Stirling engine design manual.
6. West, C. (1986). *Principles and applications of Stirling engines*. Springer.
7. Finkelstein, T. (1975). Computer analysis of Stirling engines. *Advances in Cryogenic Engineering, 20*, 269–282.
8. Babaelahi, M., & Sayyaadi, H. (2014). Simple-II: A new numerical thermal model for predicting thermal performance of Stirling engines. *Energy, 69*, 873–890.
9. Hachem, H., Gheith, R., Aloui, F., & Nasrallah, S. B. (2018). Technological challenges and optimization efforts of the Stirling machine: A review. *Energy Conversion and Management, 171*, 1365–1387.

10. Hosseinzade, H., & Sayyaadi, H. (2015). CAFS: The combined adiabatic–finite speed thermal model for simulation and optimization of Stirling engines. *Energy Conversion and Management, 91*, 32–53.
11. Babaelahi, M., & Sayyaadi, H. (2015). A new thermal model based on polytropic numerical simulation of Stirling engines. *Applied Energy, 141*, 143–159.
12. Ahmed, F., Huang, H., Ahmed, S., & Wang, X. (2020). A comprehensive review on modeling and performance optimization of Stirling engine. *International Journal of Energy Research, 44*, 6098–6127.
13. Wu, C. (1999). *Recent advances in finite-time thermodynamics*. Nova Publishers.
14. Martaj, N., Rochelle, P., Grosu, L., Bennacer, R., & Savarese, S. (2009). Numerical Study of an LTD Stirling engine with a porous regenerator. In *Conf. COMSOL.*
15. Udeh, G. T., Michailos, S., Ingham, D., Hughes, K. J., Ma, L., & Pourkashanian, M. (2020). A new non-ideal second order thermal model with additional loss effects for simulating beta Stirling engines. *Energy Conversion and Management, 206*, 112493.
16. Getie, M. Z., Lanzetta, F., Bégot, S., Admassu, B. T., & Djetel-Gothe, S. (2021). A non-ideal second order thermal model with effects of losses for simulating Beta-type Stirling refrigerating machine. *International Journal of Refrigeration, 130*, 413–423.
17. Feurer, B. (1973). Degrees of freedom in the layout of Stirling engines. *Stirling Engines, 28*, 18–37.
18. Andersen, S. K. (2006). *Numerical simulation of cyclic thermodynamic processes*. Technical University of Denmark, Department of Mechanical Engineering.
19. Toghyani, S., Kasaeian, A., Hashemabadi, S. H., & Salimi, M. (2014). Multi-objective optimization of GPU3 Stirling engine using third order analysis. *Energy Conversion and Management, 87*, 521–529.
20. Mahkamov, K. (2006). Design improvements to a biomass Stirling engine using mathematical analysis and 3D CFD modeling.
21. Abuelyamen, A., & Ben-Mansour, R. (2018). Energy efficiency comparison of Stirling engine types (α, β, and γ) using detailed CFD modeling. *International Journal of Thermal Sciences, 132*, 411–423.
22. Jan, W., & Marek, P. (2016). Mathematical modeling of the Stirling engine. *Procedia Engineering, 157*, 349–356.

Chapter 19
State of the Art of Human Energy Acquisition Technologies for Electric Power Generation: A Review

Addisu Alamirew Meku, Berihun Abebaw Mebratie, Bekalu Sintayehu Ayele, Samuel Getachew Chekol, and Addisu Derso Asemie

19.1 Introduction

Electricity is a widely used, clean, and efficient energy source, but the increasing global population has led to a rise in demand, with 2 billion people in developing countries lacking reliable access to electricity. Fossil fuels are the primary source of energy, contributing to 82.7% of total energy consumption and causing global warming and grid decarbonization [1–5]. To address these challenges, nontraditional clean energy systems such as wind, solar, and wave energy have been developed, which can provide increased environmental safety, economic security, and efficient manufacturing [6]. Renewable power plants (RPPs) now account for over a quarter of the world's total electricity-producing capacity, providing clean, safe, dependable, and affordable electricity [7]. Hybrid renewable energy systems (HRESs) have also been developed to achieve higher efficiency by combining multiple renewable energy sources or at least one renewable and conventional source. However, factors such as reliability and cost-effectiveness must be considered [3, 8–11].

As global energy consumption increases, researchers are exploring human energy acquisition technologies (HEAT) to harness energy from the human body [12–14]. HEAT refers to technologies that harness energy from human activity, such as walking, running, and cycling, and convert it into usable forms of energy, such as electricity. In recent years, there has been a growing focus on these technologies as

A. A. Meku (✉) · B. A. Mebratie · B. S. Ayele · S. G. Chekol
Faculty of Mechanical and Industrial Engineering, Bahir Dar Institute of Technology, Bahir Dar University, Bahir Dar, Ethiopia

A. D. Asemie
Department of Mechanical Engineering, Gafat Institute of Technology, Debre Tabor University, Debre Tabor, Ethiopia

K. Mequanint et al. (eds.), *Sustainable Development Research in Materials and Energy*, Green Energy and Technology,
https://doi.org/10.1007/978-3-031-69860-6_19

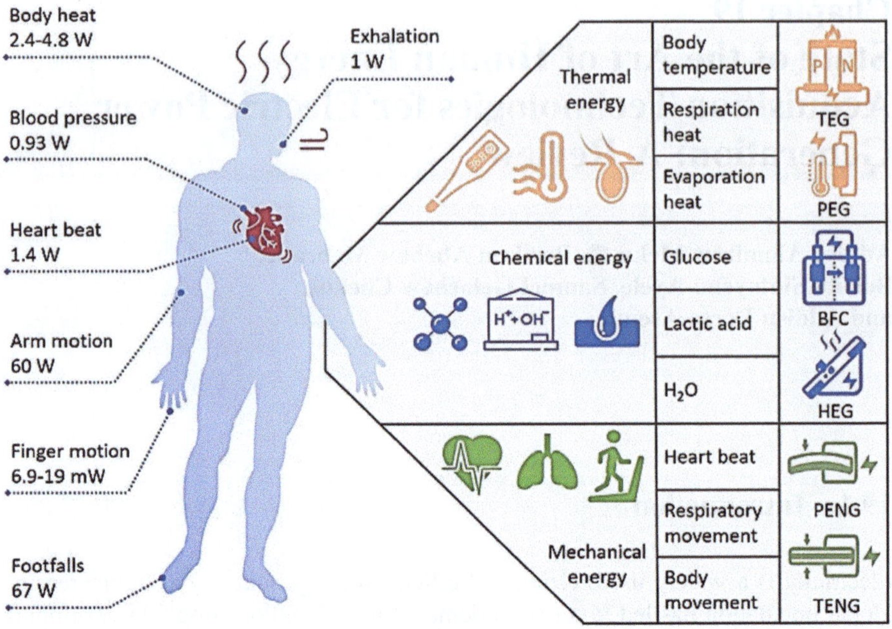

Fig. 19.1 Source and distribution of human body energy and applicable energy-harvesting technologies [20–22]

they hold immense potential in offering sustainable and eco-friendly energy sources. Notably, they have the capability to effectively mitigate greenhouse gas emissions and extend energy accessibility to underserved and remote communities.

Wearable energy generation is one area of focus in HEAT, allowing individuals to generate energy on-the-go using devices that can be worn or integrated into clothing or accessories [15, 16]. Another area of focus is on developing technologies that can convert human-powered motion, such as cycling, into usable electricity. These generators have been used in both developed and developing countries as a source of off-grid power, particularly in remote areas where access to electricity is limited [17]. In addition, technologies that can convert the energy generated by exercising in gyms into usable electricity have been developed, such as the SportsArt Eco-Power Line of fitness equipment and the ReRev system [18, 19].

The human body generates three types of energy (Fig. 19.1)—chemical, thermal, and mechanical—through various physiological processes such as metabolism, body temperature regulation, and muscle movement. These energies can be harnessed and converted into electrical energy, paving the way for a clean and renewable source of power. By exploring different sources of energy and their outputs, this study aims to shed light on the thermal and mechanical human energy acquisition technologies. While the efficiency of HEAT is relatively low compared to conventional power sources, ongoing research and development hold promise for continued improvement and expanded applications of these technologies [23].

19.2 Thermal Energy Harvesting in the Human Body

The human body has a naturally constant heat source. It constantly releases heat into the surroundings while consuming a significant quantity of energy each day to keep its body temperature constant. Hence, the heat generated by the human body can be harnessed and repurposed as a perpetual energy source. Crucially, this method of energy harvesting poses no adverse effects on the human body, making it an ideal and sustainable resource that can be continuously tapped into at any given time. Currently, the thermoelectric and pyroelectric effects, which are represented by the thermoelectric generator (TEG) and the pyroelectric generator (PEG), respectively, are the two types of energy harvesters that are primarily responsible for the extraction of thermal energy from the human body [24].

19.2.1 Thermoelectric and Pyroelectric Generators

Human energy acquisition technologies are a rapidly evolving field, with one promising area of research being the use of thermoelectric generators (TEGs). These generators can convert temperature differences into electrical power, making them ideal for wearable devices that can utilize the body heat of the wearer to generate electricity. In fact, researchers at North Carolina State University have already developed a wearable TEG that can harvest body heat to power small electronic devices, while others have integrated TEGs into clothing such as jackets and shirts to generate electricity from body heat [25, 26].

As metabolic processes inherently produce heat, there are two distinct types of thermal energy harvesters that can capture heat from the human body using different approaches and convert it into electrical energy. Enhancing the thermal contact surface and employing an efficient power management system can augment the output and conversion efficiency. Thermoelectric generators (TEGs) are presently the primary method for harvesting thermal energy from the human body. They employ an all-solid-state energy conversion approach, devoid of chemical reactions or fluid mediums. Consequently, TEGs offer noise-free, vibration-free operation and exhibit a prolonged service life during the energy generation process [27, 28].

Nevertheless, TEGs operate based on a temperature gradient and often consist of multiple sets of semiconductor thermocouples arranged in series and parallel, resulting in a complex and costly design. Conversely, pyroelectric generators (PEGs) can solely capture thermal energy that exhibits temperature variations over time, such as intermittent heat from breathing through the mouth and nose. PEGs possess a straightforward device structure and yield relatively high-output voltages. However, they lack the stability of TEGs and are susceptible to influences from high temperatures and humidity.

Notwithstanding these obstacles, TEGs offer a potential solution for powering wearable electronics. In fact, the National Science Foundation (NSF) Advanced Self-Powered Systems of Integrated Sensors and Technologies (ASSIST) initiative

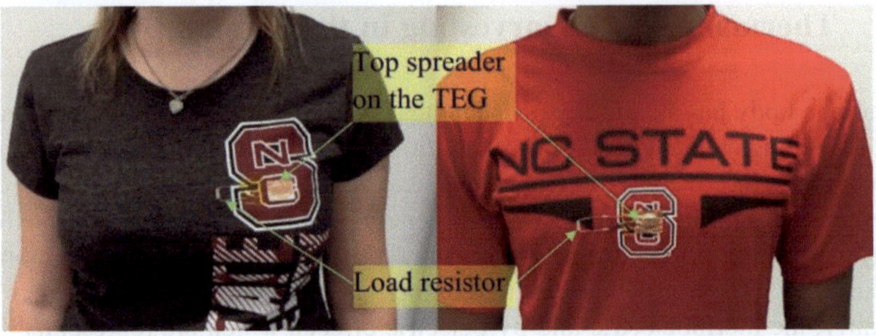

Fig. 19.2 The T-shirts TEG devices integrated [29]

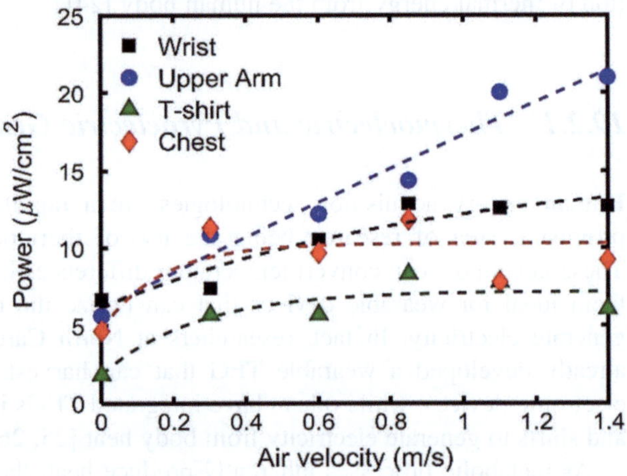

Fig. 19.3 Comparison of the TEG power on the wrist, upper arm, T-shirt, and chest [29]

is actively exploring TEGs to develop wearable, self-powered systems for health and environmental monitoring. This research aims to integrate TEG technology into wearable devices, enabling them to operate autonomously while providing valuable insights into personal well-being and environmental conditions. Figure 19.2 depicts the different sources of energy generated by the human body and their outputs, highlighting the exciting possibilities of human energy acquisition technologies as a clean and renewable source of energy [29].

TEG energy is gaining prominence due to advancements in the expanding realm of green technologies and its ability to convert low-grade or waste heat into electricity. In scenarios where there is a heat source, TEGs have the potential to significantly reduce or even eliminate the need for batteries [30–32].

The research findings indicate that the upper arm exhibits the highest power generation, while the wrist and chest produce comparatively lower power (refer to Fig. 19.3). The TEG suspended on the T-shirt demonstrates the lowest power output. The upper arm and chest locations are particularly suited for integration with electrocardiogram (EKG) sensors. When the person is stationary, both body locations yield similar power outputs. However, when the person is walking, the upper arm location generates higher power [29].

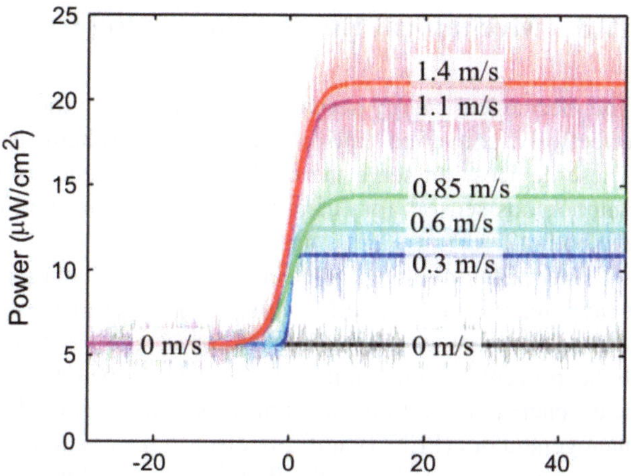

Fig. 19.4 Power generated from the TEG on the upper arm at different walking speeds [29]

One of the primary challenges in textile-based wearable electronics is ensuring washability. E-textiles with high performance should withstand repeated washing cycles while preserving their desirable properties. Given the growing interest in textile-based wearables, several recent studies have explored the development of wash-resistant E-textiles.

Figure 19.4 reveals that the output electric power can be altered based on the speed. The collected data from these experiments demonstrated a noteworthy increase in power generation with higher walking or running speeds, aligning with expectations. Running speed directly correlates with power production, resulting in an approximate power output of 18 μW/cm^2. The data trends indicate that as both body heat and airflow intensify at faster running speeds, greater power can be generated. Importantly, no significant disparity in power output was observed between male and female participants wearing the T-shirts during this experiment [29].

There is coin-size coiled-up wearable [20] and silk-fabric-based flexible [31] TEG devices that can produce electric energy. With a temperature difference of 5 K, the apparatus could only produce 0.8 W/cm^2 of power. It is anticipated that the power will be considerably lower on the body given that there is typically a 1 K temperature differential between the skin and ambient temperature. With a 50-mm-by-50-mm device area, the manufactured TEG could produce 2.1 μW of power on the human body. Once more, the highest power output on the human body was extremely low, in the region of 15 nW with a device area of 4 cm × 8 cm [33]. Through the utilization of a TEG integrated with optimized heat spreaders, the maximum output power achieved was 6 μW/cm^2 when there was no motion, and it increased to 20 μW/cm2 at a normal walking speed. Small environmental and health monitoring tools like accelerometers, ozone sensors, and electrocardiogram (ECG) sensors can be operated with the output power. These devices exhibit power consumptions of approximately 10, 150, and 50 μW, which can be supplied by the enhanced TEG device with respective areas of 0.5, 7.5, and 2.5 cm^2 [34].

19.3 Mechanical Energy Harvesting from the Human Body

The mechanical energy derived from human activities and organ movements can be effectively converted into electricity by mechanical energy-harvesting devices. These devices can serve the dual purpose of generating power and functioning as autonomous sensors for detecting human motions and physiological signals within the body [27].

The invention of the first electromagnetic induction generator, although large and heavy, represented a significant milestone as it marked the conscious conversion of human mechanical energy into electrical energy. Subsequently, radios and flashlights powered by human motion were developed. Presently, two emerging mechanical energy-harvesting technologies have gained prominence: piezoelectric nanogenerators (PENG) and triboelectric nanogenerators (TENG). These technologies are well-suited for distributed, mobile, low-frequency, and micro-energy-harvesting, and conversion applications.

Piezoelectric materials generate electricity when subjected to mechanical stress or vibration and can be incorporated into wearables, such as shoes or clothing, to capture energy from the wearer's movements, such as walking or jogging [35]. PENG is based on the displacement and capacitive current models generated inside the generator and in the external circuit, respectively [27]. In contrast, TENGs use only Al electrodes and are flexible and lightweight, making them suitable for wearables. They can generate electricity through the triboelectric effect, which is the result of friction between two dissimilar materials. When integrated into wearable devices, TENGs can convert mechanical energy from body movements, such as walking or hand gestures, into electrical power [21]. TENGs offer a sustainable solution to the finite energy supply issue by harnessing mechanical energy, making them suitable as sustainable power sources and self-powered active sensors [22, 36–46]. These devices boast several advantages, including high-output voltage, efficiency, versatility, simplicity in structural design and fabrication, low cost, stability, and robustness.

A TENG (triboelectric nanogenerator) is a device that converts mechanical energy into electric energy through the coupling of triboelectrification and electrostatic induction. It utilizes the time-varying electric field generated by surface electrostatic charges to induce the flow of electrons [47]. TENGs operate through four fundamental working modes: vertical contact-separation mode, in-plane sliding mode, single-electrode mode, and free-standing triboelectric-layer mode [48]. The initial TENG design involved stacking two thin polymer films with separate metal electrodes deposited on the top and bottom surfaces [49]. Another variant, known as a fiber-structured TENG, consists of an Al wire and a PDMS tube covering a vertically aligned nanotextured surface [50].

19.3.1 Walking/Running

As the demand for portable electricity grows, walking has become a key focus of energy collection. Energy can be harvested from both human and machine motion using wireless electronic devices such as piezoelectric, electromagnetic, and triboelectric methods, as demonstrated in different studies [51–53]. The University of Wisconsin-Madison has even developed a flooring system that generates electricity from the pressure of footfalls, which has the potential to be used in high-traffic areas like airports and shopping malls to reduce energy costs [12]. Energy acquisition technologies also have the potential to revolutionize the development of medical devices. Researchers at the University of California, Berkeley, have created a device that generates electricity from the motion of the human jaw, which can be used to power medical implants and reduce the need for battery replacements [54].

Researchers have also directed their attention toward the bending motion around the knees as an additional area of interest within the context of walking movements. Researchers at the Massachusetts Institute of Technology (MIT) developed an energy harvesting device that used a dynamo-based mechanism inside a person's shoe to harvest energy from their feet striking the ground. Additionally, a paper-thin triboelectric nanogenerator (TENG) was developed that can be inserted into a shoe's insole to capture energy from movement and track bodily activity [55]. Because walking is the most significant component of human activity, harvesting waste energy from walking has great promise for powering wearable devices [56].

As a result, researchers have developed various nanogenerators to generate relatively high electric energy from human movement. By harnessing the power of human motion, energy acquisition technologies have the potential to not only reduce energy costs but also improve the lifespan of medical implants and power wearable devices.

TENGs have the capability to harvest mechanical energy from various directions and generate a high surface charge density of 250 $\mu C/m^2$ [27]. Conversely, a wireless, fully self-powered wearable sweat analysis system, featuring a high-performance freestanding-mode TENG (FTENG), sweat sensor patch, efficient power management, and Bluetooth low-energy module integrated onto a flexible printed circuit board, can generate an impressive output power of up to 416 mW/m^2. This level of energy is sufficient to independently operate the entire system during on-body human trials, eliminating the need for any external power supply [57]. Furthermore, it is feasible to generate electric power from walking and running activities without the need for a specially designed shoe or TENG. By utilizing Al foil and electric wire (Fig. 19.5b and c), electric power can be generated from human locomotion (Fig. 19.5a) [56].

Using commercially available materials such as aluminum foil and electric wire can significantly reduce the manufacturing time and cost of raw materials [58–60]. During walking, triboelectrification occurs between the shoe's sole and an insulating surface, resulting in the transfer of electrons from the ground to the sole, leaving static electrons on the sole and positive charges on the surface due to differences in triboelectric polarities (Fig. 19.5d, I). Due to the phenomenon of

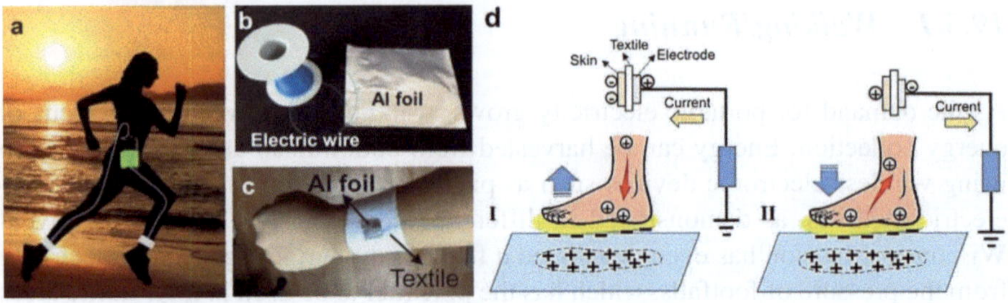

Fig. 19.5 (**a**) Schematic illustration of the human body-based triboelectric system for body motion energy harvesting. (**b**) Materials used to fabricate the triboelectric system. (**c**) Picture showing the Al electrode used for the human body based triboelectric system. (**d**) Schematic illustration of the working mechanism for the single-electrode-mode triboelectric system [56]

electrostatic induction, negative charges have the ability to transfer from the lifting foot to other regions of the body, leading to positive charging of the lifting foot. This process leads to the generation of positive charges on the aluminum electrode, thereby causing a flow of current from the electrical ground toward the electrode. When the foot makes contact with the ground, the transferred charge on the body returns to achieve electrostatic equilibrium. This process results in a charge balance and triggers a current flow in the opposite direction, from the electrical ground toward the electrode (Fig. 19.5d, II). This triboelectric system, based on the human body, is characterized by its lightweight nature, eco-friendliness, simple structure, and cost-effectiveness [56]. The energy harvested from activities such as walking and running is sufficient to illuminate a series connection of 100 commercial LEDs. Additionally, this system can serve as a monitoring sensor for detecting emergencies, monitoring patient conditions, and ensuring security.

Researchers at the University of Pennsylvania and the University of Michigan have examined energy harvesting with a backpack by leveraging a person's "bobbing" motion as they walk, converting their whole-body motion into a source of energy by allowing the backpack to move up and down on a frame and gathering energy from the linear motion [61, 62]. Similarly, a US Navy project has developed a backpack that harnesses the oscillating motion of a person's walk to generate power and reduce the need for additional 20-pound batteries to power equipment such as night-vision goggles and GPS gadgets [63, 64]. These systems generate between 4 and 7.4 W of energy, depending on their design, and the power output from different movements (shown in Table 19.1) is more than enough to power several devices simultaneously [65].

Researchers at MIT have made advancements in the development of a dynamo-based system integrated into a person's shoe has achieved a peak output of 1.61 W and an average output of 58.1 mW throughout the gait. However, it falls short of the targeted 250-mW output level. Collectively, these studies demonstrate the potential of human motion to generate significant amounts of energy, which can be harnessed to power wearable devices, monitor patient conditions, and ensure security at a low cost.

Table 19.1 Test results of the average power for different movements [65]

	Power (μW)		
Type of movement	Elbow		Knee
Squatting	–		0.136
Walking	0.363		0.216
Jogging	0.389	0.226	
Fast running	0.457	0.291	

19.3.2 Energy-Generating Gymnasium Machines (EGGMs)

A gym provides a convenient and potentially promising setting for the collection of human energy. Through the execution of periodic and repetitive movements using gym equipment with varying degrees of resistance, calories can be burned and muscular function can be improved. The requirement for activities with relatively high-resistance levels offers a suitable amount of mechanical energy that can be harnessed and converted into electricity. Additionally, the large number of people frequenting a gym enables the generation of power on a significant scale. When compared to other viable public energy collection locations (walking, running, backpack), gym utilities are an excellent fit [66]. As a moving vehicle generates electrical energy from the weight (potential energy) and motion (kinetic energy) of the vehicle [67], the electric power generation from EGGMs depends on the weight of the sportsman. EGGMs can be classified as hand and leg operations.

19.3.2.1 Hand-Operated Workouts in the Gymnasium

Various task-related factors influence human strength, including horizontal distance, height, exertion orientation, and velocity. Military research has even explored the integration of energy-harvesting technology into commercial products, such as hand-crank generators designed to provide electricity to soldiers [68]. Typically, a human can endure approximately 30 W/h by hand cranking [69]. As shown in the flow chart (Fig. 19.6), this review discusses different types of working postures in gymnasiums for both men and women.

When considering the measurement of hand strength, comparing the sitting and standing postures of men and women offers valuable insights into the task requirements and dynamics. According to recent studies, a person can safely push and pull up between 20% and 30% of his or her body weight. In both static and dynamic strength (Fig. 19.7), the horizontal distance and vertical height of exertion have a significant impact on the force exerted [70].

For achieving maximum pull strength output, a seated working position was found to be more favorable compared to a standing working position. In the standing position, the average male's pull strength was approximately 65% of the value achieved in the seated position. Conversely, women exhibited about 71% of their seated pull strength when in a standing position. As for standing push strength, men

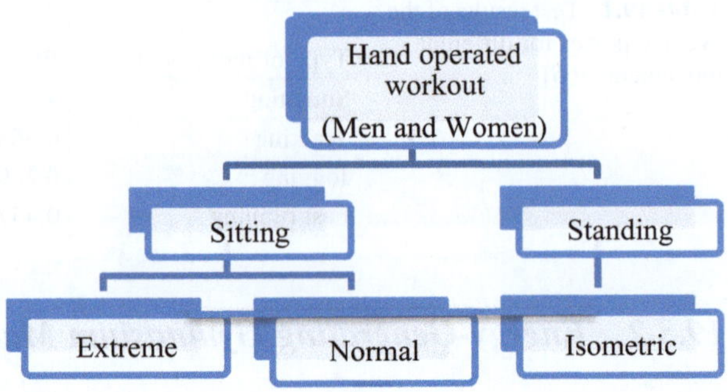

Fig. 19.6 Hand-operated workouts

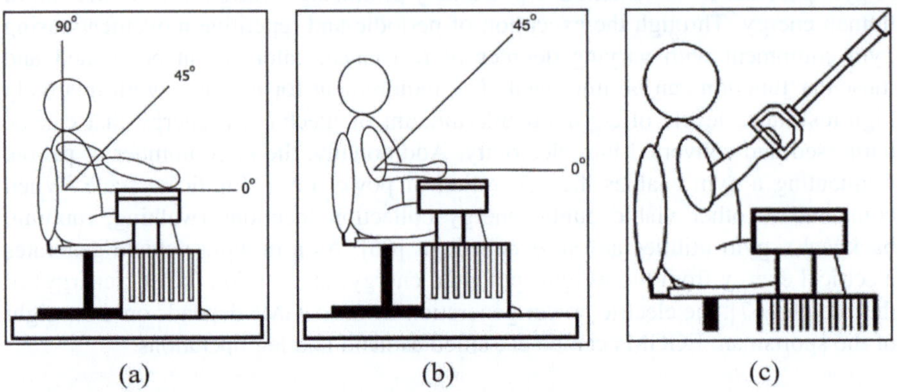

Fig. 19.7 Hand-operated workouts in the seated position: (**a**) extreme reach envelope, (**b**) normal reach envelope, and (**c**) isometric strength [70]

demonstrated approximately 88% of the seated push strength. In terms of standing pull and push strength, women outperformed around 58% and 69% of men, respectively. In the seated position, women surpassed approximately 53% and 48% of men in pull and push strength, respectively [67]. However, the extent of the gender difference in strength is influenced by the specific body parts involved in the exertion [70].

Based on those relations, the pushing (Fig. 19.8) and pulling (Fig. 19.9) strengths of men and women at the sitting and standing positions were analyzed. As such, as the weight of the athlete increased, the ability to generate electric power also increased. The average pushing and pulling strength of men and women, measured in both sitting and standing working positions, is presented in Fig. 19.10.

19.3.2.2 Feet-Operated Workouts in the Gymnasium

Human power potential studies have found that legs have four times the strength of hands [69]. There are a variety of workout devices that have a high potential for generating power while exercising. The alternator's speed and the device's age

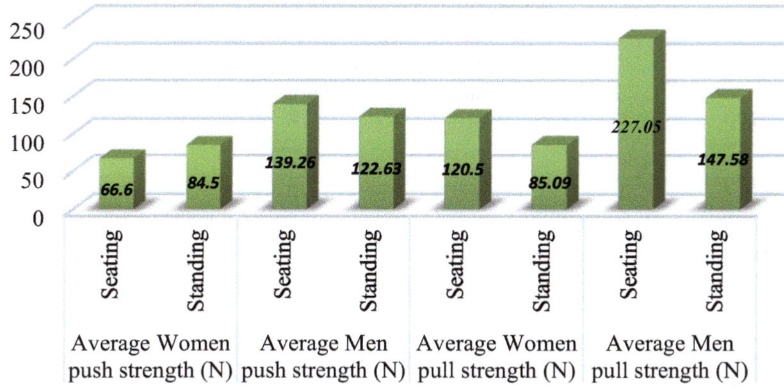

Fig. 19.8 The average strength of humans

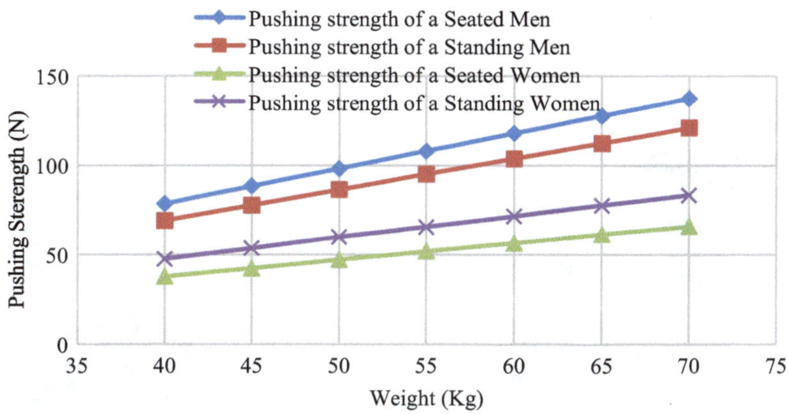

Fig. 19.9 Pushing strength in eco-gym utilities

restriction determine how much power it can generate [71]. The Great Outdoor Gym Company, based in the UK, has installed fitness machines that generate green energy at various locations such as Sir George Monoux College, Trafalgar Square, and the Green Heart in Hull City. These machines enable the generation of electrical energy through play and exercise, utilizing the concept of EGGE (energy generating gym equipment) [72]. With some adjustments, a variety of EGGEs (stationary bikes, treadmills, steppers, ellipticals, and arc trainers) can generate electricity.

Stationary bikes (Fig. 19.11a) are popular cardio exercises in gyms for both men and women. It encourages someone's nutrition and is especially excellent for cardiac regulation. In the modern gym, stationary bikes with an electrical generator, sometimes known as "eco-bikes," are employed. This foot exercise uses a dynamo-integrated charge controller to transform the mechanical rotational power of human pedaling into usable electrical energy. At the same time, health metrics such as temperature and heart rate can be improved. Treadmills (Fig. 19.11b) are

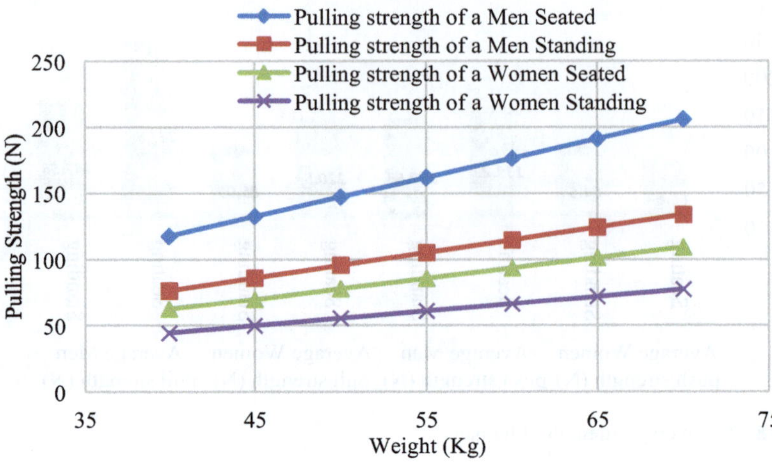

Fig. 19.10 Pulling strength in eco-gym utilities

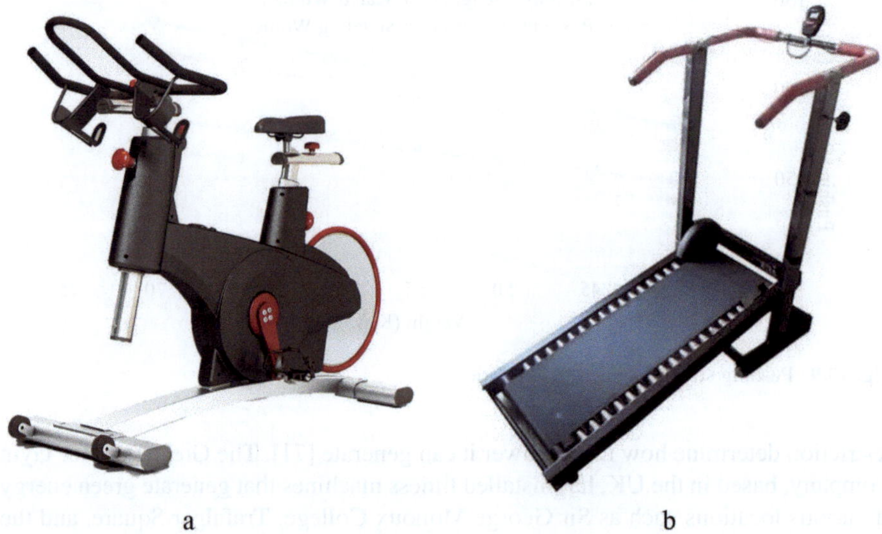

Fig. 19.11 Energy-generating gymnasium machines: (**a**) stationary bike and (**b**) manual treadmill [73]

another type of workout machine that can be used at home or in a gym. People are expected to run on a treadmill, which is made up of a path-like structure that shifts and rotates when a person walks and runs. This type of equipment can also generate electrical power.

From a single physical workout on a 10-treadmill system, 500 kW of electric power can be generated monthly [73]. Mostly, the output power depends on the walking speed of the sportsman and the type of treadmill (see Fig. 19.12).

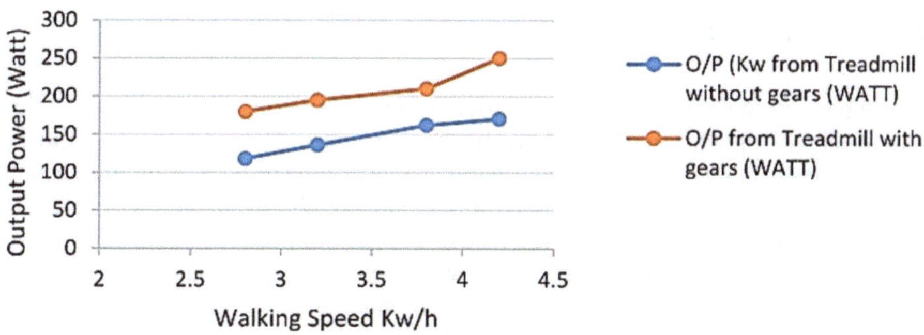

Fig. 19.12 Treadmill power generation [73]

The results obtained from the number of weight lifters who were more efficient at working in a gymnasium can produce 318.18–540 kilos of watts per hour [66]. Experimental results have demonstrated that the system is capable of generating electrical energy during cycling activities performed by young humans, achieving an average efficiency of 17% or higher [33].

19.4 Limitations of Human Energy Acquisition Technology

Human energy acquisition technologies hold immense promise as sustainable and eco-friendly sources of energy. Nonetheless, they come with their fair share of challenges and limitations. Let's explore some of these constraints that need to be addressed for the widespread adoption and success of human energy acquisition technologies.

19.4.1 Low-Power Density

One of the key challenges in harnessing human energy lies in its relatively low-power density, which refers to the limited amount of power that can be generated per unit of volume or weight. Human energy sources, such as body heat and motion, inherently produce small amounts of energy, posing a significant obstacle to generating substantial electricity. The limited temperature difference between the human body and the surrounding environment is the primary factor contributing to this phenomenon. Moreover, comparing the temperature difference inside and outside the body is not a straightforward task.

However, researchers are actively addressing these challenges with innovative approaches aimed at boosting the power density of human energy acquisition technologies. For instance, they are exploring the development of advanced materials for wearable devices that can maximize electricity generation from body motion [15]. By pushing the boundaries of material science and engineering, these efforts seek to revolutionize the potential of human energy, making it a more viable and attractive option for powering various applications.

19.4.2 Efficiency

In comparison to conventional power sources, human energy acquisition technologies often face the challenge of relatively low efficiency. The performance of these technologies is influenced by various factors, including the user's weight, movement speed, and the characteristics of the terrain. Commercially available thermoelectric materials typically exhibit efficiency ranges of approximately 5–10%. In contrast, triboelectric nanogenerators (TENGs) have shown impressive conversion efficiencies ranging from a few percent to over 50% in laboratory settings. However, when implemented in wearable applications, TENGs may experience lower practical efficiency due to factors such as mechanical losses and electrical impedance matching.

Furthermore, piezoelectric energy harvesters typically demonstrate efficiencies ranging from a few percent to around 20%. The efficiency of these harvesters is influenced by factors such as the quality of the piezoelectric material, device design, and the magnitude and frequency of applied mechanical stress. It is important to note that a significant amount of energy is lost during the conversion process. Currently, researchers are actively engaged in developing new materials for thermoelectric generators, aiming to enhance their efficiency [23, 25]. These endeavors seek to overcome the limitations and improve the overall performance of human energy acquisition technologies.

19.4.3 Power Storage

Human energy acquisition technologies often generate electricity intermittently, making it challenging to store the energy for later use. However, researchers are exploring various approaches to energy storage, such as using supercapacitors and batteries that can store energy generated by human activity [12].

19.4.4 Durability

Wearable human energy acquisition technologies, such as piezoelectric nanogenerators and thermoelectric generators, must be durable enough to withstand the wear and tear of daily use. Ensuring the durability of these devices is crucial to making them practical and cost-effective. Ongoing researches are exploring to enhance the durability of these devices by developing new materials that are more resistant to damage and wear [26].

19.4.5 Cost

Human energy acquisition technologies can be expensive to develop and manufacture, which can make them less accessible to people in developing countries or low-income communities. Researchers are developing more cost-effective materials and manufacturing processes [54].

19.5 Application Areas

Human power is already being utilized in a variety of unique applications as a renewable, eco-friendly, and cost-effective energy source that leaves no carbon footprint or electronic waste [74]. The concept of the "Internet of Things" (IoT), where interconnected devices with embedded electronics, sensors, actuators, and software communicate and interact via the Internet, has gained significant attention [75]. Wearable technology, as depicted in Fig. 19.13a, b, has become increasingly popular for personal health monitoring [76] and various electronic sensors [77]. By combining IoT and wearables, the quality of life can be improved by reducing healthcare costs, enhancing the well-being of elderly individuals, and enabling daily body workout monitoring [78]. Moreover, the generated power from human activities has been used in commercial products such as flashlights, radios, and cell phone chargers [79]. In general, it has the potential to be applied in remote healthcare monitoring, security monitoring, and energy harvesting, as indicated in Table 19.2 [56, 80–84].

19.6 Conclusion

The adaptability of recent electrical energy-based technologies has significantly impacted the advancement of modern civilization. This study reviewed strategies for acquiring human energy through thermal and mechanical means, which have not been widely used in today's culture. However, they offer a viable option for those cut off from the world due to natural disasters, military deployment, or remote locations. HEATs hold promising potential for applications in remote healthcare monitoring, security monitoring, and energy harvesting. EGGMs are particularly important for generating electric power and promoting human health, especially for athletes. Their low-cost and easy-to-implement design make them useful in developing countries with limited resources. In general, thermal and mechanical human energy harvesters can generate approximately 20 μW/cm2 and 540 kW/hr. of electricity, respectively. However, the output performance and sensitivity of recent devices still need improvement. Future studies recommended to focus on addressing practical applications of the devices in various environments.

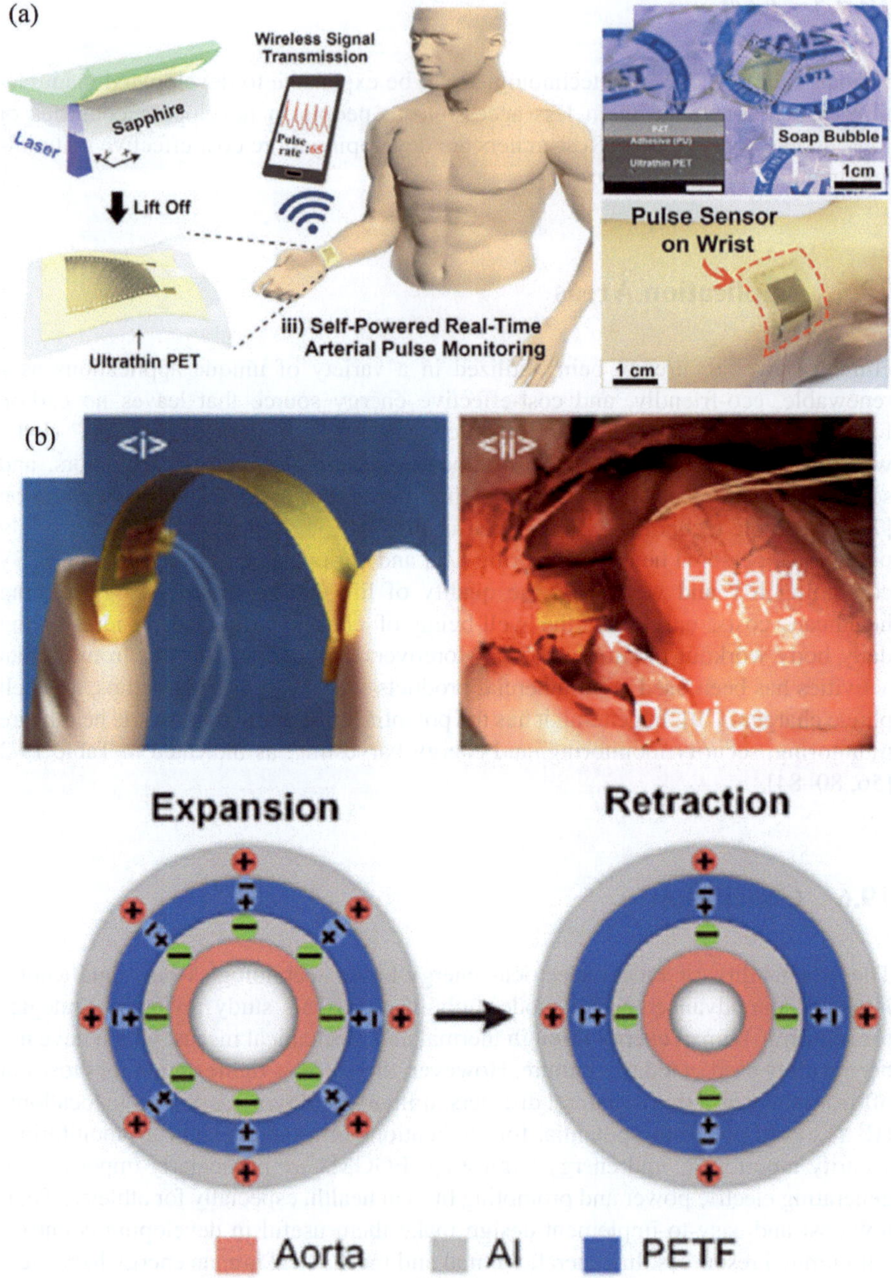

Fig. 19.13 Applications of mechanical energy-harvesting technologies. (**a**) An ultrathin epidermal piezoelectric sensor for real-time pulse monitoring. (**b**) A self-powered blood pressure monitor based on a piezoelectric thin-film wrapped around the ascending aorta of a porcine [27]

Table 19.2 Applications of human energy acquisition technology [56, 80–84]

No		Application area
1.	Wearable electronics	Smartwatches, fitness trackers, augmented reality (AR) glasses, and other wearable gadgets
2.	Healthcare and bio-medical devices	Integrated into medical devices (biosensors, health monitoring patches, and implantable devices), it can enhance patient comfort, enable continuous monitoring, and improve the overall efficiency of healthcare systems
3.	Internet of things (IoT)	It enables the seamless integration of IoT devices (environmental sensors, asset trackers, and smart clothing) into our daily lives
4.	Military and defense	It helps soldiers to utilize self-powered wearable devices for communication, surveillance, and situational awareness purposes
5.	Outdoor and sports applications	It can be useful for outdoor enthusiasts, hikers, campers, athletes, and adventure enthusiasts It can power small devices such as GPS trackers, headlamps, or emergency communication devices and ensuring a continuous power supply during outdoor activities
6.	Fashion and design	It can be seamlessly integrated into garments, allowing for self-illuminating clothing, interactive fashion, or even powering fashion accessories like handbags or jewelry
7.	Human-ma-chine interfaces	It can enable self-powered human-machine interfaces (gesture recognition systems) By capturing and converting body movements into electrical energy, these interfaces can be used in virtual reality (VR), gaming, robotics, and other interactive applications

References

1. IEA. (2020). Energy Access Outlook 2020. International Energy Agency.
2. Alizadeh, M. I., et al. (2016). Flexibility in future power systems with high renewable penetration: A review. *Renewable and Sustainable Energy Reviews, 57*, 1186–1193.
3. Husin, H., & Zaki, M. (2021). A critical review of the integration of renewable energy sources with various technologies. *Protection and Control of Modern Power Systems, 6*(1), 1–18.
4. Gan, L., Jiang, P., Lev, B., & Zhou, X. (2020). Balancing of supply and demand of renewable energy power system: A review and bibliometric analysis. *Sustainable Futures, 2*, 100013.
5. Rowejeh Saeed, S. (2020). Iba Harvest Human Kinetic Energy: A Framework for Gym's Wasted Energy Usage.
6. REN21. (2021). Renewables 2021 Global Status Report. REN21 Secretariat.
7. IRENA. (2021). Renewable Energy Statistics 2021. International Renewable Energy Agency.
8. Kaldellis, J. K., & Kondili, E. (2019). Hybrid renewable energy systems: A review of the state of the art. *Handbook of Clean Energy Systems*, 1–36.
9. Liao, H., et al. (2010). A study on the difference and inequality of world energy consumption and their changes. *China Soft Science, 10*, 6–14.
10. Getachew Bekele, G. T. (2012). Feasibility study of small hydro/PV/wind hybrid system for off-grid rural electrification in Ethiopia. *Applied Energy, 97*, 515.
11. Liang, H., et al. (2019). Fiber-based energy conversion devices for human-body energy harvesting. *Advanced Materials, 32*, e1902034. https://doi.org/10.1002/adma.201902034

12. Wang, Y., Li, X., & Chen, Z. (2018). Human energy harvesting based on wearable technology: A review. *Smart Materials and Structures, 27*(10), 103002.
13. Jansen, A., & Stevels, A. (2006). Combining eco-design and user benefits from human-powered energy systems, a win–win situation. *Journal of Cleaner Production, 14*(15–16), 1299–1306.
14. Cyders, T., & Kremer, G. G. (2008). Engineering around the world: Driving local economics in Africa with human power. In *ASME international mechanical engineering congress and exposition*.
15. Wu, J., et al. (2017). Wearable energy harvesting technologies for human-powered electronic devices: A review. *SCIENCE CHINA Technological Sciences, 60*(6), 845–859.
16. Zhang, L., et al. (2016). Recent progress on piezoelectric and triboelectric energy harvesters in biomedical systems. *Advanced Science, 3*(11), 1500286.
17. Kohler, M., Li, Y., & Gian Chandani, Y. B. (2013). Human-powered energy harvesting from cycling for wireless sensor nodes. *Journal of Microelectromechanical Systems, 22*(6), 1427–1434.
18. SportsArt. (n.d.). *Eco-power Products*. Retrieved from https://www.gosportsart.com/eco-powr-products/
19. ReRev. (n.d.). *Our products*. Retrieved from https://www.rerev.com/products/
20. Weber, J., et al. (2006). Coin-size coiled-up polymer foil thermoelectric power generator for wearable electronics. *Sensors and Actuators A: Physical, 132*, 325–330.
21. Wang, Z. L. (2017). Triboelectric Nanogenerators as new energy technology and self-powered sensors – Principles, challenges and perspectives. *Faraday Discussions, 198*, 11–30. https://doi.org/10.1039/c6fd00263k
22. Wu, C., et al. (2017). Enhanced triboelectric nanogenerators based on MoS2 monolayer nanocomposites acting as electron-acceptor layers. *ACS Nano, 11*, 8356–8363.
23. Yun, H., Kwon, J., & Kim, S. (2017). Harvesting human energy for battery-free internet of things. *Nano Energy, 37*, 136–146.
24. Siddique, A. R. M., et al. (2017). A review of the state of the science on wearable thermoelectric power generators (TEGs) and their existing challenges. *Renewable and Sustainable Energy Reviews, 73*, 730–744.
25. Kim, S. K., Lee, Y. K., & Kim, S. W. (2018). Wearable thermoelectric generator for harvesting human body heat energy. *Applied Energy, 211*, 1126–1134.
26. Kim, M. K., Kim, M. S., Lee, S., Kim, C., & Kim, Y. J. (2014). Wearable thermoelectric generator for harvesting human body heat energy. *Smart Materials and Structures, 23*(10), 105002.
27. Zou, Y., et al. (2021). Recent progress in human body energy harvesting for smart bioelectronic system. *Fundamental Research, 1*, 364–382.
28. Park, J. H., et al. (2019). Ingenious use of natural triboelectrification on the human body for versatile applications in walking energy harvesting and body action monitoring. *Nano Energy, 57*, 872–878.
29. Hyland, M., et al. (2016). Wearable thermoelectric generators for human body heat harvesting. *Applied Energy, 182*, 518–524.
30. Madan, D., et al. (2015). Printed flexible thermoelectric generators for use on low levels of waste heat. *Applied Energy, 156*, 587–592.
31. Jo, S. E., Kim, M. K., Kim, M. S., & Kim, Y. J. (2012). Flexible thermoelectric generator for human body heat energy harvesting. *Electronics Letters, 48*(16), 1015–1017.
32. Lu, Z., et al. (2016). Silk fabric-based wearable thermoelectric generator for energy harvesting from the human body. *Applied Energy, 164*, 57–63.
33. Kumar, S., & Pahuja, R. (2020). Eco-friendly power generator cum fitness analyzer. *Instrumentation Mesure Métrologie, 19*(1), 9–18.
34. Dieffenderfer, J., et al. (2016). Low power wearable systems for continuous monitoring of environment and health for chronic respiratory disease. *IEEE Journal of Biomedical and Health Informatics, 20*, 99.

35. Zhu, G., et al. (2012). Triboelectric-generator-driven pulse electrodeposition for micropatterning. *Nano Letters, 12*(9), 4960–4965. https://doi.org/10.1021/nl302879t
36. Zhang, Q., et al. (2018). Electromagnetic shielding hybrid nanogenerator for health monitoring and protection. *Advanced Functional Materials, 28*, 1703801.
37. Wu, C., et al. (2016). Wearable electricity generators fabricated utilizing transparent electronic textiles based on polyester/Ag nanowires/graphene core-shell nanocomposites. *ACS Nano, 10*, 6449–6457.
38. Wu, C., et al. (2017). Reduced graphene-oxide acting as electron-trapping sites in the friction layer for giant triboelectric enhancement. *Nano Energy, 32*, 542–550.
39. Zhang, Z., et al. (2018). Facile method and novel dielectric material using a nanoparticle doped thermoplastic elastomer composite fabric for triboelectric nanogenerator applications. *ACS Applied Materials & Interfaces, 10*, 13082–13091.
40. Yu, B., et al. (2017). High-power triboelectric nanogenerator prepared from electro spun mats with spongy parenchyma-like structure. *Nano Energy, 34*, 6975.
41. Kim, W.-G., Kim, D., Jeon, S.-B., Park, S.-J., Tcho, I.-W., Jin, I.-K., Han, J.-K., & Choi, Y.-K. (2018). Multi direction and multi amplitude triboelectric nanogenerator composed of porous conductive polymer with prolonged time of current generation. *Advanced Energy Materials, 51*, 1800654.
42. Kim, S., Gupta, M. K., Lee, K. Y., Sohn, A., Kim, T. Y., Shin, K. S., Kim, D., Kim, S. K., Lee, K. H., Shin, H. J., & Kim, D. W. (2014). Transparent flexible graphene triboelectric nanogenerators. *Advanced Materials, 26*, 3918–3925.
43. Tang, Q., Yeh, M.-H., Liu, G., Li, S., Chen, J., Bai, Y., Feng, L., Lai, M., Ho, K. C., Guo, H., & Hu, C. (2018). Whirligig-inspired triboelectric nanogenerator with ultrahigh specific output as reliable portable instant power supply for personal health monitoring devices. *Nano Energy, 47*, 74–80.
44. Wang, H. S., Jeong, C. K., Seo, M.-H., Joe, D. J., Han, J. H., Yoon, J.-B., & Lee, K. J. (2017). Performance-enhanced triboelectric nanogenerator enabled by wafer-scale nanogrates of multistep pattern downscaling. *Nano Energy, 35*, 415–423.
45. Sun, J.-G., Yang, T. N., Kuo, I.-S., Wu, J.-M., Wang, C.-Y., & Chen, L.-J. (2017). A leafmolded transparent triboelectric nanogenerator for smart multifunctional applications. *Nano Energy, 32*, 180–186.
46. Xing, F., Jie, Y., Cao, X., Li, T., & Wang, N. (2017). Natural triboelectric nanogenerator based on soles for harvesting low-frequency walking energy. *Nano Energy, 42*, 138–142.
47. Wang, Z. L. (2017). On Maxwell's displacement current for energy and sensors: The origin of nanogenerators. *Materials Today, 20*(2), 74–82.
48. Luo, J., & Wang, Z. L. (2020). Recent progress of triboelectric nanogenerators: From fundamental theory to practical applications. *EcoMat, 2*(4), e12059.
49. Fan, F. R., Tian, Z., & Wang, Z. L. (2012). Flexible triboelectric generator. *Nano Energy, 1*(2), 328–334.
50. Kim, K. N., Chun, J., Kim, J. W., et al. (2015). Highly stretchable 2D fabrics for wearable triboelectric nanogenerator under harsh environments. *ACS Nano, 9*(6), 6394–6400.
51. Priya, S., & Inman, D. J. (Eds.). (2009). *Energy harvesting technologies* (Vol. 21). Springer.
52. Prasad, C. R., Rao, P. R., Srikanth, Y., & Chakradhar, A. (2022, May). Review on energy harvesting techniques for wearable devices in wireless body area networks. In *AIP conference proceedings* (Vol. 2418, No. 1). AIP Publishing.
53. Zhou, B., et al. (2023). A review of evaluation, principles, and technology of wearable electromagnetic harvesters. *ACS Applied Electronic Materials, 5*, 4035.
54. Kim, S. K., Lee, Y. K., & Kim, S. W. (2016). A piezoelectric energy harvester using the motion of the human jaw for powering medical implants. *Applied Energy, 170*, 124–131.
55. Wu, C., Kim, T. W., Sung, S., Park, J. H., & Li, F. (2018). Ultrasoft and cuttable paper-based triboelectric nanogenerators for mechanical energy harvesting. *Nano Energy, 44*, 279–287.

56. Park, J. H., Wu, C., Sung, S., & Kim, T. W. (2019). Ingenious use of natural triboelectrification on the human body for versatile applications in walking energy harvesting and body action monitoring. *Nano Energy, 57*, 872–878.
57. Song, Y., Min, J. H., Yu, Y., et al. (2020). Wireless battery-free wearable sweat sensor powered by human motion. *Science Advances, 6*(40), eaay9842.
58. Wu, C., Park, J. H., Sung, S., Koo, B., Lee, Y. H., & Kim, T. W. (2018). Integrable card-type triboelectric nanogenerators assembled by using less problematic, readily available materials. *Nano Energy, 51*, 383–390.
59. Wu, C., Park, J. H., Koo, B., Chen, X., Wang, Z. L., & Kim, T. W. (2018). Capsule triboelectric nanogenerators: Toward optional 3D integration for high output and efficient energy harvesting from broadband-amplitude vibrations. *ACS Nano, 12*, 9947–9957.
60. Mallineni, S. S. K., Behlow, H., Dong, Y., Bhattacharya, S., Rao, A. M., & Podila, R. (2017). Facile and robust triboelectric nanogenerators assembled using off-the-shelf materials. *Nano Energy, 35*, 263–270.
61. Lester, G. (2008). *Shoe leather as a renewable resource: Penn biologists invent power-generating backpack*. Office of University Communications – University of Pennsylvania.
62. Kuo, A. D. (2005). Harvesting energy by improving the economy of human walking. *Science, 309*, 1686–1687.
63. Rome, L. C., Flynn, L., Goldman, E. M., & Yoo, T. D. (2005). Generating electricity while walking with loads. *Science, 309*(5741), 1725–1728.
64. Roach, J. (2005). New *Backpack generates its own electricity*. National Geographic News.
65. Li, K., He, Q., Wang, J., Zhou, Z., & Li, X. (2018). Wearable energy harvesters generating electricity from low-frequency human limb movement. *Microsystems & Nanoengineering, 4*, 24. https://doi.org/10.1038/s41378-018-0024-3
66. Haji, M. N., Li, K., & Agogino, A. M. (2010). Human power generation in fitness facilities. In *Proceedings of the ASME 2010 4th international conference on energy sustainability*.
67. Jordehi, A. R. (2018). How to deal with uncertainties in electric power systems? A review. *Renewable and Sustainable Energy Reviews, 96*, 145–155.
68. Moyers, W. L., Coombe, H. S., & Hartman, A. (2004). Harvesting energy with hand-crank generators to support dismounted soldier missions.
69. Dean, T. (2008). *The human-powered home: Choosing muscles over motors* (p. 64). New Society Publishers.
70. Das, B., & Wang, Y. (2004). Isometric pull-push strengths in a workspace: 1. Strength profiles. *International Journal of Occupational Safety and Ergonomics, 10*(1), 43–58.
71. Sunil, H. V., et al. (2019). Power generation using gym equipment.
72. The Great Outdoor Gym Company (GOGC). (2018). Green Energy.
73. Mustafa, F. F., Oh, O., Daham, S. R., & Sabry, A. H. (2020). New treadmill design for energy harvesting. *Journal of Mechanical Engineering Research and Developments, 43*.
74. Avani. (2021). *Avani eco power*. Retrieved from https://www.avani.eco/avanieco-power.html
75. Atzori, L., Iera, A., & Morabito, G. (2010). The internet of things: A survey. *Computer Networks, 54*(15), 2787–2805.
76. Misra, V., Bozkurt, A., Calhoun, B., Jackson, T., Jur, J. S., Lach, J., et al. (2015). Flexible technologies for self-powered wearable health and environmental sensing. *Proceedings of the IEEE, 103*(4), 665.
77. Castano, L. M., & Flatau, A. B. (2014). Smart fabric sensors and e-textile technologies: A review. *Smart Materials and Structures, 23*(5), 053001.
78. Hiremath, S., Yang, G., & Mankodiya, K. (2014). Wearable internet of things: Concept, architectural components and promises for person-centered healthcare. In *2014 EAI 4th international conference on wireless mobile communication and healthcare (Mobihealth)*, IEEE.
79. Merchant, B. (2008). *Hand crank power, common electronics power*. Planet Green.
80. Goktas, M., & Elitaş, M. (2020). Wearable energy harvesting technologies for smart textiles: A comprehensive review. *Sensors and Actuators A: Physical, 310*, 111832. https://doi.org/10.1016/j.sna.2020.111832

81. Paradiso, R., & Loriga, G. (2013). Energy harvesting wearable electronics for wireless monitoring of physiological signals. *IEEE Transactions on Information Technology in Biomedicine, 17*(3), 392–399. https://doi.org/10.1109/TITB.2013.2257951
82. Khan, U., Ghosh, A., & Hussain, M. M. (2018). Wearable energy harvesting: Perspectives, challenges, and future directions. *Applied Physics Reviews, 5*(3), 031104. https://doi.org/10.1063/1.5022560
83. Gupta, S., & Jha, R. M. (2020). Wearable energy harvesting: A review on recent developments in materials, technologies, and applications. *Journal of Energy Storage, 32*, 101903. https://doi.org/10.1016/j.est.2020.101903
84. Kim, D. H., Lu, N., Ma, R., Kim, Y. S., Kim, R. H., Wang, S., et al. (2011). Epidermal electronics. *Science, 333*(6044), 838–843. https://doi.org/10.1126/science.1206157

91. Paradiso, R., & Longa, O. (2013). Energy harvesting wearable electronics for wireless monitoring of physiological signals. *IEEE Transactions on Information Technology in Biomedicine, 17*(3), 392–399. https://doi.org/10.1109/TITB.2013.2257951
92. Khan, U., Ghosh, A., & Hossain, M. M. (2013). Wearable energy harvesting: Perspectives, challenges and future directions. *Applied Physics Reviews, 5*(3), 031101. https://doi.org/10.1063/1.5022560
93. Gupta, S., & Jha, R. M. (2020). Wearable energy harvesting: A review on recent developments in materials, technologies, and applications. *Journal of Energy Storage, 32*, 101903. https://doi.org/10.1016/j.est.2020.101903
94. Kim, D. H., Lu, N., Ma, R., Kim, Y. S., Kim, R. H., Wang, S., et al. (2011). Epidermal electronics. *Science, 333*(6044), 838–843. https://doi.org/10.1126/science.1206157

Chapter 20
Influence of S/N Ratio on Biogas Desulfurization Performance Using Conventional Anoxic Bioreactor (ABR) and Membrane Bioreactor (AMBR)

Ebrahim Tilahun

20.1 Introduction

Biogas can play a major role in the developing market for renewable energy, and it is expected that the power production capacities from commercial biogas facilities in the world will be more than doubled over the next decade [1]. Consequently, the preparation and development of biogas production have been extensively studied in recent times [2–4]. The created biogas largely comprises of methane (CH_4) and carbon dioxide (CO_2) but also trace amount of hydrogen sulfide (H_2S) and other impurities [5, 6]. CH_4 is the combustible constituent of biogas, which can be used for mutable energy sources [7, 8]. However, the utilization of biogas as an energy source is limited greatly due to the existence of high concentrations of H_2S in the biogas. H_2S in the biogas not only cause corrosion of pipelines and instruments but also could cause harm to the human health. Therefore, it is essentially important to minimize H_2S from biogas before being utilized as energy source.

Biogas desulfurization methods mainly include physicochemical and biological processes. Physicochemical methods have been widely used for the removal of H_2S from both liquid and gas phases. However, these methods have problems, such as generation of secondary pollutant, high operation cost, and requiring frequent regeneration. Compared with physicochemical methods, biodesulfurization can oxidize H_2S to either elemental sulfur or sulfate by microbial metabolism, which is environmentally friendly with several benefits, such as low-energy requirement and less secondary pollutant and operation cost. Biodesulfurization can be an aerobic or anoxic process. Presently, aerobic biogas desulfurization are mostly used, which

E. Tilahun (✉)
Center for Renewable Energy, Addis Ababa University, Addis Ababa, Ethiopia
e-mail: ebrahim.tilahun@aait.edu.et

K. Mequanint et al. (eds.), *Sustainable Development Research in Materials and Energy*, Green Energy and Technology,
https://doi.org/10.1007/978-3-031-69860-6_20

requires a stoichiometric oxygen level depending on the inlet concentrations of H_2S in biogas. Moreover, due to safety concerns and dilution effects, it is also essential to control the oxygen/air dosage in order to avoid reaching high concentrations of oxygen and nitrogen content in biogas, respectively. However, these drawbacks might be avoided by using NO_3^-/NO_2^- as electron acceptors in the process of biodesulfurization. Hence, biogas desulfurization could be coupled with wastewater denitrification simultaneously.

Within the scope of this study, an innovative approach is developed to reduce the need for external carbon source and increase biogas quality, in combination with autotrophic denitrification. The main purpose of this research was to compare the potential of conventional anoxic bioreactor and anoxic membrane bioreactor performance for simultaneous biogas desulfurization and autotrophic denitrification. The effects of S/N molar ratio and biogas flowrate on both bioreactors are investigated.

20.2 Materials and Methods

20.2.1 Anoxic Bioreactor

Within the scope of this study, a laboratory-scale anoxic bioreactors were established. A cylindrical shape bioreactor with dimensions of 0.12 m diameter and 0.20 m depth is fabricated from a plexiglass material. Tubular PDMS membrane with an internal diameter of 7 mm, a wall thickness of 1 mm, a length of 3.25 m, and an active surface area of 0.0.092 m^2 is used. The liquid in the bioreactor is continuously stirred with a magnetic stirrer to achieve uniform mixing. During the experiments, conductivity, ORP, and pH values of the liquid phases were monitored online and recorded instantly using a digital multimeter (Multi 9430, WTW GmbH, Germany). The operating temperature was kept under mesophilic conditions (21 ± 1°C) using an electric heating blanket. The pH in the bioreactor was adjusted at 7.0 by adding a 1 N NaOH solution using a dosing pump. The stock alkaline solution was prepared by using distilled water as required. The schematic of the anoxic bioreactor was show in Fig. 20.1. During the study, the experiments were carried out in a fume hood to avoid the leak of the H_2S gas in to the other worker in the laboratory. A simulated biogas mixture, containing of 60% (v/v) CH_4, 39% (v/v) CO_2, and 1% (v/v) H_2S, was supplied with mass flow controllers and continuously fed into the bioreactor. The biogas flowrate was measured by digital gas counters (MGC, Ritter, Germany). The process consists of three basic steps, gas separation, gas absorption, and biological degradation. The presence of membrane supports diffusing of H_2S and immediately scrubbed into the alkaline liquid medium.

Finally, the sulfide-oxidizing bacteria (SOB), in particular *Thiobacillus* spp., biologically oxidized scrubbed H_2S to elemental sulfur and/or sulfate; at the same time, the nitrate/nitrite in the synthetic wastewater reduced to nitrogen gas. The ratio between the available electron acceptor and electron donor, (S/N) ratio, is the key parameter determining the formation of elemental sulfur or/and sulfate.

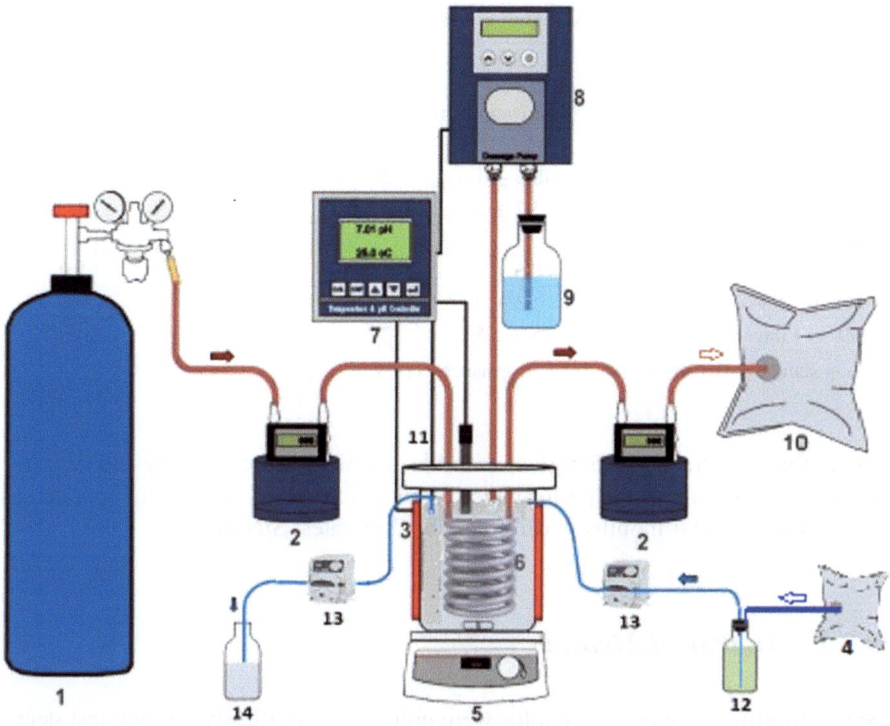

Fig. 20.1 Schematic representation of anoxic bioreactor: (1) biogas cylinder, (2) milligas counter, (3) heating blanket, (4) N_2-containing gas balloon, (5) magnetic stirrer, (6) folded tubular membrane, (7) pH transmitter and temperature sensor, (8) NaOH dosage pump, (9) NaOH solution, (10) effluent biogas balloon, (11) multiprobe meter, (12) bottle containing the feed liquid medium, (13) peristaltic pump, and (14) bottle containing the effluent liquid medium

20.2.2 Operational Conditions, Feeding Composition, and Inoculums

The operational conditions of both bioreactors were indicated in Table 20.1.

The bioreactor was inoculated with active sludge taken from laboratory-scale anoxic bioreactor treating thiosulfate-containing wastewater. The experiment was conducted in a continuous gas and liquid flow, under anoxic conditions by using nitrate/nitrite as an electron acceptor. During desulfurization, most of H_2S and some CO_2 have permeated through the membrane and dissociated into sulfide (H_2S_{aq}, HS^-, S^{2-}) and bicarbonate ion (HCO_3^-). The scrubbed sulfide and bicarbonate were used as electron donor and inorganic carbon source, respectively. A stock solution of 1000 mg NO_3^-/L was prepared from sodium nitrate ($NaNO_3$). The composition of the medium: 57 mg/l of K_2HPO_4, 10 mg/l of $MgCl_2.6H_2O$, 10 mg/l of $CaCl_2.2H_2O$,

Table 20.1 Operational conditions in both bioreactors

Period (days)	Biogas flowrate (l/d)	H_2S loading rate (mg/m^2.h)	Liquid flow rate (l/d)	NO_3^- loading rate (mg/l.d)	S/N molar ratio
ABR					
I (1–22 days)	13	48.5 ± 1	2	85	1.9
II (23–43 days)	23	92 ± 2	2	85	3.0
AMBR					
III (46–75 days)	13	48.5 ± 1	2	85	1.9
IV (76–98 days)	23	92 ± 2	2	85	3.0
V (99–114 days)	35	135 ± 2.5	2	85	4.0

Mean ± standard derivation at stable performance period

3 mg/l of yeast extract, and 5 mg/l of ascorbic acid were prepared daily and kept in a refrigerator. To evaluate the performance of both bioreactors, the system operated as a whole for around 4 months in different gas flow rates, S/N ratio.

20.2.3 Analytical Methods

For the purpose of analysis, samples were collected periodically and the test determinations performed in duplicate. Sulfide was analyzed spectrometrically (WTW PhotoLab 6100VIS) following the method described by Cord-Ruwisch [9]. Sulfate was measured according to the *Standard Methods for the Examination of Water and Wastewater* [10]. High-performance liquid chromatography (HPLC) was used for the determination of nitrate/nitrite concentration in the supernatant of the reactor. For gas analysis gas chromatography was used (Shimadzu GC-2014ATF) equipped with TCD. The temperatures of injection, column, and detector are 150, 40, and 150 °C, respectively.

20.3 Results and Discussions

20.3.1 Conventional Anoxic Bioreactor Process Performance

The conventional anoxic bioreactor (control) experiments run for about 1.5 months. The control experiment was performed without installing membrane in the system; hence, the control experiment was considered as conventional bioreactor in which the biogas directly contacted with the alkaline liquid medium. So that, H_2S was trapped and easily solubilized in slightly basic liquid medium inoculated with sulfide-oxidizing bacteria (SOB). The liquid medium in the bioreactor comprises

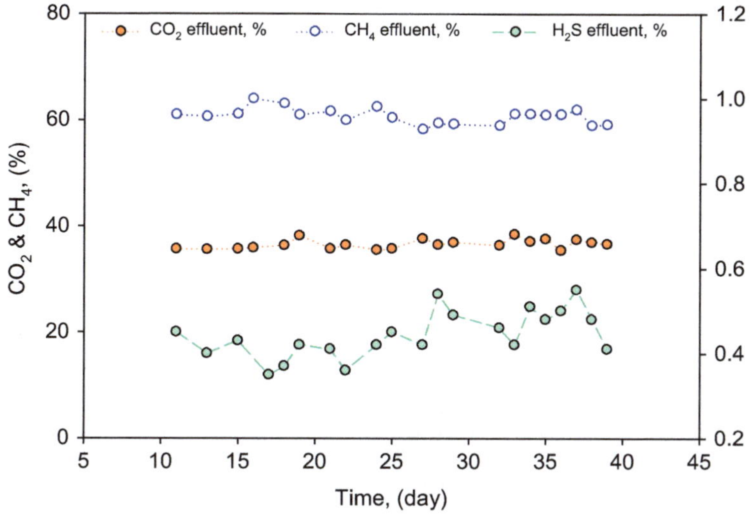

Fig. 20.2 The effluent biogas contents in the anoxic bioreactor without membrane

suspended active microorganism that can biochemically oxidize scrubbed sulfide into sulfur or/and sulfate depending on the supply of electron acceptor/donor.

In this conventional test, to retain great dissociation and scrubbing of H_2S, the pH of the liquid medium was kept about 7.0. The results revealed that the average gas phase H_2S removal efficiency of the anoxic bioreactor was nearly 62 ± 2% and 54 ± 3% at a gas flowrate of 13 L/d and 23 L/d, respectively. The high concentration of CO_2 present in biogas makes the satisfactory removal of H_2S difficult; the CO_2 also reacts with sodium hydroxide and uses it up quickly. One more barrier that still limits the H_2S removal was uneven distribution of H_2S within the conventional bioreactor. Additional restraining factor was the gas diffuser used during operation of the bioreactor might also result in lower gas transferring capacity. As shown in Fig. 20.2, throughout the whole conventional experiment, the concentration of H_2S in the exit was significantly greater than 500 ppmv which can cause corrosion in the internal combustion engines [11]. In the case of biogas combustion, it could emit SO_2 to the environment [12].

Figure 20.2 also represents the CO_2 contents at the outlet of the conventional experiments. It can be seen that the CO_2 content reduce to some extent only; this might happen owing to its direct contact between biogas and the liquid medium. The scrubbed CO_2 could be used as inorganic carbon source for the SOB. On the other hand, the CH_4 content in the treated biogas, almost similar with the methane content in fed biogas even at lower gas flowrate tested, remained nearly 62 ± 2% (Fig. 20.2). The poor CH_4 enrichment was probably caused by minimal dilution of the biogas with nitrogen gas created during the denitrification process, since CH_4 is only sparingly soluble in water and not degraded by the microorganism in the bioreactor [13].

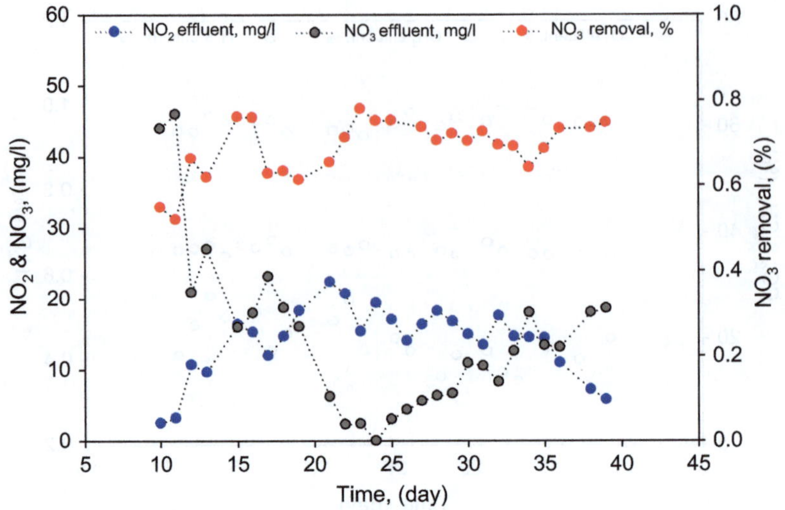

Fig. 20.3 The nitrate removal efficiency of NO_3^- and NO_2^- effluents in the conventional ABR

Alike the earlier literature by More et al. [14], the anoxic bioreactor had been fed with nitrate in order to avoid the accumulation of nitrite due to its inhibitory effect on sulfide oxidation. On the other hand, the nitrate removal efficiencies raised from $54 \pm 2\%$ to $68 \pm 3\%$ in parallel with the hoisted gas flowrate (Fig. 20.3). The lower removal efficiency might be the result deficient electron donor in addition to the limited bacterial biodegradable capacity in this particular test. Fernández et al. [15] report on the absence of an inhibitory effect on a biotrickling filter when nitrite accumulated up to 0.45–0.79 gN-NO_2^-/l. Thus, in this study, no inhibitory effect on the activities of bacteria was observed.

As shown in Fig. 20.4, gas flowrate of 13 L/d, with S/N ratio of 1.9, nearly generated 58 ± 2.5 mg SO_4/l.d, while at 23 L/d gas flowrate (S/N > 2.5), just about 69 ± 3 mg of SO_4/l.d were generated. Less sulfate production appeared even at high S/N ratio, which might favor other intermediate compound generation rather than sulfate (Fig. 20.5).

As indicated in Fig. 20.4, at lower gas flow rates (13 L/d), roughly 0.18 g NaOH was added; by contrast at higher gas flowrate (23 L/d), the amount of NaOH supplied was only 0.22 g. This might happen because of hydroxide ion generation owning to elemental sulfur production during the process. In general, the conventional anoxic bioreactor tested here was not efficient for both autotrophic denitrification and biogas desulfurization. Therefore, to increase the system efficiency, it is reasonable to adapt membrane-based biogas upgrading technology as an alternative to the conventional processes examined in Sect. 20.3.2.

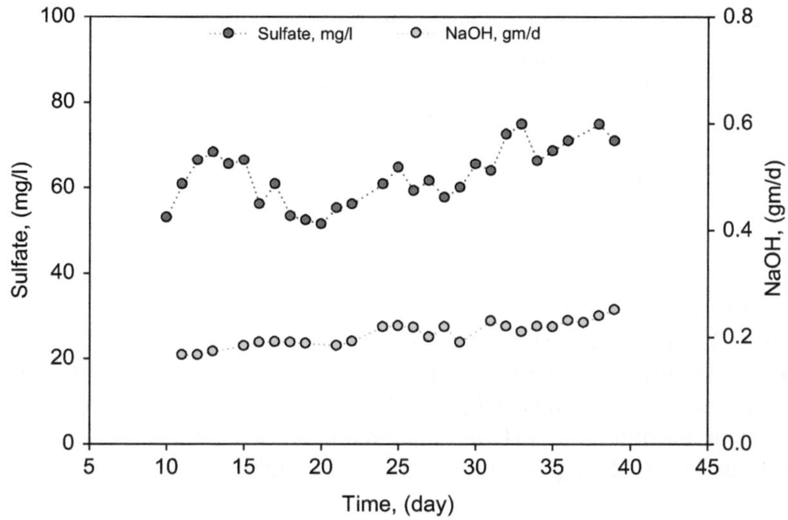

Fig. 20.4 Generation of sulfate and NaOH consumption in the conventional ABR experiments

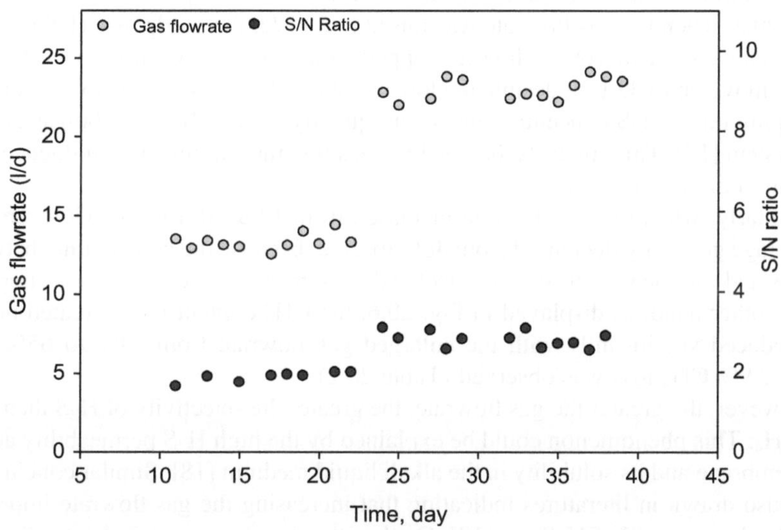

Fig. 20.5 The S/N ratio and gas flowrate tested in the conventional ABR experiments

20.3.2 Membrane-Supported Anoxic Bioreactor Process Performance

In this section in order to avoid the drawback of the conventional method and enhance the process efficiency, dense PDMS membrane had been inserted into the bioreactor. Therefore to evaluate the effectiveness of the membrane-supported

Table 20.2 The effects of different gas flowrate on the AMBR

	Gas absorption efficiencies (%)			Gas selectivity		Gas flux ($g/m^2.d$)		
Gas flow rate (l/d)	H_2S	CO_2	CH_4	H_2S/CO_2	H_2S/CH_4	H_2S	CO_2	CH_4
13	98.9	46.3	2.92	2.14	34	2.13	50	1.75
23	95.6	32.4	2.22	2.94	43	3.4	58	2.21
35	87.0	21.3	1.34	4.08	65	4.97	61	2.14

bioreactor for concurrent desulfurization and denitrification, the biogas flowrate was raised from 13 to 35 L/d, whereas the nitrate loading was kept nearly 85 mg/l.d; with these effects of different S/N ratio on the process, end products were tested. The pH values were also kept constant at 7 ± 0.1. In earlier literatures, Fernández et al. [16] reported that the optimum pH of nitrate-reducing and sulfide-oxidizing culture ranged from 7.0 to 7.5, and the microbial activity inhibited at a lower or higher pH value.

According to the results of AMBR at the gas flowrate of about 13 L/d, the outlet H_2S concentrations of the biogas was below 200 ppmv; besides nearly all H_2S gas diffused through the membrane was absorbed in the liquid phase. As indicated in Table 20.2, when the gas flowrate was raised to 23 L/d, the gas0phase H_2S removal efficiencies closed to 96%, while it dropped to about 87% when the gas flowrate further move up to 35 L/d; despite the low variation, this corresponds to greater than 500-ppmv outlet H_2S concentrations. Consequently, it can affect the biogas utilization system [17]. This suggests that the high gas flowrate significantly influenced the separation of H_2S from biogas.

Similarly, when the gas flowrate increased from 13 to 35 L/d, the CO_2 removal percentage gradually decreased from 46% to 21%, because the contact time between the gas and the membrane steadily declined as higher volume of gas was supplied. On the other hand, as displayed in Fig. 20.6, the CH_4 content in the treated biogas was reduced significantly with the enlarged gas flowrate from 74% to 65%, and nearly 2.9% CH_4 loss was observed (Table 20.2).

However, the greater the gas flowrate, the greater the selectivity of H_2S than CO_2 and CH_4. This phenomenon could be explained by the high H_2S permeability across the membrane and its solubility in the alkali liquid medium [18]. Similar conclusions were also drawn in literatures indicating that increasing the gas flowrate improved the selective removal of H_2S gas [19]. On the other hand, increasing the gas flowrate enhanced H_2S and CO_2 flux from 2.13 to 4.97 $g/m^2.d$ and from 49 to 61 $g/m^2.d$, respectively (Table 20.2). Therefore, gas separation using dense membrane could remove substantial H_2S and marginal amount of CO_2 in one step if sufficient driving force for permeation is provided [20].

The result also indicated that as the gas flowrate was raised, the NaOH consumption also increased from 0.22 g/d to 0.33 g/d, owing to more acid gases absorption (Table 20.3). This result revealed that the current system favor high H_2S removal capacity, but CO_2 removal was kept low, because H_2S could be easily diffused

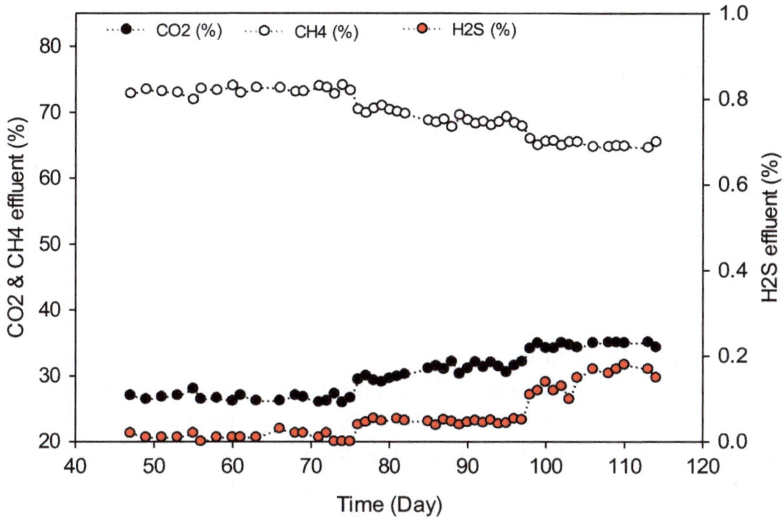

Fig. 20.6 The effect of gas flowrate on the effluent H_2S, CO_2, and CH_4 contents in AMBR test

Table 20.3 Performance of the AMBR with different S/N ratios

Period (days)	S/N ratio	Dissolved sulfide (HS^-) oxidation efficiency (%)	Nitrate removal efficiency (%)	NaOH consumption (g/d)
III (46–75 days)	1.9	98 ± 0.5	80 ± 5	0.22 ± 0.02
IV (76–97 days)	3.0	96 ± 0.7	91 ± 2	0.28 ± 0.02
V (98–114 days)	4.0	94 ± 1.1	97 ± 3	0.33 ± 0.03

Mean ± standard derivation at stable performance period

through the membrane and used as electron source for SOB in the process of desulfurization and denitrification.

As illustrated in Table 20.3, the removal pattern of nitrate and sulfide at a gas flowrate of 13 L/d (S/N ratios 1.9) was tested. In the first few days, between 46 and 50, nearly all H_2S diffused through the membrane, dissolved in the liquid medium, and eliminated by SOB, whereas the nitrate removal efficiency was less 75%; the reason might be due to the oxygen entrance into the anoxic bioreactor while inserting the membrane into liquid medium. However, after day 50, the nitrate removal efficiency raised above 84%. In the membrane-supported experiment, the amount of sulfate generated was about 95 mg/L, which is higher compared to the conventional test. However, this value was below the expected theoretical sulfate production probably because of partial sulfide oxidation that boosts intermediate sulfur compounds formation. In addition, the development of whitish layer on the external surface of the membrane observed. This was indicative of biological elemental sulfur generation and its accumulation within the bioreactor [21].

On day 76, the biogas flowrate went up to 23 L/d, (S/N ratio 3); as a result 96% of dissolved sulfide and 90% of nitrate removal were achieved. Here, even though the S/N ratio is greater than 2.5, complete denitrification was not attained because the constrained growth rate of the autotrophic bacteria in the bioreactor might not be enough for the entire nitrate degradation.

As demonstrated in Table 20.3, on day 98, the biogas flowrate raised to 35 L/d (S/N ratio 4); consequently, 94% sulfide were dissolved and nearly 100% of nitrate eliminated. Under excess sulfide loading, declining of sulfate production had been seen, meaning it is possible to control the oxidation products by varying the supplied sulfur to nitrogen molar ratio. This indicates that the biological elemental sulfur generated and attached on the external surface of the membrane and suspended in the liquid phase was attributed to a high sulfide loading [22]. Similar behavior previously reported by Almenglo et al. [23] stated that when limited amount of nitrate is supplied, the sulfate production was always lower than the values predicted by the theoretical calculations. In the earlier literature, Cai et al. [24] also evaluated the optimized S/N ratio as 2.5 for preferably generation of elemental sulfur and nitrogen gas. Hence, it can be conclude that the high S/N molar ratio had a significant negative effect on the sulfate creation and favors biological elemental sulfur generation, but not on the H_2S gas removal efficiency. Hence, the result of the current anoxic membrane-supported bioreactor behavior was in agreement with previous reports. This demonstrated that the membrane-supported anoxic bioreactor tested in this study had better capacity for biogas desulfurization and autotrophic denitrification simultaneously, which is very promising to apply in pilot-scale experiments.

20.4 Conclusions

The results show that the conventional and membrane-supported anoxic bioreactor could realize the biogas desulfurization and wastewater denitrification simultaneously, and the AMBR had better and more stable performance in the S/N ratios tested in this study. The shift of S/N ratios did not influence the sulfide oxidation significantly; almost all diffused and dissolved H_2S were removed, while the desulfurization end products and nitrate removal were affected evidently. These results also indicate that H_2S gas diffusion through PDMS membrane could be the limiting step (mass transfer limitation); however, the biological sulfide oxidation in liquid phase was relatively fast. With the increase of S/N ratios from 1.9 to 4, the formation of sulfate was decreased, and denitrification performance was improved markedly. Higher S/N ratio led to larger elemental sulfur generation, which favored in saving of alkaline consumption by the bioreactor. This work revealed that, by combining membrane-supported autotrophic denitrification with desulfurization, not only allows selective H_2S removal from biogas but also improves the nitrate removal from wastewater without a need of external carbon source and dilution effects.

Therefore, the combined AMBR process developed in this study is an easy-to-operate and better alternative than conventional anoxic bioreactor for biogas desulfurization and autotrophic denitrification.

References

1. Sun, Q., Li, H., Yan, J., Liu, L., Yu, Z., & Yu, X. (2015). Selection of appropriate biogas upgrading technology-a review of biogas cleaning, upgrading and utilisation. *Renewable and Sustainable Energy Reviews, 51*, 521–532. https://doi.org/10.1016/j.rser.2015.06.029
2. Bohutskyi, P., & Bouwer, E. (2013). Biogas production from algae and cyanobacteria through anaerobic digestion: A review, analysis, and research needs. In *Advanced biofuels and bioproducts* (pp. 873–975). https://doi.org/10.1017/CBO9781107415324.004
3. Esen, M., & Yuksel, T. (2013). Experimental evaluation of using various renewable energy sources for heating a greenhouse. *Energy and Buildings, 65*, 340–351. https://doi.org/10.1016/j.enbuild.2013.06.018
4. Mussgnug, J. H., Klassen, V., Schlüter, A., & Kruse, O. (2010). Microalgae as substrates for fermentative biogas production in a combined biorefinery concept. *Journal of Biotechnology, 150*, 51–56. https://doi.org/10.1016/j.jbiotec.2010.07.030
5. Frigon, J. C., & Guiot, S. R. (2010). Biomethane production from starch and lignocellulosic crops: A comparative review. *Biofuels, Bioproducts and Biorefining, 4*, 447–458. https://doi.org/10.1002/bbb.229
6. Truong, L. V. A., & Abatzoglou, N. (2005). A H_2S reactive adsorption process for the purification of biogas prior to its use as a bioenergy vector. *Biomass and Bioenergy, 29*, 142–151. https://doi.org/10.1016/j.biombioe.2005.03.001
7. Khanal, S. K. (2009). *Anaerobic biotechnology for bioenergy production: Principles and applications*. Willey and Blackwell. https://doi.org/10.1002/9780813804545
8. Lantz, M., Svensson, M., Björnsson, L., & Börjesson, P. (2007). The prospects for an expansion of biogas systems in Sweden – Incentives, barriers and potentials. *Energy Policy, 35*, 1819–1829. https://doi.org/10.1016/j.enpol.2006.05.018
9. Cord-Ruwisch, R. (1985). A quick method for the determination of dissolved and precipitated sulfides in cultures of sulfate-reducing bacteria. *Journal of Microbiological Methods, 4*, 33–36. https://doi.org/10.1016/0167-7012(85)90005-3
10. Lenore, S. C., Arnold, E. G., & Andrew, D. E. (2012). *APHA/AWWA/WEF: Standard methods for the examination of water and wastewater* (p. 541). American Public Health Association.
11. Fortuny, M., Gamisans, X., Deshusses, M. A., Lafuente, J., Casas, C., & Gabriel, D. (2011). Operational aspects of the desulfurization process of energy gases mimics in biotrickling filters. *Water Research, 45*, 5665–5674. https://doi.org/10.1016/j.watres.2011.08.029
12. Su, J. J., & Chen, Y. J. (2015). Monitoring of sulfur dioxide emission resulting from biogas utilization on commercial pig farms in Taiwan. *Environmental Monitoring and Assessment, 187*(1), 4109. https://doi.org/10.1007/s10661-014-4109-7
13. Nikiema, J., Bibeau, L., Lavoie, J., Brzezinski, R., Vigneux, J., & Heitz, M. (2005). Biofiltration of methane: An experimental study. *Chemical Engineering Journal, 113*, 111–117. https://doi.org/10.1016/j.cej.2005.04.005
14. Mora, M., Fernandez, M., Gomez, J. M., Cantero, D., Lafuente, J., Gamisans, X., & Gabriel, D. (2015). Kinetic and stoichiometric characterization of anoxic sulfide oxidation by SO-NR mixed cultures from anoxic biotrickling filters. *Applied Microbiology and Biotechnology, 99*, 77–87.
15. Fernández, M., Ramírez, M., Pérez, R. M., Gómez, J. M., & Cantero, D. (2013). Hydrogen sulphide removal from biogas by an anoxic biotrickling filter packed with pall rings. *Chemical Engineering Journal, 225*, 456–463. https://doi.org/10.1016/j.cej.2013.04.020

16. Fernández, M., Ramírez, M., Gómez, J. M., & Cantero, D. (2014). Biogas biodesulfurization in an anoxic biotrickling filter packed with open-pore polyurethane foam. *Journal of Hazardous Materials, 264*, 529–535.
17. Ramos, I., & Fdz-Polanco, M. (2014). Microaerobic control of biogas sulphide content during sewage sludge digestion by using biogas production and hydrogen sulphide concentration. *Chemical Engineering Journal, 250*, 303–311.
18. Kennedy, N., Zhao, Q. B., Ma, J., Chen, S., & Frear, C. (2015). The selective removal of H_2S over CO_2 from biogas in a bubble column using pretreated digester effluent. *Separation and Purification Technology, 144*, 240–247. https://doi.org/10.1016/j.seppur.2015.02.046
19. Lu, J. G., Zheng, Y. F., & He, D. L. (2006). Selective absorption of H2S from gas mixtures into aqueous solutions of blended amines of methyldiethanolamine and 2-tertiarybutylamino-2-ethoxyethanol in a packed column. *Separation and Purification Technology, 52*, 209–217. https://doi.org/10.1016/j.seppur.2006.04.003
20. Harasimowicz, M., Orluk, P., Zakrzewska-Trznadel, G., Chmielewski, A. G. (2007). Application of polyimide membranes for biogas purification and enrichment. In: Selected papers of the proceedings of the 5th European meeting on chemical industry and environment, EMChIE 2006 Held in Vienna, Austria, 3–5 May 2006, vol. 144(3), pp. 698–702. https://doi.org/10.1016/j.jhazmat.2007.01.098.
21. Li, X., Jiang, X., Zhou, Q., & Jiang, W. (2016). Effect of S/N ratio on the removal of hydrogen sulfide from biogas in anoxic bioreactors. *Applied Biochemistry and Biotechnology, 180*, 930–944. https://doi.org/10.1007/s12010-016-2143-3
22. Zhou, Q., Liang, H., Yang, S., & Jiang, X. (2015). The removal of hydrogen sulfide from biogas in a microaerobic biotrickling filter using polypropylene carrier as packing material. *Applied Biochemistry and Biotechnology, 175*, 3763–3777.
23. Almenglo, F., Ramírez, M., Gómez, J. M., & Cantero, D. (2016). Operational conditions for start-up and nitrate-feeding in an anoxic biotrickling filtration process at pilot scale. *Chemical Engineering Journal, 285*, 83–91. https://doi.org/10.1016/j.cej.2015.09.094
24. Cai, J., Zheng, P., & Mahmood, Q. (2008). Effect of sulfide to nitrate ratios on the simultaneous anaerobic sulfide and nitrate removal. *Bioresource Technology, 99*, 5520–5527. https://doi.org/10.1016/j.biortech.2007.10.053

Index

K. Mequanint et al. (eds.), *Sustainable Development Research in Materials and Energy*, Green Energy and Technology,
https://doi.org/10.1007/978-3-031-69860-6

MIX
Papier aus verantwortungsvollen Quellen
Paper from responsible sources
FSC® C105338

If you have any concerns about our products,
you can contact us on
ProductSafety@springernature.com

In case Publisher is established outside the EU,
the EU authorized representative is:
Springer Nature Customer Service Center GmbH
Europaplatz 3, 69115 Heidelberg, Germany

Printed by Libri Plureos GmbH
in Hamburg, Germany